FOR TRISH

SPIRIT IN EXILE

EXILE

PETER PORTER

AND HIS POETRY

BRUCE BENNETT

OXFORD

OXFORD UNIVERSITY PRESS AUSTRALIA

Oxford   New York   Toronto
Delhi   Bombay   Calcutta   Madras   Karachi
Petaling Jaya   Singapore   Hong Kong   Tokyo
Nairobi   Dar es Salaam   Cape Town
Melbourne   Auckland

and associated companies in
Berlin   Ibadan

OXFORD is a trade mark of Oxford University Press

National Library of Australia
Cataloguing-in-Publication data:

Bennett, Bruce, 1941–
   Spirit in exile, Peter Porter and his poetry.

   Bibliography.
   Includes index.
   ISBN 0 19 554970 8.

   1. Porter, Peter, 1929–   . 2. Poets, Australian – 20th
   century – Biography.   3. Australian poetry – 20th century.
   I. Title.

A821.3

Designed by Marion Buckton
Typeset by Setrite Typesetters Ltd, Hong Kong
Printed by Condor Production Ltd, Hong Kong
Published by Oxford University Press,
253 Normanby Road, South Melbourne, Australia

CONTENTS

	ACKNOWLEDGEMENTS	vii
	A NOTE ON TEXTS AND ABBREVIATIONS	ix
	INTRODUCTION	xi
1	THE BRIGHT LOCKED WORLD	1
2	EACH UNSHINING HOUR	14
3	SUMMER HERMIT	29
4	AT HOME, AWAY	43
5	CARNIVAL WINTER	60
6	THE STRETCH MARKS ON HISTORY	92
7	FUEL FOR THE DARK	117
8	THE EASIEST ROOM IN HELL	149
9	THE LYING ART	177
10	BEYOND APOCALYPSE	197
11	POSSIBLE WORLDS	237
	NOTES	260
	BIBLIOGRAPHY	278
	INDEX	287

ACKNOWLEDGEMENTS

This book could not have been completed to deadline without a 75th Anniversary Research Award from the University of Western Australia, which relieved me of teaching and administrative duties in the second half of 1989. I' thank Bob White and Susan Miller, in particular, for temporarily reducing the load of administrative responsibilities which rapidly overtake a university teacher. The professional skill and patience of Pauline Dugmore, who typeset this manuscript at the University of Western Australia, have been outstanding. Bruce McClintock helped with library searches. I wish to thank Sandra McComb for showing faith in this project and Yvonne White, Peter Rose and Bronwyn Collie who have given constructive and encouraging advice. I would also like to thank Robert Gray and Bill Grono, who read and commented on the manuscript.

Since 1981, when I first met and interviewed Peter Porter in London, he has been generous in meeting my requests for information and material which would help my understanding of his work and its relation to his times and experience. I thank him for his permission to quote freely from his published and unpublished writings. Permission to quote from Porter's work has been granted by André Deutsch (for extracts from *Mars*), Oxford University Press (Oxford) and Secker and Warburg. *London Magazine*, ABC Radio Helicon and *Westerly* have granted permission to quote from articles, talks and interviews. Al Alvarez, Gavin Ewart, John Lucas and Anthony Thwaite have granted permission to quote from articles and reviews, Peter Edwards from correspondence, Peter Spearritt from an interview with Porter and Charles Osborne from *Giving It Away: The Memoirs of an Uncivil Servant* as acknowledged in the notes. I have special thanks for Arthur Boyd, for permission to reproduce *Picture on the Wall, Shoalhaven 1979−80* on the cover of this book and to quote from his letter to me.

Librarians at the Australian National Library in Canberra (where the majority of Porter's notebooks and manuscripts are held) and at the British Museum have been helpful in my research. John Kaiser, librarian at Pennsylvania State University, provided me with a pre-publication copy of his typescript bibliography, which has subsequently been published as *Peter Porter: A Bibliography 1954−1986* (Mansell Publishing, London and New York, 1990),

Those people whom I interviewed in order to better understand Porter and his context were generous with their time and thoughts.

These include Fleur Adcock, Christine Berg, Alan Brownjohn, Geoffrey Burgon, Roger Covell, Trevor Cox, Gavin Ewart, Ian Hamilton, Sally McInerney, Blake Morrison, Jill Neville, Diana Phillips, Susanna Roper, Jacqueline Simms and Anthony Thwaite. Informal talks with many other people have contributed to the discussion in this book.

It would be difficult to thank sufficiently Trish, Cathy and Michael Bennett, for adapting to the state of exile which the completion of this book has required.

A NOTE ON TEXTS AND ABBREVIATIONS

Page references for quotations are taken from the editions listed in the Bibliography, unless otherwise stated. For ease of reference, quotations of published poems which appeared in Peter Porter's *Collected Poems* (1983) refer to this source. The page reference in the *Collected Poems* is preceded by the volume in which the poem first appeared. Quotations from poems in volumes published after 1983 refer to those volumes alone.

The following abbreviations have been used for volumes of poetry by Peter Porter:

CP	*Collected Poems* (1983)
OBTB	*Once Bitten, Twice Bitten* (1961)
1962/3	Poems in *Penguin Modern Poets 2* (1962) and *A Group Anthology* (1963)
PAAM	*Poems Ancient and Modern* (1964)
APF	*A Porter Folio* (1969)
TLOE	*The Last of England* (1970)
PTTC	*Preaching to the Converted* (1972)
AM	*After Martial* (1972)
LIACC	*Living in a Calm Country* (1975)
TCOS	*The Cost of Seriousness* (1978)
ES	*English Subtitles* (1981)
FF	*Fast Forward* (1984)
TAO	*The Automatic Oracle* (1987)
PW	*Possible Worlds* (1989)
APS	*A Porter Selected* (1989)

One's life is not a case
except of course it is.

Les A. Murray, 'Three Poems in Memory of My Mother'

It is not just curiosity that makes an expatriate, there must be
something that happens in the very soul of them.

George Johnston, *My Brother Jack*

And all my calling cannot bring her back
To this real house, she in so much of it.

Douglas Dunn, 'Pretended Homes'

I have nothing to say about this garden
I do not want to be here, I can't explain
What happened. I merely opened a usual door
And found this.

Fleur Adcock, 'Unexpected Visit'

INTRODUCTION

◆

The search for the Common Reader, like that for the Bunyip, has been largely abandoned. The reader of this book — a study of a major Australian writer who has lived in London since 1951 — is likely to be uncommon, since the audience for serious poetry is unfortunately restricted. One of the purposes of this book is to show that the readership for contemporary poetry, and for Peter Porter's in particular, should be wider, for the best poetry of our times offers adventures in language, thought and feeling provided by no other literary genre with such compact intensity.

Peter Porter's reputation as a poet began in London in the early 1960s with the publication of his first book *Once Bitten, Twice Bitten* (1961), followed by his inclusion in *Penguin Modern Poets 2* (1962), and in various anthologies, including the revised edition of Al Alvarez's influential selection, *The New Poetry* (1966).

The first reviewers and commentators tended to focus on Porter's critiques of contemporary British society and presented him as another 'angry young man', a leading figure in the Group, whose flair derived in some way from his Australian origins ('the bruiser from Australia', one of the critics called him), if they knew his background at all. Porter's fellow Australian expatriate in London, Clive James, described the British view of his compatriot thus:

> Before the age of thirty, Porter simply drops off the map: by British standards he's almost completely untraceable, whether to school, university or class. When he does get on the map, he can be only approximately identified, as the man from nowhere who talks with a slight Movement accent in Group company.[1]

James's mockery of a restrictive, 'little England' view of outsiders fails to do justice to those British reviewers and commentators who recognized Porter's talent early. Uncertainty has persisted about when Porter put his own stamp on contemporary poetry, when a distinctive voice became evident, and implanted itself in readers' minds. A tentative consensus has emerged that this occurred around 1970, following publication of Porter's fourth volume, *The Last of England* — the first of his books published with Oxford University Press (the previous three were from the small Scorpion Press). But Porter's now recognizable voice had its earlier manifestations, as this study shows.

An important element in Porter's poetry, early and late, is a sense of exile, of separation not only from the country of his birth and upbringing but from any sense of 'home' — one of the words most highly charged with ambivalent feeling in his vocabulary. This sense of exile is more deeply embedded and more persistent in Porter's writings than the often superficial posturing of his contemporaries for whom Colin Wilson's *The Outsider* (1956) was a bible, for whom 'alienation' was a fashionable prerequisite to acceptance. Porter's poetry increasingly involved dramatizing migratory selves, whose relationship to any 'home' was deeply problematic.

The most significant year in Porter's personal and poetic fortunes was 1974. This was the year of his first return to Australia in twenty years, of a significant love affair with an Australian woman, and of the death of his English wife Jannice. Laments and love poems are joint legacies of this year, together with a deepening appreciation of the complex relationship between pain and art. A profound questioning of origins and allegiances also followed. Increasingly, from the mid-1970s, Porter returned to Australia on writing or teaching assignments, and the country of his birth was revitalized in his imagination. Yet just as British readers were only vaguely aware of Porter's Australian origins and background, most Australians were ignorant of his work in Britain and of its importance. This was due partly to a resistance towards complication from outside influences during a period of conscious national revival of the arts in Australia in the 1970s, together with a resentment in some quarters of those who had 'made it' overseas. But it revealed also an unfortunate lack of belief in the 'home-grown' product, which was remedied to some extent in the 1980s with a recognition of the achievements of expatriate writers such as Christina Stead, Sumner Locke Elliott, Shirley Hazzard, Randolph Stow, Janette Turner Hospital, Porter and others. Despite such signs of cultural maturation, however, the fate of expatriate authors still often fell between nations, never to be taken up fully by the publishers and readers either of their adopted country or of the country which had once been, and often remained psychologically, home.

As a poet, Porter's reputation has oscillated between that of social poet and elegist. This study shows that any such strict division is false, that an important personal and elegiac dimension both preceded and followed Porter's ebullient satiric engagement with London's 'swinging sixties' when the 'social poet' stereotype became entrenched. The ground bass of feeling in his work reached its apogee in Porter's seventh volume, *The Cost of Seriousness* (1978), in which a suite of poems addressed his feelings about the death of his wife. Subsequent volumes also faced this experience, both as an occasion of pain and loss and as a basis for questioning the limits of poetry as consolation and communication.

The publication of Porter's *Collected Poems* in 1983 brought him the Duff Cooper Memorial Prize. Reviewers and critics now placed Porter with the major poets writing in English, including Philip Larkin, Ted Hughes and Seamus Heaney. He was compared with John Ashbery in America and considered alongside A. D. Hope, Judith Wright and Les A. Murray in Australia. He was also perceived as a fit successor to W. H. Auden and Wallace Stevens, two major influences on his poetic thought and practice. On Porter's appearance in *Penguin Modern Poets 2* (1962), Stephen Spender is said to have asked, 'Who is Peter Porter?' Upon publication of the *Collected Poems* twenty-one years later, Spender wrote of Porter as 'an immensely fertile, lively, informed, honest and penetrating mind' whose brilliant use of language has 'an intoxicating effect' on the reader.[2] When Porter was awarded the Whitbread Prize for poetry for his tenth volume, *The Automatic Oracle* (1987), his co-prize winner in the prose fiction section was Salman Rushdie, for his novel *The Satanic Verses*. But in Australia, which had entered Porter's thought and verse increasingly since the mid-1970s, he was still largely unknown, or, if known, considered as a figure remote from the interests and concerns of Australians. The Australian Literature Society awarded Porter the Gold Medal for Literature in 1990; this may represent a turning point in Australians' recognition of this outstanding contemporary author.

My own first encounters with Porter's poetry may be indicative of the piecemeal and chancy way in which other Australians have come to his work. Certain popular anthology poems such as 'Your Attention Please' and 'Phar Lap in the Melbourne Museum' had caught my attention, and I had been impressed with a selection of Porter's poems from the 1960s which appeared in Alexander Craig's Australian anthology, *Twelve Poets 1950–1970* (1971). But it wasn't until I went to England on study leave in 1981 that I obtained my first volume of Porter's poems, *English Subtitles* (1981). This was Porter's eighth volume. I had not previously appreciated the skill and the subdued, but nonetheless powerful emotional force of his work, nor his wit and his seriousness. Reading back through Porter's work was an education of the mind and the emotions. Why had I been so unaware of this major talent from my own country? Was it because he had left Australia in 1951 and was now, in a period of nationalistic self-assertion, an outcast? I decided to find out more, wrote to Porter and interviewed him for the literary magazine *Westerly*.[3] When I returned to Australia, we corresponded and I met him subsequently at readings in Australia and elsewhere. In 1987, Porter spent two months as a colleague and writer-in-residence at the University of Western Australia. I found his company enormously stimulating and his conversation, like his poetry, was studded with brilliant and provocative observations. I was impressed with Porter's intellectual openness, his capacity to entertain alternative hypotheses, and to doubt with passion.

The title of this book, *Spirit in Exile*, is meant to indicate not only a major theme in Porter's writings (exile from home, school, country, self) but also the spirited quality of his response to this condition. Exile appears in many forms in Porter's writings. A poem of the late 1980s, 'Stratagems of the Spirit' (*PW* 55–6) indicates the sense of incompleteness, of separation from perfection, which stimulates, in Porter's case, a variety of procedures to negotiate the limbo in which the spirit is felt to be lost. Whether this sense of exile is projected in dramatic monologues, or through literary figures such as Byron and Christopher Smart, or figures from biblical and classical myth such as Cain and Orpheus, or in the fates of exiled painters and musicians, Porter's discontent smoulders, flickers or flares, lighting up a personal 'map of loss'. Irony, humour and ridicule are as much part of this process as tones of tragedy or pathos. In more directly autobiographical ways too, as in the image of a husband grieving for his dead wife, or a lover separated from his loved one, this sense of exile is evoked.

Spirit in Exile takes up a suggestion by Anthony Thwaite, to read Porter's poems as 'discontinuous parts of a huge verse novel, a *roman fleuve* of our time'.[4] The central figure in this 'river novel' or saga is my version of 'Peter Porter'. In this account, Porter comprehends more than his own individuality. The historical Porter would seem to approve of this presentation of the individual self in a wider context, for he has commented that the chief value of the ego/self is that it represents 'the only filter through which things much more wonderful than ourselves can pass'.[5] He has found the pronoun 'I' less justifiable than its fellow-pronoun 'me', which was notable for receiving 'exalting messages from the world beyond it — in dreams, in other people's art and artefacts, in the assertion of its birthright in feeling and knowing'. In short, he was not just himself; he was part of 'a world-nexus which is extraordinary ... a given not a chosen domain'.[6] Taken too far, this view could suggest a crude determinism. This book does not accept the implications of such determinism and presents Porter not as a passive receptor, but as an active thinker in his writing, a resistance fighter against unthinking conformism, a thorn in the side of tyrants and uncaring philistines. In Porter's hands, then, a poetry which involves dissembling and betrayal can be transformed into 'a nearly truthful art' ('The Lying Art', *TCOS*, *CP* 243–4).

The personal allegory which runs through Porter's writings does not display the unbending faith which enables the pilgrim to remain on the path to salvation. Porter's stratagem reveals a greater awareness of hell or its correlatives than of heaven or paradise, and his journey is plagued by guilt and doubts. Since allegory has been observed to arise chiefly in periods of loss, when 'a once powerful theological, political or familial authority is threatened with effacement',[7] the story of Peter Porter might be linked with two major losses of the post-second world war years in western industrialized countries: the loss of a sense of the past, and the loss of religious faith. Porter writes from within these gaps which have opened up so sharply in contemporary metropolitan societies, calling attention to the human needs which they make apparent rather than proposing solutions. As one who has tasted religion and journeyed widely through past ages in his reading, Porter is qualified to criticize the limitations of modernism and 'future shock', but he is fascinated and attracted also by the rush of modernity. A sceptical humanist, who bears the traces of a residual Christianity, an Anglo-Australian Protestantism, Porter conveys in some of his poetry a near-religious nostalgia for a barely apprehended but haunting notion of perfection, presented most often as gardens which are reclamations of an Eden. In the sense that his search sporadically uncovers the numinous in everyday reality, his work reflects something of a religious sensibility, but hopes of perfection are generally dashed. More characteristic of his work is the pessimism he expressed in a free rendering of Leopardi: 'To human kind fate has allocated/only dying: scorn Nature then,/the brutal power which rules for misery,/and the vanity of everything that is' (*FF* 50). But the extremist's rhetoric here tantalizes the reader with a lingering hope, revealing one of the spirit's most cunning stratagems: to hold despair at bay by dramatic exaggeration, a performative act which defines the speaker's presence and his needs, thereby making a claim for attention to fellow humans and, if they are listening, to the lost God or gods.

Like religion, Porter's social and political thinking emerges in his verse in fragments, images, aphorisms, and through a range of personae. Indeed, Porter's poetry and *ex cathedra* statements continually reject the notion that abstract theories can have an explanatory power superior to the plural, arbitrary insights of dreams and everyday observations, tempered by wide reading and art; he has ridiculed the view that 'it's the shape the theory makes that counts' ('Disc Horse', *TAO* 49−50). On social questions, Porter emerges as a resistance fighter against the purveyors of utopias and the theories they build. His outlook develops beyond the idealistic Fabian socialism (via George Bernard Shaw) to which he had felt drawn in his teenage years in Brisbane, towards an unillusioned socialism which accepts the necessity of a competitive market, but not the dominance of a mercantile ethos. Like Czechoslovak playwright-president, Vaclav Havel, Porter has come to see socialist and other utopias as dangerous, and bound for disillusionment. Havel has observed a direct link between beautiful utopias and the concentration camp: 'What is a concentration camp, after all, but an attempt by Utopians to dispose of those elements which don't fit their Utopia?'.[8] Like Havel, when he was an adversarial author (before he took over the presidency of his country), Porter might be criticized for not grasping all the moral nettles that he plants, but his principal concern has been to shake readers into an awareness of their unresolved problems. The allure of cities and declining empires for Porter seems related to a sense that the best that can be rescued from the chaotic present is an enlightened humanism which incorporates a tolerance of cosmopolitan diversity.

Like his acceptance of the business world and the realities of trade, Porter's verse demonstrates an intellectual accommodation of the changing balance of global power in the post-second world war years. An ambivalence about the British Empire and its fate is evident in poems such as 'Seaside Resort' (*PTTC*, *CP* 163−5). Porter's intellectual and imaginative engagement with Europe (especially Italy and Germany) was in advance of many in Britain. Lacking the credentials, or the beliefs, to be a bardic nationalist (like Les A. Murray in Australia, for instance), Porter has nevertheless emerged as a 'seer and spokesman' for a more complex international and metropolitan intelligentsia, for whom migration across borders is a way of life.

In looking for shapes and patterns in the writings and life to date of Peter Porter (he turned sixty in 1989), I have been interested in what James Olney called 'metaphors of self'[9] and their relation to significant events in Porter's life experience. Through a variety of dramatic personae, including an increasingly recognizable first-person speaker, Porter suggests for contemplation and pleasure an extensive allegory of life, which he has surveyed in one of his later poems, 'River Run' (*PW* 7−8). Deriving in part from his perceptions of the Shoalhaven River in southern New South Wales, which he first saw when he visited his artistic collaborator, Arthur Boyd, at Riversdale in 1975, Porter's allegory liberally employs local details of the Australian setting. In writing an allegory, Porter went beyond experiments in his third collaborative volume with Boyd, *Narcissus* (1984), though 'River Run' had in common with that book the desire of its author to 'pin down pictures of the self/I know are true of everywhere'.[10] The point made strongly in 'River Run' is that, just as we humans have no memory of where we come from, neither can we have knowledge of where we are going: our vantage points on experience are few and problematical, and we are inevitably part of the flow towards an unknowable end. The poem obviously alludes to the notion of the Heraclitean flux.

Like T. S. Eliot's Mississippi in *The Dry Salvages*, Porter's allegorical Shoalhaven in 'River Run' carries all kinds of impurities. Literal and symbolic details combine to suggest an actual river which begins among twigs and gullies, with a bird (of annunciation?) hovering, and ends among the 'fish heads and the floating debris/of the wharfs', suggesting on one level the mouth of the Shoalhaven near Nowra, and on another the inescapable physicality of death. Always mindful of ends, Porter offers no answers, only questions: 'The river has no start,/ How could it bring us to a proper end?'

The stages of Porter's allegorical river offer a collage of impressions from childhood to late middle-age and, more speculatively, beyond this to 'the ghostly dock of dreams', where we each take nightly boats 'upon a different thread of delta', listening in the dark, dreaming, and hoping 'never to reach the pounding Heads'. Porter's river of life begins, not with the historical chronology of the first day of a child's life, but with its first expression of emotion, signified by crying. Like magic, these tears gather and roll forward to 'A pool, a pearl, another pearl, a pool'. From the 'playground kills' and 'wet despair' of childhood, Porter moves to adolescence ('A time of waterfalls, of leaps round rocks') and reminders of nature red in tooth and claw, in the image of 'blood smears of bottlebrush and banksia'. The flatter terrain of early middle-age does not omit the cruelties either, or the politics of bedroom and boardroom betrayals:

> Hard-working days of green ambition,
> the river and the self are broadening
> among short-lived crimes, a gaudy flap of parrots.

Changing gear, out of Sturm und Drang days, 'As suddenly as afternoon the surge becomes/a modified achievement'. Some of the more voluptuous pleasures of slowing down are reflected in the river's mature phase, in a slower modulation towards late middle-age:

> Wide as a yawn, the slow-coach river now
> bastes in itself and boils the leaves
> upon its surface: it is going home.

In spite of its inevitability, the last phase, moving towards death, still holds terror and uncertainty, and a turmoil churns below the surface. Porter's allegory here, as in previous stanzas, is a compound of personal experience and observation, reflecting a self which, in spite of a sense of exile, retains significant links with its surrounding world.

Like Arthur Boyd's painting *Picture on the Wall, Shoalhaven 1979–80*, which is reproduced on the cover of this book, Porter's river holds ambiguous possibilities. To one side, a direct view, through the window, records a calmness and symmetry in nature; but beside this a picture within the picture introduces desire and fear, and a wholly different perspective becomes apparent. Ursula Hoff has glossed the difference between Boyd's two images of the river by pointing out that the framed picture shows 'an imaginary view stylized in memory of Baldovinetti's *Madonna and Child* in the Louvre, thus connecting the past with the present and associating Boyd's Australian home with his summer holidays in Tuscany'.[11] Beside the peaceful translucent river of direct vision, Boyd's 'created' artefact has darker, more pessimistic overtones, and an atomic mushroom cloud rises in the distance. Such dark visions are also seen to cloud Peter Porter's artistic progress along his river.

1

THE BRIGHT LOCKED WORLD

♦

How, and where, does a sense of exile begin? For Peter Porter, who was to become Australia's leading author-in-exile in London, it began in earnest in a suburb of Brisbane in 1938, when he was a nine year old boy — his mother died, without warning, of a burst gall bladder. Later, in his early fifties, the development of a philosophic cast of mind enabled Porter to write that 'Pain is the one immortal gift'.[1] But as a child no such props were available to resist the raw onslaught of emotion. From the enclosed and protected environment of a home with a quiet and somewhat timorous father and an ebullient, life-loving mother, the only child found himself suddenly thrust into what seemed a loveless world. On the verandah of his Brisbane home he had a premonition of death. Masochistic fantasies and apprehensions of mortality haunted him. Subsequently, a paradise would be constructed from fragments of his early childhood. For the young child, however, time was arrested at the moment of his mother's death, and the world seemed a bleak and bitter place for many years afterwards.[2]

In the construction of a mythology of self in his published poems, reviews, interviews and articles, Porter's mother's death is of paramount significance. It is the central image in a mosaic of loss. From the age of nine he recalls that he felt 'locked out of Paradise'.[3] This sense of expulsion provides the emotional impetus for a mythology which mirrors in some respects the Christian narrative of a fall from grace, with its many literary analogues. In most of Porter's work, hell is more vividly realized than Eden or Paradise, though it should also be said that the temporary imaginings, the fleeting images of a state of grace, are consequently more expressive and poignant.

The historical Marion Main married William Porter in 1922, when she was thirty-two and he was thirty-seven. She had worked as a nurse before her marriage. After some five miscarriages she gave birth to her only surviving child, Peter Neville Frederick Porter, on 16 February 1929. She is unlikely to have imagined that this son would become one of the major English-speaking poets of the twentieth century, or that he would owe so much to her. The reinvention of the mother-figure in memory and imagination assumes special iconic force for the young writer in his twenties, and at different phases later in his writing career. In finely-etched images rendered in prose and in verse, Porter describes this primal source of emotional sustenance:

She was the oldest of eight children, and the most loved, as well as the most ebullient in personality. All her life she masked a restless melancholy under a party-going extravagance. Her vision of existence was a florid excess of light and laughter followed by darkness and dreams. She was openly, even determinedly, vulgar — but this was a reaction to the increasing gentility of Australian suburban life. She was no intellectual and did not care for reading, but she loved company and holidays. She was the firstborn's mixture of rapture and grace, but she was overweight and radiated doubt.[4]

One of the few photographs of Peter Porter smiling shows him as an eight year old holding the hand of a large, soft-faced woman who is also smiling.[5] Their smiles are not assertive, but confident, sure. The 'restless melancholy' masked beneath a 'party-going extravagance' is perhaps the most obvious inheritance of the son from his mother. The poise and play of the phrase 'light and laughter' against 'darkness and dreams' seems also self-interested. But it is where these images of the lost mother are most at odds with the public 'man of letters' persona that they become most revealing. The mother is projected as anti-intellectual and 'determinedly vulgar'. While Porter's poetry is often loaded with cultural referents, it also demonstrates a relish in attacking social and intellectual pretensions with rude energy. These are the son's equivalents of the mother's behaviour, who could refer to chamber music as 'people pissing into pots'.[6] So are the often rapid transitions between formal and colloquial speech which British critics have sometimes put down to Porter's 'Australian' origins.

Marion Main also transmitted to her son an incipient sense of exile. Although born in Adelaide in 1890, she moved with her parents, brothers and sisters to Sydney, where she was brought up, first in Randwick and then at Woolwich. By the time she moved to Brisbane and married, Marion Main was a convinced Sydneysider. Brisbane seemed a form of exile to her, and her son subsequently accepted that Sydney was 'the real Australia — the authentic Babylon'.[7] The laughter, parties and drinking at his Sydney grandparents' home soothed him to sleep during his holidays far more easily than the bareness and silence of the Brisbane house in Annerley where he spent most of his childhood years.

Although Porter has not to date written a book on the houses of his childhood, as fellow Brisbane writer David Malouf did in *12 Edmonstone Street* (1985), these have figured significantly in his autobiographical commentaries, interviews and poems. The wooden house on stumps at 51 Junction Terrace, Annerley, in East Brisbane (called 'The Nook') is a place without magic, the lost place in which the boy is remembered, through the mythologizing filter of later years, as living chiefly 'Under the House':

... where lawnmower, wash tubs, copper, rain tank and the thousands of pieces of *disjecta membra* of suburban living collected. I always associated under the house with illicit activities: in early childhood, lighting fires and sharpening pencils with my father's cut-throat razors. Later on, when I came back to live in the house after nine years at a boarding school, I used the darker recesses for those desperate, solitary sexual manoeuvres there seemed to be no way of avoiding.[8]

The boy's bedroom in this house is inhabited by shadows and night terrors. Nearby is a quarry, to which the boy sometimes retreats, and the Chinese market gardens. Reflecting on his childhood terrors, the older Porter cannot account for them in terms of any lack of parental love, but sees them instead as 'spirits of the house':

When I imagine the psychopomp, arriving to collect my soul for the underworld, I see myself back in a room of this child-hood house, when at last the door really will open and usher in the remorseless assessor.[9]

The sense of judgement expressed here, and the terror of inchoate death, find different correlatives at different stages in Porter's writing career, but are constantly in his personality.

Following Porter's recurrent visits to Australia after 1974, it is Sydney rather than Brisbane which provides him with the central images of his childhood;[10] these derive from Christmas and other holidays with his mother at his maternal grandfather's house at Woolwich. Such early plantings of imagery assume primal significance and are closely linked with the mother. Porter has evoked the house and seascape in prose as well as in verse:

My grandfather's land went down to the water, and there he built a single storey brick house, complete with a deep cellar (something a boy from burning Brisbane and its wooden houses on stilts found eerily fascinating) and spacious, pleasantly laid-out garden. The path down to the harbour was terraced and cut deeply into the soft blue rock. Ivy and other creepers and perennials grew at each turn as you made your way from the lawn above to the swimming pool with its coarse buffalo grass and adjacent boatshed. All you had to do in those days before pollution if you wanted a swimming pool was run wooden stakes into the harbour and drape narrow-mesh net around them. At high tide you had a feasible body of water to swim in. You could manage a fair number of strokes before you bumped up against the shark-defying net. At low tide the baths were a beautiful if ragged rock pool made up of muddy flats and luminously green seaweed with bladder wrack and stranded jellyfish. I loved to sit on the grass sward beside the boat ramp on a summer's morning as the Lady Chelmsford, the diesel ferry with the larger funnel, went by, heading up the river. Her sister ships, also named after former Governors' Ladies had shorter stacks. There was a haunted house at Onions Point, which I would never venture beyond. The river was often crowded with sailing and rowing boats, prawn fishermen at night with lanterns in their boats, and on special days the Showboat with music blaring from her saloons.[11]

Significantly, these memories have entered Porter's dreams and he has associated them with music. He has identified this music as Schubert's song 'Die Götter Griechenlands', in which a lost fair world is summoned up in Schiller's words, 'Schöne Welt, wo bist du?' as the music modulates to A major.[12] The landscape and waterscape of Porter's grandfather's house and the road along the Lane Cove river become entwined in a key stanza in the autobiographical poem 'A Christmas Recalled', written in London in 1959 and published in *Once Bitten, Twice Bitten* (1961):

Summer was December and the water sounds
Of the presiding Harbour. Ferries named
For Governors' Ladies wallowed round
The river bends, past one haunted house, by
A thousand boatsheds, past the water games
Of Spartan Girls Schools, regatta crowded bays,
Resting heavily at barnacled, bituminous quays.

The poem as a whole is less carnivalesque than this. It evokes images of the child, apart from the mother and her family at the house by the harbour, overhearing the adults talking and drinking, their noise 'a secure lullaby,' their drinking and talk anaesthetizing his fears. What intervenes is an uncle's mention of the forthcoming war ('We've got about two or three/Years before the Japs come'), and premonitions of death and destruction are aroused. The boy escapes the house and sits crying under the pepperina tree which, like the house, the boat-shed, the ferries and the harbour, subsequently assumes iconic signifi-cance among the voyages and returns of Porter's work.

Later poems in Porter's *oeuvre* take up memories of the Woolwich house and its associations, including 'Landscape with Orpheus' (*ES, CP* 309−10), 'Essay on Dreams' (*TAO* 38) and a stanza in 'The Story of U' (*TAO* 62−3), a poem which derives from Mallarmé's sonnet, 'Une dentelle s'abolit':

This is the house they made for you
With water-steps and angled palms,

A cellar where your tears came true
And terrors took you in their arms:
Down by the water a boatshed
Collected the dynastic dead
Who heard cicadas keeping on
Their etching of a single song.

The mood is higher pitched than in 'A Christmas Recalled', and mythic
associations have thickened the texture, so that the Lane Cove river
dissolves into an image of the Styx. The cellar in the house at Woolwich is
an ante-chamber to death, recalling the 'Under the House' area in
Brisbane, where the boy cries as he has beside the pepperina tree.
Instead of the mother's loving embrace, 'terrors took you in their arms'.
The 'single song' of the cicadas is one of grief and incipient loss. Here,
among images of the holiday house by the harbour, the boy-man
commences a career of exile in a fallen world, a pattern of dislocation
and apartness. Such a sense of self might have culminated in the
production of a Romantic, but Porter resisted neo-Romanticism in
favour of more ironic, and sceptical, modes of address, thus accepting
the mixed blessings of the person suspended outside all places of final
belonging.

One of Porter's contributions to the study of childhood is his
recognition of its denial of simple innocence to the mind looking back
from adulthood. 'Ghosts' (1962/3, CP 42−3), a series of directly autobiographical vignettes, is influenced in its ideas by Ibsen's play
Ghosts, which Porter saw in London in 1960, and in tone by Strindberg and Hardy. The poem evokes not the grand holiday home at
Woolwich but the interior of the weatherboard house in Brisbane.
While the poem contains no direct reference to Ibsen's play, apart from
the title, it implicitly polarizes Brisbane and Sydney as places of
provincial enclosure and metropolitan freedom respectively, as Ibsen
does with his Norway versus Paris opposition in Ghosts. In Porter's
poem the central figure is the autobiographical speaker's mother,
who comes from a more energetic, mercurial family in lively arriviste
Sydney, but has married into what the speaker recalls as the deadened,
colonial-Victorian city of Brisbane. The figure of the mother is rendered
with a dramatic naturalist's specificity:

A large woman in a kimono, her flesh
Already sweating in the poulticing heat
Of afternoon − just from her bath, she stands,
Propping her foot on a chair of faded pink,
Preparing to cut her corns. The sun
Simmers through the pimply glass − as if
Inside a light bulb, the room is lit with heat.

The watcher of this scene is a 'little feminine boy/In striped shirt, Tootal tie, thick woollen socks,/His garters down'. The only child's inheritance from this mother, his twenty years older self muses, in a rapid transition, is 'her party melancholy and a body/Thickening like hers'.

The mother's imputed viewpoint controls the tone of 'Ghosts'. By moving to Brisbane and marrying into the remnants of an 'Old Colonial Family', her perspectives are narrowed like Ibsen's returnees to the darkened interiors of Norwegian houses, after the cafés and restaurants of cities further south. The boats on Brisbane's Moreton Bay provide compensating images of release and pleasure, as do the ferries at Lane Cove, but the poet seems cut off, financially and temperamentally, from the pleasurable physical adventure which these craft might offer:

> Who goes for weekends down the Bay
> In thirty footers to St Helena, Peel and Jumpin' Pin?
> No yachts stand off the Old People's Home,
> Out past the crab-pot buoys and floating mangrove fruit.

What then is the cause of such disillusion and joylessness? Naturalism in drama of the Ibsen kind customarily seeks causes. Porter's poem ends with an attempt at explanation:

> I was born late in a late marriage. Psychiatrists
> Say it makes no difference — but now I think
> Of what was never said in a tropical house
> Of five miscarriages. If the words were said
> They'd start the deaths up that I left for dead.

Like Ibsen's play this ending implicitly accepts the power of heredity as an explanation of human behaviour, while also doubting such determinism. The deep disquiet in the poem can have no simple physical or psychological explanation, but an influence is established, a set of questions raised. The poet's task here, it seems, is to raise questions rather than answer them, to doubt psychiatrists and theorists of human behaviour. It is a customary refusal of this author to be browbeaten by authorities or to accept facile conclusions and generalities. At the same time, the slangy facetiousness in the speaker's last line opens the way for an interpretation of his present state of exile as an over-hasty retreat from the buried fears of his Brisbane home. The poem raises questions of psychoanalytical accountability and personal responsibility which are explored from a variety of angles in later work.

The figure of the mother in the years of childhood is presented as one of fullness and generosity; the years of adolescence that follow are by contrast an emotional wasteland. In 'Two Merits of Sunshine' (*PAAM, CP* 56−7) Porter creates counterpointed voices which explore the mother-fixation from different angles, somewhat in the manner of Harold Pinter. The first voice is that of the younger self, and recalls the child's perceptions at his Brisbane home in precise and vivid detail:

> When you were about seven you were playing
> Among the fallen fleshy palmnuts
> On the pink cement path running through buffalo grass.

The mother returns from shopping 'powdered in a talc of sweat' and smiles 'a brown and gold mouthful/Of smile, a donation of her stones/ Of sunburned fat'. The mother is large, generous, outgoing, though she is also one who doubts and worries. She gives her son a wind-up toy, but he breaks the spring by over-winding it. The voices of suburban order, friends and neighbours, seem to mock him: 'you had broken your mother's heart'. However the mother's image remains full and generous, 'under a spinnaker of laughter'; though in retrospect the speaker now knows 'she was bellying out to death'. The concluding lines given to the first voice offer an apparent finality:

> Having defined love you never needed it again,
> Or its fat victim in a flowered dress,
> Or a garden to tease perfection in.

If the first voice recalls principally the 'teasing perfection' of the garden of love which his parents, and in particular his mother provided, and which the son blames himself for bringing to an end, the second voice seems to further fracture any simple notion of identity. It represents the forces of repression. Although the speaker moves to distance himself with irony or disdain from the loss of the mother, he remains totally implicated:

> I now read fashion magazines and
> Deplore Mother's Day. There are no
> Lovable fat women in Heaven, nothing crooked
> Is made straight and no rough places plain.

The inversion of the biblical psalmist's prophecy in the last lines,[13] following the pathos of the speaker's transparent attempt to distance himself from the pain of his loss, is both ironic and poignant. The outlines of Porter's personal allegory are already evident: it will be one in which distant and barely attainable hints of perfection occasionally appear — in gardens, seascapes, music, paintings or people — setting themselves against the ordinary course of life, which is characterized by crooked paths, rough places, and the persistence of pain. But elements of Eden will continue to taunt his human consciousness.

While the mother is the main figure in the personal allegory of

Porter's writings, the father has an important complementary role. Their place together in the son's early childhood is recalled in images of the garden and house in Annerley. Here, if anywhere in Porter's writings, is a coherent vision of an Australian Eden:

> Our family seemed to communicate best in the garden. I would follow my father around as he weeded, mowed the lawn, planted out his seedlings and wove webs of cotton above the beds to daunt the sparrows. They were not put off, of course. I would watch while he made fires and clipped the hedges. My mother would make forays into the garden, to bring tea to my father, to help him smoke out wasps and to talk over the side fence to the neighbours. She would also stand on the back landing and shout the names of horses she wanted to back to the woman next door, who would then ring them through to our SP bookmaker. We had no phone. It is not much to remember of a close-knit family life, yet it suffices. In Eden, Adam and Eve, after naming the animals, may have had little more to say to each other. Loquacity, even chat came with the fall. We required few furnishings and few points of reference beyond nature to keep ourselves well in our islanded stockade.[14]

These images of harmony in casual Australian backyard activity are remarkable in the writings of a man for whom dislocation and creative discontent are the more usual states of mind, though the passage also contains hints of a dangerous isolation. In the mid to late 1970s, however, after a period of twenty years absence from Australia between 1954 and 1974, Porter's writings became imbued with a new spirit of reconciliation, and also a recognition that his early view of Australia had been skewed towards a projection onto this country of 'the monsters of my daily life'.[15] While hints of the early perfection are contained in a number of poems set in gardens, the archetypal return occurs in 'An Australian Garden' (*LIACC*, *CP* 208–10), in which the speaker and his lover enact a union among the hybrid blooms of a Sydney garden.

The reconciliation between Porter and Australia is most evident also, however, in the changed imagery of his father. The most genial of this imagery links the father with the mother and garden. In 1977 the father is projected, semi-comically, as an artist *manqué*, perhaps a poet himself, part-British, part-Australian:

> My father loved [the garden] with the concern of a Capability Brown correcting God's poor sense of design. His way of gardening would hardly recommend itself to the 'natives only' gardens of modern Australia. He saw our suburban terrain as

wild land requiring to be tamed. So he kept only three or four trees ... There were garden beds of all shapes — rectangles, circles, parabolas, parallelograms and chains of mere holes, in which my father sowed a rotation of flower crops. He favoured iceland poppies, cinnerarias, stocks, asters and dahlias — all imported flowers bought from nurserymen, nothing native to Australia ... My father's greatest love was reserved for roses — he was always arriving home with new cuttings wrapped in hessian and slightly dampened sawdust. When he reaches Heaven, he will undoubtedly approach the throne down an avenue of rose bushes.[16]

This is a paean to the father who, after the death of his wife, when himself cast down and out of work, cast forth his young son (on the advice, the son avers, of rich friends and relations[17]) onto the waters of the Anglican private schools of Queensland, where he was miserably unhappy. School was 'hell'; or, in later formulations, 'Auschwitz'.[18] The father who drove him out, and for whom a repressed anger must have been felt, nevertheless is understood in later years as:

... a man not fully at home in the fallen world, [though he was] most at ease in his garden. From it he drove, angel-like, most of the native plants and shrubs ...[19]

From this garden too, the only son was expelled to a series of exiles at various schools, followed by short-lived 'apprenticeships' as a reporter and then warehouseman in Brisbane, before departing by ship, in 1951, for London and the life of an Australian expatriate.

The relationship of son to father is touched on in the poem 'Syrup of Figs Will Cast Out Fear' (*OBTB, CP* 8). Porter recalls his parents as hypochondriacs with a special concern for their bowels.[20] 'Syrup of Figs' was one of the crude remedies in the medicine cabinet and its comic inadequacy to the child's problem of fear is evident enough. The poem, written in London in 1960, ends with the stanza:

I know from my Father I was a sullen child
Scolding my indulgent parents' love.
Today in bed, the hegemony of smiles
Raises the prison cot bars — I might prove
Lucky in the future dark, but not reconciled.

This poem might be compared with Bruce Dawe's 'Condolences of the Season', which was written in 1964.[21] Whereas Porter's poem evokes an unresponsive child and tackles somewhat bluntly the idea that the child is father to the man, Dawe's poem is bathed in the tribal warmth of the extended family and of a special alliance between father and son. Porter's fate is to break with family; Dawe's to remain within its fold, mildly rebuking, though enjoying, its ritual togetherness. The isolation of the exiled figure and his need for reconciliation are not pressing issues for Dawe; for Porter they mean everything.

Memories and images of the father are an important element in the efforts toward self-definition which recur throughout Porter's writing. In 1982, not long after William Porter died at a Brisbane nursing home, three months short of his ninety-seventh birthday, his son commented that he had inherited from his father the sense of being 'ill-at-ease in my homeland'.[22]

> My father was a decent, timid man with no great expectations of life ... He was a gentile in the rag trade — he spent the whole of his working life selling the products of Manchester — sheets, pillowcases, towels and the like — to Australians. There's another thing about him. He was the only one of all his acquaintances who didn't volunteer for the first World War; and therefore, given the logic of the times, he was almost the only one who survived. My mother's brothers and my father's only brother were all killed in the war.[23]

In this assessment, the father is far from being strong and purposeful, although there are hints of a resistance to war which Porter was himself to demonstrate later in his Campaign Against Nuclear Disarmament marches in London. Strangely, Porter expresses no anger at his father's decision to send him off to boarding school:

> He did his best. In fact, he made great sacrifices. But what he didn't do was to keep the family home and me living in it, which was what he should have done ... I got a series of psychosomatic illnesses (mastoids, pneumonia, every childish epidemic there was). Then, against the advice of his smart friends, he sent me to Toowoomba Prep and Grammar Schools — where I got dramatically better.[24]

Tensions are more evident in images of father and son in 'Ghosts' (*1962/3*, *CP* 42–3). Presented here are images of Toowoomba Grammar School on Sunday, parents' visiting day. The son is taken by his father for 'milk shakes and a demure high tea' at the Canberra Temperance Hotel. He sulks, and adopts a priggish apartness. But the older self recognizes, in retrospect, that the father's fear of the small child matches his own. This awareness of the son's resumption of his father's fears is perhaps a necessary phase in the acceptance of personal responsibility in the adult person. It marks a stage in Porter's personal mythology which cannot be signified accurately as made of a series of Freudian or Piagetian stages. Rather, the allegory of Porter's life shows a continual sifting and reordering of his dreams, memories and observations, by a sceptical, restless intelligence for which living is a continual critical inquiry, the only final resolution being death.

Although less immediate than relations with his mother and father, Porter's understanding of his wider family is also utilized in literary constructions of himself. Beyond this, his notion of family is closely linked to perceptions of the country into which he was born. In a period when family history is apparently becoming one of the most popular prose forms in Australia, one might point out that Porter's principal concern is not historical accuracy. Nevertheless, he withdrew his most detailed poem on family history, 'Five Generations', from publication in his *Collected Poems* partly because of inaccurate information it contains. Its 'cavalier indifference to fact', which Porter says in the preface to the volume persuaded him to omit the poem may, however, be of less concern to readers who are principally interested in the *impulses* which contribute to personal constructions of meaning, rather than in their factual notation.

One of the temptations for a family chronicler in the 1960s was to emulate the American poet, Robert Lowell, whose *Life Studies* appeared in 1959. Dorothy Hewett did this in swashbuckling style with her 'Legend of the Green Country', in the book *Windmill Country* (1965). Porter attempts it in 'Five Generations'. There is hyperbolic verve in his verse portrait of his great-grandfather, Robert Porter, and of his father, William:

> He built Brisbane's gaol and deep-water pier
> and was hated by a little boy, my father.
> When Robert cried the flood of '93 began;
> the little boy whistling at his birds
> lived with a mad god . . .

The contrast between 'big' grandparents and a father brought down to size is Lowellian; so too is the melodramatic rendering of extreme emotion, and the hints at madness. But Lowell's father, a naval officer, and his land-owning grandfather, were ruling class Bostonians. Even in a poem such as 'Five Generations', which gestures at extremes, Porter's impulse is to place himself among 'ordinary' people. Hence the central assertion of that poem:

> Our family's modern,
> moving out of historic darkness
> in a diaspora of the undistinguished.

That resonant last phrase is characteristic of Porter, in the stylish way it disports the costume of ordinariness. While asserting the 'undistinguished' nature of his family, Porter's use of the word 'diaspora' hints at a grandeur and dignity hidden amongst lost and ordinary experience. The desire for an 'ultra' element in experience finds other modes of representation in later work, without the Lowellian influence.

References to family history provide a frame within which auto-biographical investigations of one's past can proceed. In the poem 'Forefathers' View of Failure' (*OBTB*, *CP* 3) Porter paints one area of his self-portrait, that deriving from the Scottish ancestors on his mother's side, quite different 'builders' from those on the Porter side of the family. The 'undistinguished' forefathers arrived in Australia, and the heritage they offered was what historian Manning Clark summed up as British philistinism with strong lashes of puritanism:[25]

> Men with religion as their best technique,
> Who built bush churches six days a week,
> Stencilled failure's index on their brains.
> Whisky laced the mucous of their heads,
> Flushed their pores, narrow-bored their veins,
> But they were building still on their death-beds
> Having no life but the marking-time of work,
> Sleeping collapsed outside despair and talk.

Porter has commented on Mark Main, his maternal grandfather, who built the stone house at Woolwich near Sydney Harbour, which Porter visited as a child.[26] The main models here, though, are the Porter males, with their tradition of building. Even so, the poem is far from being straightforward family history; it exemplifies the poet's straining in the 1950s and 1960s for generalizable individuals, situations and incidents. In this respect, 'Forefathers' View of Failure' is similar to the Theophrastian portraits in 'A Farewell to Theophrastus' (*APF*, *CP* 116–18), in which the 'characters' of the Greek philosopher and pupil of Aristotle are counterpointed with characterizations of various types of human failing evident in London's 'swinging sixties'.

Like one of the greatest of Australian expatriate authors, Henry Handel Richardson, Porter is much more engaged, artistically and emotionally, in the experience of failure than in popular myths of success and achievement.[27] 'Forefathers' View of Failure' projects a view of colonial Australia similar to those parts of *The Fortunes of Richard Mahony* which question the meanings of success and failure. For both Porter and Richardson, the blind compulsion to build, among individuals who have 'no life but the marking-time of work', is a kind of failure seldom recognized as such in colonial societies, where building may become an end in itself. Drunkenness in such societies may be a way of blotting out a deep sense of insecurity and failure. The life of the mind or spirit, exemplified in protests, or ideals of the millennium, is expunged from the experience of these colonists. The couplet at the end of Porter's second stanza sums it up:

These narrow fates had a viciousness
They drank for, but no vicariousness.

The jarring rhyme of 'viciousness' and 'vicariousness' here draws
attention to the distinction between 'narrow fates' and those that could
be broadened by imagination; but it also links them by association.
'Viciousness' connotes a depravity of imagination rather than of fact:
these colonial Australian men are far from vice-ridden in the terms of
their society. However they are deprived of those influences which
allow a person to understand or sympathize with the experience of
others. Hence 'vicarious' is a very positive word in Porter's canon; and
it connotes in this poem a capacity to imagine beyond the confines of a
utilitarian, masculinist way of living based on physical achievements
to the 'sophistication' of new ideas, the capacity to doubt conventions,
and to protest against them.

The implied outlook of the poet in 'Forefathers' View of Failure'
is, in a sense, protestant. Yet protestant colonial Australia is also a
target of the poem's critique in these tightly corseted verses:

The weatherboard churches bleached white
As the calcimined crosses round them invite,
Like the War Memorial with ten names,
Eyes up to plain Heaven.

Such archetypes of Australian country towns have become the
territory of a number of Australian poets in the 1970s and 1980s —
most notably, perhaps, Les A. Murray and Geoff Page. Porter's contribution in 'Forefathers' View of Failure' is to set a certain con-
straining world view arising from a selective reading of his family tree
against his own, evolving set of metropolitan values in the early 1960s:

In this new land the transplanted grasses root,
Waving as sulkily as through old falling soot.

Australia is not in this view a new land of hope and opportunity
so much as a repository of British industry and vice. At this stage,
Porter's historicist critique remains somewhat fuzzy, but poems such
as 'Sydney Cove, 1788' (*PAAM, CP* 50−1) and the later 'Spiderwise'
(*TAO* 56−9) give it more bite. Suffice to say here that Porter uses
family legend, not to create liberationist expectations about migration
from one country to another, but to reinforce a view of the fallen
nature of human beings, for whom the anatomy of their failures may be
the most illuminating of explorations.

2

EACH UNSHINING HOUR

◆

The Education of Young Peter is a significant phase in the personal allegory which runs randomly through Porter's poems and commentaries. As with his use of family history, autobiographical detail is used selectively, to highlight personal concerns and to comment on social situations and ideas.

Porter's references to education show that he is not a thinker in the Wordsworthian tradition, didactically concerned to present his education as an exemplary 'growth of the poet's mind'. Although his poems, articles, reviews and interviews are often critical of the institutions of education, they do not imply a belief in 'natural' learning such as we find in Rousseau or Tolstoy. On the contrary, Porter retains a belief in a relatively formal, classically oriented education in the humanities. Nevertheless, he is not inflexible in the face of changing conditions. Although due in part to financial constraints, his two daughters, Katherine and Jane, attended state schools in London, culminating in the co-educational Holland Park Comprehensive School, of which Porter has spoken appreciatively, in terms of the preparation it provided for living in a multicultural society, if not for the rigour of its arts and humanities curriculum.

Porter's own schooling commenced in 1934 at the Junction Park State School off Ipswich Road, Annerley, in Brisbane. The Porters' house in Junction Terrace was opposite the school; the protected little middle-class boy had only to cross the road to go to his classes. The suburb of Annerley was not, however, uniformly middle-class and one of the sharp memories Porter later recalled was of the penury and general squalor of the home of a fellow pupil whose father drove a steam roller for the Council.[1] Young Peter also attended Anglican Sunday School in Annerley in 1933 and 1934, although his parents were only nominal Christians; the older Porter has called his father a 'Baptist atheist' and his mother a 'Presbyterian atheist'.[2] The Bible, the Book of Common Prayer and Hymns Ancient and Modern were to become important reference points in his writing.

After his mother's death, the nine year old Porter stayed at his grandfather's house in Norman Crescent, Norman Park. He attended the local state school, and was badly bullied.[3] After attempting to run away from school, he was sent as a boarder to the Church of England Grammar School in Brisbane. Porter's later views on his first years of boarding school as a ten and eleven year old at 'Churchie' are expressed with characteristic verve, mordant wit and a degree of hyperbole:

> After my mother died there was no longer a house. What we
> did is that my father rented the house and sent me off to
> boarding school, and that was the great change in my life. Not
> only did I lose my mother, but I was also taken out of this
> admittedly over-protected atmosphere — because I was an
> only child — and exposed to the full horror of the Church of
> England Grammar School in Brisbane, which was an appalling
> institution modelled, miles after the event, on English public
> school practice of the most barbarous and outdated sort. It
> was quite literally not very far removed from Auschwitz . . .
> In fact, I'm sure some people did not survive it — they are
> probably buried in the grounds somewhere.[4]

While boarding at the Church of England Grammar School Porter
was a regular visitor to the Matron and sick bay — 'I am a castaway
in a world/Ruled by Matron — she is the right hand/That makes
my world left-handed' ('The View from Misfortune's Back', *PAAM*, *CP*
57–9). The headmaster, Canon Morris, was a 'snobbish sadist' according to Porter. When the 'new boy' contracted double pneumonia in
1940 and the school sick bay was not equipped to look after him
properly, Canon Morris showed his colours: he would allow William
Porter to take his son to hospital in Brisbane only if he were conveyed
in a taxi, since the visit of an ambulance would provide bad publicity
for the school.[5] Such were the ways the school ruled.

From the personal hell of 'Churchie', Porter was transferred in
1941 to the less prestigious but more humane environment of Toowoomba Church of England Boys' Preparatory School, 135 kilometres
west of Brisbane in the foothills of the Great Dividing Range. Porter's
schooling was again disrupted in 1942 when, in fear of a Japanese
invasion, the girls at St. Hilda's, Southport, were evacuated inland to
Toowoomba, while the boys were sent to the girls' school on the coast.
Although Porter came to feel more relaxed at Toowoomba Grammar
School, as he moved through the school to complete his secondary
education in 1946, the sense of being an outsider remained. One
former student recalls that Porter sometimes adopted the role of 'court
jester', making sarcastic comments at others who aspired to be the
intelligentsia in the school.[6] His personal insecurity never led to bullying however, perhaps because he had endured this himself. At
Toowoomba, as at other boarding schools, bullying appears to have
been a favourite pastime, no doubt exacerbated by the additional value
given to the virtues of masculine toughness at a time when boys were
still expected to go into the armed forces when they turned eighteen.
Porter missed out on the war by a couple of years, but observed the
war games at his school, and images that connected with his own

personal and sexual uncertainty remained in his mind. Indeed, the shadow of war hangs over much of his work. A specific memory which stimulated imagined fears is of the venereal disease clinic in Brisbane:

> As we had all shuddered at the letters VD at school (so much more terrifying in this starkness than if spelled out), I was greatly taken by a poster in the window showing three Japanese aircraft being brought down in flames by long-barrelled anti-aircraft guns. Along their wings were printed the words Syphilis, Gonorrhoea and (rather mysteriously) Chancre. The guns were labelled Prophylaxis, Precaution and Early Treatment.[7]

But Porter was never in this firing line, either, being too timid to try out Brisbane's Albert Street brothels, or those across the river by the Cremorne Theatre, and a tortured sexuality persisted into his twenties.

The figurehead and chief transmitter of Australian public school values was the headmaster who, following the British model (which some Australian teachers viewed even less critically than their English counterparts) inculcated virtues of patriotism, cleanliness and godly living. Porter has been highly critical of his two headmasters, Canon Morris at Brisbane's Church of England Grammar School and Henry Emmanuel Roberts at Toowoomba Grammar School. Looking back from the late 1980s, Porter described Canon Morris as a 'truly vile example of the Anglican hearty, for whom the playing-field was the only true *campus martius*'.[8] Henry Roberts, the original of the figure in Porter's poem 'Mr Roberts' (*OBTB, CP* 5) was, by comparison, more 'original and decent', though his masters through this war time period were 'a weird enough bunch — drug-takers, alcoholics, nymphomaniacs and straight-forwardly unqualified dullards'. According to another of Porter's fellow students, Peter Edwards, the original Mr Roberts was unaffectionately known by the boys as 'Poo-Bah' or 'Poob' for short, and he inspired:

> ... fear and loathing — almost Gothic horror — in everybody except a few sycophantic prefects-in-the-making. Rightly or wrongly, we were all convinced that he wielded the cane with sadistic relish. As tyrants always are, he was universally credited with eyes in the back of his head, and he seemed to haunt dark corners and hunt down minor malefactors with Javert-like pertinacity and blood-lust.[9]

Porter's impulse in the poem 'Mr Roberts' is less *ad hominem* than this, though it shares something of the myth-making capacities of the schoolboy who feels himself tyrannized. A retrospective prose portrait by Porter complements that of Edwards:

Despite this emphasis on sports (all were compulsory except gymnastics — thank God for that much remission!) Roberts was good at none himself, except athletics, though he had grotesquely small legs for his enormous trunk and torso. I believe now that it was ambition rather than conviction which made him so assiduous an advocate of the public school creed. He genuinely loved Latin and would quote Horace to us to our bewilderment. And he had an interesting personal library which we could use as seniors. There I first encountered the Decameron and Graves's Claudius novels. I'd have learned more Latin if I hadn't hated him so much. He was a terrifying man in aspect and treated his staff with a ferocity as great as that he used on the students ... My misfortune was that he took an immediate and radical dislike to me, and he was too childish a man to see anything disproportionate in the opposition of an all-powerful headmaster and a naively rebellious but totally ineffectual schoolboy.[10]

Porter's 'hatred' of this figure is modified in the satiric elements of his verse portrait to include observations which touch upon broader ideological concerns:

This pedagogue pushed: he owned them for four years.
A Rugby field was the Republic's mould

'Rugby' here suggests both the sport and the ethos of British Public Schools, as symbolized in Thomas Arnold's famous Rugby School in England. Appropriately, Matthew Arnold, the headmaster's son, was one of the authors studied in a Toowoomba literature programme top heavy with British Victorians.

It is interesting that Porter mistakenly assumed that both headmasters were English — they were actually native Australians. The error reveals an early antipathy towards certain English ruling class values which would increase rather than diminish when he lived in England, especially as he observed the residue of these values operating in Margaret Thatcher's Conservative Governments through the 1980s. But the error is revealing in another way; as Porter himself has noticed, it was a simplification, a stereotyping, of differences between Australia and England:

It is curious now to think, after years of living in England and after the many wrinkles of antinomous feeling which the sense of exile, if only symbolic, has wrought on me, that I believed both men to be Englishmen. This was because they embodied the stale Anglo-Saxon public school ethic, which is a parody of *mens sana in corpore sano*, and strove to convey an im-

pression of Oxford and Home Counties patriotic heartiness
which was probably old-fashioned already in England. But
then the values Australian public schools conveyed were
curiously brutalised, or perhaps I should say, made ridiculous
by being taken to extremes. This is what 'provincial' as a
pejorative term really means.[11]

This statement indicates a significant point in Porter's dichotomy
between 'provincial' and 'metropolitan' cultural values. It also indicates
the close connection which Porter makes between the institutions
which have influenced him personally and his broader thinking about
issues and problems. In much the same way as Porter's reading of his
family tree identifies the importation of philistine values from the
British Isles, so does his representation of Australian public schools
show the perverse ends towards which British models of education
may be put. As we will see, however, he typically (especially after the
1960s) presents the recipients of such values as being personally responsible for their behaviour: behaviour not simply determined by circumstances. In this at least, the protestant ethic of individual responsibility is maintained, even while its historical associations with capitalism and the work ethic are interrogated.

'Mr Roberts' was written in London in 1958−9, three years after
Porter had begun attending weekly meetings of 'the Group' under
Philip Hobsbaum's and Edward Lucie-Smith's guidance. The significance of these meetings on Porter's work is discussed in Chapter 4. It
is perhaps sufficient to note here that other 'Group' poets such as
Martin Bell, George MacBeth and Alan Brownjohn also drew upon
their schooldays experience, each in his individual way.[12] The notebook version of 'Mr Roberts', now held in the National Library of
Australia in Canberra, shows that it was not born fully formed, but was
always close to completeness, and in this reveals the author's fluency
in composition. The only earlier poem which deals with public school
life is a section in an unpublished, semi-autobiographical piece written
in the early 1950s, 'Anson Jones', in which the protagonist:

> . . . played the British games.
> He played up and played the great inane.
> On Eton's playing fields, he was informed
> Mind and beauty are greatly scorned.

With its loosely rhymed couplets and ragged metrical arrangement, the
poem is both less focussed and less technically adroit than the later
'Mr Roberts', which partakes of the critical realism of many of Porter's
best poems of the late 1950s and early 1960s.

Porter establishes Mr Roberts's grandiose image of himself in an
opening line of stately iambs — 'He was the great Consul and his

teacher's gown'. But the image is immediately thrown into comic
disarray in the second line, when the teacher 'Out-toga'd the forum of
his Latin Class'. Thus the tone is set for bathos and farce, though Porter
never allows the poem to slip into inconsequential humour. His end is
more ambitious than this, and what begins as a portrait of a pompous
headmaster ends by implying a more widespread condition of excessive pride and self-importance, and its consequences. A deft switch of
viewpoint in the third line of the first stanza reveals the headmaster's
devastating perception of his students:

> His eyes translated what they rested on:
> Boys of the pudding world, unstoical faces,
> Ears beyond the ablative, all thickened glass.
> Staring into them he might have drowned
> In such transparency ...

'Ears beyond the ablative' is one of those brilliantly arresting and
memorable phrases with which Porter's work is peppered, and which,
when examined more closely, are usually of pin-point relevance. Here,
for instance, the boys' abstraction is expressed by their lack of interest
in the one Latin case which expresses a direction from a place or time;
their lostness is, as it were, doubly verified. Yet the reader is aware
that this is only one side of the equation, that of the teacher-authority
figure. This bifocal view gives the poem its special piquancy. While it
has its amusing moments, 'Mr Roberts' stops far short of the facetiousness which fellow Australian expatriates Clive James or Barry
Humphries, or Englishman Gavin Ewart might have given it. Porter's
humour holds in store its serious, almost stately-serious purpose; a
purpose which respects the Roman Republican virtues, if not their
degenerate manifestations in the English or Australian provinces. He
steps back from all-out humour as from total satiric demolition because
he sees more than one side to a case, and is unwilling to relinquish a
hold on truth in favour of wholehearted exaggeration.

The ambition of this poem is apparent also in its early mingling of
past, present and future. The self which inhabits the verse is relatively
distanced; an 'objective' manner is established. The school is shown to
be isolated and lost in the past, a point which the poet makes dramatically in an image of the father (notated as 'your father') visiting the
school on Speech Day and stepping from an 'otiose car' (which seems
superfluous in this ancient setting) and feeling the shock of 'an ancient
fright, a gladiator's jar'. In such a setting, the past continually erupts
into the present. Instead of exploring the personal impact of the headmaster or school, or the neuroticism it may have induced in the writer
and others, Porter uses a technique common in his work of the late
1950s and early 1960s: he adduces brief 'case studies' — in this
instance, of fellow victims who suicided or went mad. In a later piece,

'Futurity', he takes up the story of the suicide.[13] While actual objects or places at Toowoomba Grammar School are part of the poem's furniture (bike shed, weather vane, bell tower) these are given a classical extension and symbolic resonance as part of Tartarus, the Roman and Greek hell over which this headmaster presides. Yet in spite of its evidently diabolical influence, the school is not wracked by revolution; and the concluding line wryly announces the grim comedy of this generalizable instance of institutional endurance: 'The Old School waits and presages each unshining hour'. 'Mr Roberts' stands as a monument to the all-too-successful suppression by tyrants and their institutions of 'that soft thing the self'.

'That soft thing' speaks out more directly in Porter's dramatic monologue, 'The View from Misfortune's Back' (*PAAM, CP* 57—9). Though taking form as a schoolboy's prayer, it would be a mistake here, as it would be elsewhere, to read the poem as a direct transcription of personal experience. Porter's pervasive habit of fictionalizing his actual experience is signalled in his changing the relatively posh boarding school he attended to an orphanage. In this way, painful experience is distanced and the imagination is given scope. Again, a sense of humour is evident, and a distancing irony which allows the reader to both comprehend the child's point of view and to perceive it in a broader context:

> May God who is good make the porridge
> Have more sugar today. Also as
> Privilege let me sit next to Kleinschmidt,
> And if I whisper don't let the patrol
> Monitor hear me. I put water
> On my hair, I washed my face, I owned
> Up when the chess set disappeared.

A breathless, almost tearfully pleading voice with its intonational shifts is signalled here by the chopped lines in a virtuoso use of free verse. As in the better known 'Your Attention Please', rhetorical control is achieved through a voice well-suited to the occasion; and in each case, much more is revealed than the 'speaker' can supposedly be aware of. Yet in neither poem is Porter mainly concerned with the consistency of this voice or its authenticity. At certain points in 'The View from Misfortune's Back', for instance, a voice-over by an older, more knowledgeable speaker intrudes, to give resonance to the boy's plight.

After confessing some of his sins, including joining in the bullying of another boy and not putting his hand down the lavatory bowl to save the picture of his mother, the boy comically, but poignantly, confuses his image of God with that of R. M. Ballantyne, author of the boys' adventure story *The Coral Island*: both claim to be authors of 'the

natural way to live', and the boys' adventure story takes precedence in
the speaker's imagination. Then a more authoritative voice intervenes,
re-examining the sick boy's hatred of Matron and his literature-inspired
sense of being a castaway. Similarly, later in the poem, an older
person's view is present in the lines:

> ...When all our sharp faces
> Go into the world, I want to love
> The softness of people, the sheer chance
> Which makes them lovable...

And again in:

> I have only this black orphanage
> And the map of the world with all its red
> To make me real...

The irony here of the British Empire being protector and guarantor
of the boy in his 'black orphanage' gains additional force when considered beside Porter's other poems which reflect on his schooling and
that of others. The poem ends gloomily, and with an assertion of
psychological determinism which Porter modifies in his later thinking:

> Misfortune leaves its bite mark and you
> Are never free of anything you did.

Although the bite marks of misfortune do reappear at intervals through
Porter's writings, and the limits of personal freedom are underlined,
Dorothy Green is justified in her linking of 'The View from Misfortune's
Back' with the later poem 'What I Have Written I Have Written' (*ES*,
CP 286–7), thereby demonstrating a development from the self which
looks for sources of salvation or blame externally to one which accepts
full responsibility for itself and its actions.[14] A change in Porter's
thinking is evident during the 1970s, from the perception of individuals
as victims of circumstance to the acceptance of full personal responsibility. This view is strongly represented in the volume *The Cost of
Seriousness* (1978).

A further aspect of education is explored in the poem 'Eat Early
Earthapples' (*1962/3*, *CP* 44). The title, from *Finnegans Wake*, seems to
urge an early escape from Eden into sinful pleasure and sensuality (in
a Joycean manner), but the poem itself exposes an ironic, questioning
side to this proposition. The adolescent, prep school persona is tortured
and troubled:

> ...There were
> So many ways of losing a troublesome innocence
> But so many ways of keeping it too.

Porter's deft exposure of the double-bind of his introverted and self-critical younger self is a prelude to later wrestles with his personae over the uses and abuses of sophistication, and the associated literary problem of plain versus baroque language. In this poem the speaker finds himself deficient beside the more outgoing, audacious and physically mature country boys:

> ...who made fifty pounds
> Over the holidays selling kangaroo hides
> They'd skinned and pegged out themselves
> On their fathers' stations. Many shaved, several
> Slept with the maids — one I remember
> Running his hand up the Irish maid's leg
> At breakfast not ten feet away
> From the Headmaster's enormous armature of head.

'That soft thing the self' is here differentiated from those who do manly, daring things in the face of authority, however the recalled envy is counterbalanced by knowledge that these boys are now 'thick men ... with kids and problems' and that:

> There is no way back into their wormy Eden,
> Ripe with girls, esplanaded with sex

either for the speaker or for them. Rather than presenting himself as pleased, superior and free beside these hypothesized school companions, the older self is principally concerned to face the 'baffled fears' which dominated his earlier years.

The chief unifying device in 'Eat Early Earthapples' is the train journey,[15] which Porter uses later in some 'travel' poems — most successfully, perhaps, in 'On the Train between Wellington and Shrewsbury' (*PTTC, CP* 171−3) and 'Good Vibes' (*LIACC, CP* 202). What distinguishes 'Eat Early Earthapples' is that it brings together emotionally charged schoolday images of fear, failure and exclusion in a poignant and witty reminder of the vulnerability of youth. But it is not a poem which exudes calmness after a storm. The recalled self of the adolescent oscillates between extremes which still seem alive in the older speaker: puritanical at heart, and timorous in action, he also knows with a force greater than that in his unreflective, pleasure-seeking cohorts, the forbidden attractions of the underside of life. The boy with his head in a book has arrived there before them. Porter is not kind to his younger self here, yet he allows him at the end of the poem a degree of unsentimental sympathy:

> The boy with something wrong reading a book
> While the smut-skeined train goes homeward
> Carrying the practised to the sensual city.

The perception of the boy as having 'something wrong' — the imprecise phrase harbours a diffuse menace — is thought to be the view of the other boys, while the 'smut-skeined train' seems an older self's encapsulating memory of the sense of sin engendered by these train journeys. The 'sensual city' will not be Brisbane for this speaker but London, where other intellectually inclined provincials also set out to lose their innocence.

If the image of the bookish boy is rendered ironically and mainly from the outside in 'Eat Early Earthapples', with hints of the bitterness of exclusion felt by the child, the poem 'Reading MND in Form 4B' (*1962/3, CP* 45) presents a lyrical celebration of the magic world that books may provide — in this case, the text of Shakespeare's *A Midsummer Night's Dream*. This poem was written for inclusion in a book of poems to honour Shakespeare, published by the Trustees and Guardians of Shakespeare's birthplace at Stratford-upon-Avon.[16] The poem as it first appeared did not contain its lyrical concluding passage, which was reclaimed for the version in the *Collected Poems*. Whether the suppression was his first editor's decision, or Porter's own, the passage is finely appropriate in a poem which begins with a situation of teacher control, even domination, in a classroom where the boys are required to learn passages by heart. The themes of bullying and group terror that are important elements in 'Mr Roberts' and 'Eat Early Earthapples' recur here, but the poem ends with this coda:

> Elsewhere there is war, here
> It is early in an old morning, there is pollen
> In the air, eucalyptus slipping past
> The chalk and dusters — new feelings
> In the oldest continent, a northern race
> Living in the south. It is late indeed:
> Jack shall have Jill, all shall be well,
> Long past long standing eternity,
> Eastern Standard Time.

Australia is presented as indisputably European ('a northern race/Living in the south'), whose chief guarantee of continuity lies in its access to major creative geniuses such as Shakespeare.

Although Porter's work contains many other echoes of Shakespeare, there is only one other piece to date which is directly *about* the writer considered by Porter to be 'the supreme example of mental good sense and instinctive human understanding'.[17] That other poem, 'Exit, Pursued by a Bear' (*LIACC, CP* 233—6), is a more restless, ironic rendering of abuse of Shakespeare by the admen and others who 'ransack' the 'genius of the universe'. The mood of romantic reconciliation which inhabits the concluding passage in the later version of 'Reading MND in Form 4B' is relatively rare in Porter's work, although not as rare as some commentators have suggested. When such lyrical notes do occur, they momentarily light up the surrounding darkness; or, to use a metaphor which Porter himself has used, they 'transport' the spirit to other realms. Some artists in Porter's wide canon, Shakespeare being the outstanding example, have the capacity to transport the spirit in this way, but others such as Gertrude Stein, in Porter's view, remain at the mercy of earth-bound 'transportation'.[18]

Whether Porter's marriage to Shirley Jannice Henry in 1961 and the birth of their first daughter Katherine in 1962 contributed to his receptiveness to Shakespearean romance at the time he composed 'Reading MND in Form 4B', or other factors were at work, the outcome was a poem which celebrates the reconciliations which poetry and the imagination may achieve. Michael Hulse has observed that Porter's Australian poems 'give us the most warmly sympathetic of Porter's responses'.[19] There is evidence to support this view, as well as evidence which lends support to the counter-assertion that at certain times during his 'exile' in London, especially between 1951 and 1974, Porter used Australia as the place on his mental map where things went fundamentally 'wrong'. Historical context is important here. 'Reading MND in form 4B' significantly locates its vision of harmony against the writer's indirect experience of war in 1944. (Porter's correction of the year from '1942' in the *Collected Poems* to '1944' in *A Porter Selected* is instructive, indicating not only his wish for accuracy to autobiographical detail but also an increasing sense of this author's accountability to his text as personal history.) Like a number of Judith Wright's poems of the 1940s, mention of the war here gives an antithetical force to images of harmony and love. But the 'new feelings/In the oldest continent' are not stimulated by a mood of fear or apprehension, as are those induced by the departure of loved friends or relations for the war zones in Wright's 'The Company of Lovers' or 'The Trains'. Rather, the theme of war is deftly introduced through the adolescent boys' perceptions of their female teachers' sexual escapades with American officers on leave in Brisbane. The effect is to maintain an authenticity to the adolescent point of view.

In 'Reading *MND* in Form 4B', Porter recognizes both the sexual potency implicit in Shakespeare's text and the ways in which adolescent lust may suffuse all perceptions:

> Queen Titania, unaware of Oberon,
> Is sleeping on a bank. Her fairy watch
> Sings over her a lullaby,
> The warm snakes hatch out in her dream.
> Miss Manning is too fat for love,
> We cannot imagine her like Miss Holden
> Booking for weekends at the seaside
> With officers on leave...

War in the school is simulated in the locker rooms or on the
sporting field which Porter's persona abhors:

Out on their asphalt the teams for Saturday
Wait, annunciations in purple ink,
Torments in locker rooms, nothing to hope for
But sleep, the reasonable view of magic.

'Annunciations in purple ink' is a witty collocation — lists of sporting
terms made strange by close observation and biblical allusion. What is
conveyed through the poem is a feeling of exclusion from ordinary
sporting pleasures, the fears and torments of school for a bookish boy,
and the intoxicating power of Shakespearean language which may
bring two hemispheres together in the music of language. Life may be
hell for the schoolboy, his fat teacher may insulate her charges from
Shakespeare's genius, but a vision of pure harmony may also be snatched
from these unlikely circumstances. Porter indicates too the insecure
adolescent's sudden shifts of mood, and something of his need for a
transcendence of the confining circumstances of school life. Following
the reference to Miss Manning's capacity to insulate her charges from
Shakespeare's genius, the last line of the second passage breaks free
with the exact words of Oberon's command of music and magic in Act
IV, scene i: 'Rock the ground whereon these sleepers be'. It is as if the
child's subconscious is summoning the power of language to his aid.
The adult, through the child's perceptions, reaffirms this power.

Like many of Porter's poems, 'Reading MND in Form 4B' contains
an argument. In this poem it is an argument about the power of certain
artists to overcome national differences and distance. Like Joseph
Furphy, the Australian nationalist author from whom he differs in
most other respects, Porter's imagination is suffused with Shakespeare,
from whom he can quote almost at will.[20] An education system which
can at least give scope to such major artists as Shakespeare is therefore
to be praised, in spite of its other faults.

Porter has freely admitted the value his career as a writer owes to
his school education in English literature. Since he did not attend
university, in spite of qualifying to do so, Porter's estimate of the
worth of his formal education at high school assumes significance:

Australia in those days [the 1940s] had a fairly rigorous approach to English literature on the principle that they didn't
try to lead you by some Gradus ad Parnassum system of easier
leading to harder works: they chucked you in at the deep end.
I'm very grateful because I think, even if one is baffled by
good writing one needs to know what good writing is before
one tries to appreciate literature, let alone to compose it. I
think the thing that staggered me most as a child was to hear

ordinary people around one and to suddenly realise that this
was the same language used by William Shakespeare. The
great thing about literature is that it is based on the ordinary
language which we live and speak. For some reason when I
was at school, apart from Shakespeare, the emphasis was
almost entirely on the Victorians: so I was extremely well
grounded in Tennyson, Arnold and Browning when still at
school, but hardly at all in what now seems to me the greatest
of all centuries for English poetry, the seventeenth century.[21]

While Porter celebrates here the relationship between the language
of ordinary speech and literary language, he does not relate this to
literature written by Australians. The reason for this is simple: until
1955 there was no fully fledged Australian literature course at an
Australian university, hence no teachers were educated to any significant
extent in the literature of Australia.[22] The 'cultural cringe' phenomenon,
a description coined by Arthur Phillips in 1950,[23] was perhaps most
marked in Australia's private schools, where ethos and curriculum
were based upon British 'public' school models. Ironically, the first
pupil enrolled at Toowoomba is said to have been A. G. Stephens, who
became editor of the Red Page of the *Bulletin* magazine and one of
Australia's most famous literary critics.[24] A later student was Eric
Partridge, the well-known lexicographer, who attended the school
before the first world war. During Porter's years there from 1943 to
1946, however, few strong movements of republican or nationalist
sentiment existed in Australia. Porter's fellow Queenslander, Rhodes
Scholar P. R. 'Inky' Stephensen had published his polemical call for
the recognition and assertion of an Australian identity in *The
Foundations of Culture in Australia* in 1936.[25] From 1944, H. M.
Green and Dorothy Green were preparing material for their *History of
Australian Literature* and Manning Clark was working from nationalist
perspectives on his *History of Australia*. Nonetheless, Porter's almost
exclusive education in British literature, without serious attention to
Australian, American or other literatures, was the typical emphasis of
English curricula around the country. Although highly critical of his
schooling in other respects, Porter has remained uncritical of its literary
emphasis, except for its stress on nineteenth rather than seventeenth
century writers, on the grounds of the immense range and depth of this
literature and the flair and genius of certain authors.

Ancient history, at which Porter excelled, led him to the Roman
and Greek classics. He was already a voracious and eccentric reader,
and took in *Robbery Under Arms* and *We of the Never Never* along
with such Australians as Ion Idriess, Frank Clune and George Johnston,
though Lawson, Furphy and Richardson did not come his way at
school or from local lending libraries.[26] His education in Australian

literature would occur later, and, especially in its contemporary manifestations, would receive a strong boost during return visits to Australia in 1974 and after.

If Porter's secondary schooling provided him with his only formal literary training, it also gave him his first chance to publish. The *Toowoomba Grammar School Magazine* for 1945 and 1946 shows Peter Porter, as fifth former and then as sixth former, winning the Alan Dunn Memorial Prize. In 1946 he also won the Eric Partridge Essay Prize for a composition written under examination conditions on the topic 'Great Men I Should Have Liked to Meet'.[27] What is remarkable about this achievement is the writer's great fluency in writing an essay of 1200 words in one hour which encapsulated his interest in Themistocles, Hannibal, St Francis of Assisi, Voltaire, Dostoevsky and Oscar Wilde. The belletristic style and occasional pomposity of tone were legacies of the literary criticism of the 1920s and 1930s, but there were signs also of personal insight. The seventeen year old was interested, for instance, in the fate of exile suffered in one way or another by each of his putative dinner companions — who, it should be noted, ranged well beyond the British literary 'greats'.

The young Porter also evinced interest in modes of resistance to prevailing orthodoxy, ranging from Hannibal's military revolts to St Francis's meekness and gentleness. But it is the literary figures which evoke special insights, and the young Porter shows an amazing capacity to imagine these figures as individuals with whom he can converse. He is attracted to Dostoevsky's 'characteristic slow melancholy', to Wilde's witty conversation and to Voltaire's realization that the best weapon against the tyrants around him was laughter: '"If we cannot defeat them by fighting we must make them see how ridiculous they are" was his maxim for the people, and it took great courage to laugh at these powerful princes who were dull and snobbish and entirely devoid of humour'.

When set beside the other prize-winning writings of his age group at the grammar school, Porter's essays, short dramas and verse seem precocious, not only in technique but in their unsentimental evocation of a wide-ranging European past, in a period before Australians began to take countries of the Asia-Pacific into serious account. The sentimental notion of a distant England idealized in Australian public schools is evident in the essay of one of Porter's fellow students, whose thoughts swell out beyond the dusty farms of Queensland to his vision of an English country village:

> Hours spent by the forge of the traditional village smithy would bring back to life scenes conjured up in the story books of my childhood. There, with the children, I could marvel at the sparks and the roar of the bellows. As the farmers passed

I could read in their faces that solemnity learnt as behind
plough and horse they trod contemplating the rich loam tumbling to earth.

In such an English country village I would find the life
that is England; the life I have been taught to love, yet never
known; the life that gives to its inhabitants the love of hedges
and primroses and things eternal. These men were great, for
they were the backbone of a nation great among nations.[28]

This is clearly not Porter, whose dreams have never been so pastoral,
or sentimentally 'English'; as he has commented, with some exaggeration, his schooling gave him a 'loathing of the Anglo-Saxons [which]
could not have been greater if I had been Friedrich Nietzsche'.[29]

Porter by now was beginning to take a wider interest in Europe.
His essay on Richelieu, his two dramatic interludes, 'Santa Lucia' and
'The Secret House' (the latter inspired by Wilde's 'The Ballad of
Reading Gaol'), and his trilogy of verses relating to Napoleon at different
stages of his career show a breadth of reading and an intellectual
ambition well beyond that of his classmates at Toowoomba, and
no doubt most of his contemporaries elsewhere in Australia. In his
intellectual adventuring during his school years, Porter was broadening
his interests well beyond the scope of the literary curriculum. In doing
so he was laying a foundation for his as yet unforeseen role as commentator and critic in London, the undisputed literary metropolis for
Australians at that time. He was also preparing himself for a lifetime
career of largely self-taught responses to, and analyses of, cultural and
social issues outside the walls of formal educational establishments.

3

SUMMER HERMIT

◆

Peter Porter's autobiographical narratives, whether in prose, verse or recorded conversation, locate themselves principally in cities: London, Vienna, Rome, Sydney and Brisbane. The culture of individuals and human communities has always interested him more than physical nature. In Porter's mythology of self, Brisbane is the fallen city, in whose suburbs, after the death of his mother, he learnt the lineaments of alienation, loss and even despair. After leaving Toowoomba Grammar School in December 1946, Porter as a young man returned to Brisbane with his father. They moved into the house in Ipswich Road, Annerley, which had been rented out after Marion Porter's death. Peter Porter remained there until January 1951; these years seem to have reinforced his sense of exile, making the word 'home' more problematic than ever.

In any literal sense of the term Porter was no exile in Brisbane. His father's family had lived there since the 1860s. Moreover, whatever he thought of his schooling, in society's eyes he had been privileged to attend two of the major public schools in Queensland. But the experiences of his early life, allied with a volatile temperament with tendencies towards melancholy, led him to retain a separateness, if not an aloofness from the life of his native city.

Porter's career as a reporter for the Brisbane *Courier Mail* was a short one. It lasted from February 1947 to April 1948, when he was sacked, not for reprehensible behaviour but because of his reticence in chasing stories. After being unemployed for six weeks he obtained work at the wholesale textile warehouse of Gollin and Company in Creek Street, Brisbane. But a repetition of his father's career held no interest for Porter, who had begun to read intensely in contemporary poetry, listen to music, and educate himself in art. He began to write poetry in large quantities and to consider being a writer. He completed several plays and poem cycles. Porter, Roger Covell (whom Porter had met at the *Courier Mail*) and Brian Carne (a friend from Toowoomba Grammar School days) all decided to become great writers. On 2 January 1951, Porter and Brian Carne set sail on the *Otranto* from Brisbane for England, where they had arranged to meet Covell, who was working on music and theatre projects in London.

What kind of city, and country, was Porter leaving behind in 1951? Fellow author David Malouf described Brisbane as:

... a cultural backwater. Still recovering from two wars and
the Great Depression, still lost in a dream of empire and half-
living elsewhere, deeply puritanical and conservative, bel-
licosely unintellectual, it was the dull and decent place that
some Australians now look back to with nostalgic regret:
before migration and the various Libs; most of all, before
television, talk, Vietnam, cheap airfares and the [European]
Common Market changed us for ever.[1]

Closer to the years of post-war 'reconstruction', another Queenslander and temporarily expatriate author, Brian Penton, had foreseen
the perils of continued Australian isolation from world concerns and
the challenges which this posed for Australians:

Australia is cast in the role of the buffer state at the storm
centre of the post-war world. Here the Empire meets Asia, and
Russia meets America in a maelstrom of competing interests.
If we want peace we must reconcile these interests. In this
reconciliation Australia could play a very important part. Her
fears would awaken a new alarm throughout the western world.
Her tolerance and eagerness to co-operate with Asia, to fit
herself into a system which asks for no guns and battle fleets
to defend enclosed privilege, would encourage those who are
trying to give the world a genuine international settlement, to
beat down fear, and confound the reactionary. Australia may
never be great enough to set the world on fire, but in the years
to come she will have a splendid opportunity to help humanity
save itself from the flames.[2]

Penton had worked as a journalist in London in the 1930s and was
P. R. Stephensen's successor as business manager of the Fanfrolico
Press before returning to Sydney and, from 1941, editing the *Daily
Telegraph*. His energetic iconoclasm, and challenge to Australians to
take their role in the world seriously, hardly rippled the surface of life
in Australia's other provincial capitals. Australians generally preferred
to remain isolated from the great challenge of international involvement
which the war had thrown up.

The need to escape Brisbane, for Porter and Malouf as well as for
other would-be authors or intellectuals, seems based primarily on
personal needs, which expand into a sometimes lop-sided vision of
Australia. Both writers have evoked the houses they left, and the
significance of family in their reconstructions of self. Malouf's parents'
house in South Brisbane, to which he devotes such poetic attention in
his prose work *12 Edmonstone Street* (1985) was, like William Porter's
house in Annerley, a weatherboard house on stilts. Malouf's father,
whose family had come from Lebanon in the 1880s, was ashamed of

his house. Upward mobility was expressed, then, in brick and tiles:

> Weatherboard was too close to beginnings, to a dependence
> on what was merely local and near to hand rather than expens-
> ively imported. It was native, provincial, poverty-stricken —
> poor white. Real cities, as everyone knows, are made to last.
> They have foundations set firm in the earth. Weatherboard
> cities float above it on blocks or stumps.[3]

Malouf's reconstruction in the 1980s of his childhood rejects the
brick house to which his father aspired and to which the family
eventually moved in 1947; to the older Malouf the latter house seemed
'stuffily and pretentiously over-furnished and depressingly modern'.[4]
Although Porter sees similarities in the 'comfortably philistine' outlooks
of both families, and suggests that this provides 'not a bad start' for the
artist,[5] some differences are also revealing. Whereas Malouf's family
moved up and on in the context of Brisbane's (and middle Australia's)
culture of success, William Porter kept the weatherboard home in
Annerley after his wife died, rented it out and returned there with his
son. These events form a foundation for Porter's tale of his family's
fortunes as one of decline, in which he is the inheritor of little except a
sense of failure, and a loser in the world's success stakes. Such images,
in Porter's verse and prose, run counter to those of modern advertising,
but have a long tradition in Australian folklore and culture, ranging
from Steele Rudd's Dad and Dave stories to Peter Carey's novel
Illywhacker, in which a sardonic irony accompanies tales of loss,
failure or disappointment, and survival is the only game.[6] Elsewhere,
Carey has commented on an Australia where people 'don't dare believe
in anything too large because they will only be disappointed in the
end. The dreamers in our country all perish'.[7]

The notion of a 'house of Porter', in any nineteenth-century novel-
istic sense, is undermined at its wooden foundations. The white ants
are always at their work and Porter is ever aware of them. He has
presented himself and his father's house as part of 'shabby genteel'
Australia, not quite 'destitute enough to make exciting reading':

> We were on the Ipswich Road, an unlovely ribbon of shops,
> factories and hospitals winding out of Brisbane on the south
> side of the river ... Imagine a primitive interpretation of Le
> Corbusier's ideas, carried out in wood and painted in garish
> or depressing colours. Our house was only about five feet off
> the ground in front, but at least fifteen at the back, the ground
> sloped so steeply. It was mounted on wooden piles, each
> topped with a metal hat and coated in creosote to deter the
> white ants.[8]

Porter's better-off aunts lived in Clayfield and Hamilton, where Malouf's father had moved 'upwards' in 1947. These 'smarter' suburbs on the north side of the river constituted 'a world of confidence mined by neurosis, an Anglo-Saxon outpost of Vienna, where Freud's Wolf Man would have been quite at home, where poverty known to the Legacy Club was infinitely more humiliating than the appalling neglect of Naples'.[9]

However, Porter's grandfather's house in Norman Park, an old suburb in East Brisbane, provided a *locus classicus* for the imagination to expand in, a place where a sense of grandeur, however faded, might find sufficient earth. During his school years and after, the grandson would catch the Balmoral tram to this spacious colonial house with its tennis court and two acres of garden. Here was evidence of a past:

> The old box trees had been planted by my father's grandfather, and were over a hundred feet high. They had been used at the turn of the century by ships navigating the twists of the Brisbane river. There were numerous other trees — silky oaks, Moreton Bay figs, assorted wattles and angophoras. Flowers abounded, and my father distressed me by telling me that beneath several of the garden beds were the bodies of his pet dogs. As a youth and in his twenties and thirties, he had kept a variety of dogs as well as an aviary of bush birds he had trapped himself.[10]

Characteristically, here, the attention shifts from a scene of stability and order to considerations of death, burial and disintegration. The grandfather's house was also near the East Brisbane cemetery — although his mother had been cremated, her young son's morbid cast of mind led to an early fascination with graveyards. The cemetery was memorable for the constant buzzing of bees which fed on the nectar of the flowers brought to the graves;[11] Porter had this in mind when he referred, in a later poem, to 'the honey fields of death'.

One of Australia's most celebrated expatriates, Germaine Greer, has commented that Australians are destructive of their roots. In an interview in America she commented that it would be impossible to produce a television series like *Roots* in Australia because 'nobody gives a shit'.[12] The enormous and burgeoning interest in family and local histories in the 1980s in Australia, culminating in the various celebrations in 1988, the bicentennial year of European settlement, suggest otherwise. Greer's recent autobiographical search for her father in *Daddy, We Hardly Knew You* (1989), and Robert Hughes's search for Australia's (and his own) roots in the experience of the convicts, in *The Fatal Shore* (1987), indicate the importance which the reclamation of an Australian past can assume for writers and artists who voluntarily exile themselves to another country. Some of Porter's reclamations of family were discussed in Chapter 1. A fuller understanding of his images of Brisbane, however, and his perceived relationship to this city before he left it for London, requires further consideration of his patrilineal inheritance, which linked him by blood line to this European settlement since the 1860s. The process through this line is interpreted by Porter to show a diminution of fortunes, from the great-grandfather who built the deep water pier at Cleveland, and allegedly the Boggo Road Gaol, to his father and himself. Porter concedes that it is this view of decline and fall, arising from a set of familial circumstances, which he, presumably unconsciously, imposed on the city of Brisbane — a picture of 'all Fall and no Eden'.[13]

The mother's side of the family provided Porter with quite a different pattern of loss. Marion Main's parents had emigrated to Australia from a poor background in Glasgow. The grandfather, Mark Main, was a businessman in Adelaide before moving to Sydney, to the brick house with the dark and daunting cellar where the young Peter Porter spent Christmas holidays.

As autobiographical commentator, Porter overlooks the parable of his grandparents' material success in favour of the story of his mother's two handsome brothers, both killed in France in the First World War, and whose loss might have accounted in part for the melancholic streak Porter observed in his mother.[14] The poem 'Somme and Flanders' (*1962/3*, *CP* 40) links these two men with an uncle on Porter's father's side (his father's only brother), who also died in the 1914—18 war, making a trio with whom the young writer, with his anti-war views, communes:

> Who am I to speak up for the long dead?
> Three uncles I never knew say I'm right.
> Their tongues are speaking in my head,
> I'm related to their flesh by fright.

The exposure of fear, vulnerability and uncertainty in the speaker here are elements in a persona which Porter later tunes more finely and for a variety of effects. In this poem, a past consisting of only photographs and memories, mostly suppressed, is revivified. Reading his way back into that past in the Harmsworth picture books of the war and elsewhere was an essential task for a young poet who already felt his role was to show the interfusion of past and present. But why choose these subjects? The impulse towards mercantile success provided by Mark Main and his family, though not derided by Porter in the literary manner of an aesthete, has never spoken as compellingly to him as other kinds of ambition; or indeed the example of the three uncles who died in France:

> One image haunts us who have read of death
> In Auschwitz in our time — it is just light,

Shivering men breathing rum crouch beneath
The sandbag parapet — left to right

The line goes up and over the top,
Serious in gas masks, bayonets fixed,
Slowly forward — the swearing shells have stopped —
Somewhere ahead of them death's stop-watch ticks.

The image here of men moving in unison to their inevitable deaths
sounds a warning against blind obedience to national 'duty'. Such
thoughtless patriotism is also probed and questioned in Porter's later
work.

Other matters were being fought over closer to hand, in Brisbane,
in the post-war years, which contributed to Queensland's reputation as
a cultural desert, from which many of those with artistic inclinations
found it necessary to flee. A key figure in these battles was Johannes
(now 'Sir Joh') Bjelke-Petersen, later to be the State's longest-serving
Premier, from 1968 to 1987. Bjelke-Petersen's anti-intellectual pos-
turings set a tone of antipathy to most artistic pretensions and liberal
thought in Queensland and provided Queensland-based poet Bruce
Dawe with plenty of ammunition for his entertaining satire. The
Protestant work ethic espoused by Bjelke-Petersen — 'hard work =
success = salvation' — was a formula which flourished with electoral
success, even after its hypocrisy was exposed by the press and public
inquiries. Such success was found to be dependent on gerrymandering
and a relatively supine media corps in Queensland, who Bjelke-Petersen
often referred to as his 'chooks' needing to be fed. Social commentator
Hugh Lunn later described Bjelke-Petersen's reign as 'anti-socialist,
pro-private enterprise, anti-worker dissent, and pro-rural'.[15] Although
Porter had no special interest in party politics at this time, he was
certainly aware of a general climate of anti-intellectualism and particularly of the raw rural fundamentalism which Bjelke-Petersen
represented; he had encountered this among his fellow boarders at
Toowoomba, many of whom were the sons of farmers.

Porter was also aware of the sectarian divisions between Catholics
and Protestants in Queensland politics in the 1930s, which culminated
in accusations in 1937 that the Labor government was favouring Catholic
schools. A breakaway Protestant Labor Party formed the same year to
contest the 1938 state election.[16] If the poem 'Eat Early Earthapples'
(1962/3, CP 44) provides insight into Porter's rejection of rural fundamentalism, 'Who Gets the Pope's Nose?' (OBTB, CP 32−3) indicates
scepticism about Roman Catholicism. During his years in Brisbane,
Porter would also have been aware that the Labor Party, which
controlled Queensland politics through the years of his boyhood, dif-
fered very little in its authoritarian conservatism and primary concern
to stay in power from its Country Party opponents. Porter's evolving

preferences for socialism, metropolitan values and a kind of doubting
theism owe a good deal to these years in Brisbane, when he lived in a
personal 'nest of doubts'.

The difficulty for a socially awkward and reticent young man
whose chief interests were books and music was to find like-minded
company. At the *Courier Mail* he was fortunate to meet and begin an
enduring friendship with Roger Covell, whose musical interests and
achievements would later earn him the Chair of Music at the University
of New South Wales. The Brisbane desert had other oases. Porter
found regular refuge in the Ballad Bookshop, run by Charles Osborne,
two years older than himself. Osborne's musical and literary tastes
were precocious; he, too, learned quickly that Brisbane was too small
for his interests. In 1950, when twenty-three, he moved to Melbourne,
preparatory to departing for London. Osborne's reasons for leaving
Brisbane were clear:

> ... with the exception of books, most of the things I cared
> about were not to be found there. There was a certain amount
> of music making, it is true, but how much more of it there was
> in London. When I left Brisbane, I had not heard a live per-
> formance of a Mahler or a Bruckner symphony. The local
> orchestra was mediocre: visiting conductors occasionally suc-
> ceeded in making it play well, and Otto Klemperer made
> it sound like the Vienna Philharmonic. But Klemperer was
> Klemperer ... In theatre in the 1940s, the situation was
> hardly better: a steady diet of recent West End or Broadway
> comedies ... I wanted to see operas, real plays, real operettas
> for that matter, and to look at paintings and baroque churches,
> and even see some real modern architecture.[17]

Surprisingly, perhaps, Porter did not follow up the interests which he
shared with Osborne by joining the 'Barjai' literary group of which the
latter was a leading member. In his own words, Porter was 'filling
bucket loads of pages', writing 'piles of poetry, plays.'[18] He has
commented:

> I was just the right age to have partaken of this strangely
> anachronistic ganging-up, but years in Toowoomba and my
> own diffidence kept me from joining Barrie Reid, Charles
> Osborne, Barbara Patterson, Laurie Collinson, Laurie Hope,
> Cecil Knopke and the rest ... the once or twice I met Charles
> Osborne I found him dauntingly sophisticated ... Charles
> was always a man of great courage, and he led the way in
> defying Brisbane's puritanism and philistinism. He cultivated
> a sort of cheerful decadence ...[19]

But there was another reason. The 'Barjai' group, whose title came from an Aboriginal word meaning 'today', was predominantly homosexual, and although Porter wasn't at that stage 'any kind of sexual', his 'basic idea was to be heterosexual'.[20]

Nevertheless, the high camp, left-wing iconoclasm of the 'Barjai' group would surely have had more appeal for Porter than the somewhat older 'Meanjin' group (whose Aboriginal title meant 'meeting place'), led by Clem Christesen. This group had a more earnest, idealistic and pietistic mien before it uprooted itself for the more fertile cultural soil of Melbourne in 1945, when Porter was only sixteen and still at Toowoomba. The *Meanjin* transplantation can be seen as a foreshadowing of other cultural departures from Brisbane for larger metropolitan centres. Besides losing members to other cities *Barjai*, which had published twenty-three issues since 1943, gave up the ghost in 1947 when it lost its financial patron. By the time Peter Porter was ready to benefit from Brisbane's literary magazines and their associated communities of writers and critics they had departed, or were departing, for greener pastures. One contemporary noted that young intellectuals 'all got out at the first opportunity, initially to Sydney or Melbourne, subsequently in many cases to London or New York'.[21] While it is not quite accurate to present Porter as a pioneer expatriate, in P. D. Edwards's words, 'blazing a trail that everyone was taking within a few years,' Porter's decision to leave Queensland for London in 1951 did epitomize a recurrent problem: 'that all [Queensland's] bright young people either want or have to get out, and hardly ever come back'.[22]

Most of the 'bucket loads' of verse which Porter wrote between 1947 and 1951, after leaving school at Toowoomba and before his departure for England, have been lost. Those poems which remain in the author's possession show a young man's preoccupations with unfulfilled sexual desire, Christian belief, and a sense of sin. Reason and feeling are in combat, without resolution. Occasionally, a timorous self masks its uncertainty with a brash assertion or fine phrase before retreating behind more aggressive literary figures such as Blake, Donne or Strindberg. An ambitious poem written in 1949 sets a Voltairean rationality against the kind of religious expression which Porter absorbed at his non-denominational but essentially Protestant/Anglican schools. Titled 'Under the Circumstances', this poem mimics the form of certain eighteenth and nineteenth-century non-conformist hymns, such as those of Isaac Watts, to ridicule the traditional, public school-derived images of God: 'O circumscribed, athletic God/Calm fertile father of our tribe ...'. Unfortunately, the inspired opening lines are not sustained, but an early interest in donning verse forms for ironic or satiric purposes is evident.

A longer poem of the Brisbane years, 'Visions for Judgement', dated early 1950, alludes only fleetingly to Byron's 'The Vision of Judgment', but is symptomatic of Porter's intentions and problems as a writer at this time. First, it exemplifies early attempts at developing a persona, which are not sustained in this poem; and second, it indicates ways in which the Brisbane environment was compressed and simplified in the service of this persona. The opening lines of the poem are bravely, even romantically assertive in a Byronesque manner:

> I am the summer hermit
> In the sunburnt cave.

This is the first example of the authentic Porter voice. The poem epitomizes Porter's preference for interiors over outdoor settings, with the notable exception of certain gardens and seascapes. The troglodyte speaker is entranced, dulled by the summer heat but receptive to dreams and visions — speleology will follow:

> The summer brain is silent,
> Giving itself in great peace
> To the dreams it knows, and
> The revelations it absorbs.

This mood of quiescence is closer to some of Randolph Stow's lyrics than most of Porter's later work. Australian poet Robert Gray's comment about Brisbane summers makes a similar point: 'The place seems to want to steam from its inhabitants, through the long summers, all of their introspection.'[23] Taking up elements of the opening refrain again, Porter's speaker varies and extends his self-revelation:

> I am my own lone hermit
> Terrified by love, as though
> Tomorrow were life.

Although the speaker here cannot be equated in any simplistic sense with the author Peter Porter, it does indicate a young man's attempt to express his sense of alienation in this summer place, where, instead of liberation, he felt contained in a circle of hell.

Autobiographical commentaries suggest there were some compensations, however, for the absent life of the spirit in this Brisbane which Porter was inhabiting. The city represented a part of the world where things habitually went wrong, but it also provided a temporary resting place:

> Brisbane was a place I cruised in half asleep or tranced with incapacity. I was sure of a number of things: that I had sighted reality in the poetry of Shakespeare, Browning and Auden, and in the music of Mozart and Schubert; that I should never master the art of making love to a girl; that I would devote my

life somehow or other to writing, though I doubted my talent
for it; that I was doomed to be an air plant, déraciné everywhere and with no proper home.[24]

This was the state of suspended animation in which Porter was able to
say, in all sincerity, that a Mozart sonata was more real to him than a
Brisbane tram. An obliviousness to material reality was accompanied
by a wish to live alone in the world of language, or beautiful music, to
be wrapped up in it and transported to other realms. When he returned
with his father to Annerley in 1948, they were soon receiving notes in
the letterbox from neighbours who complained about the loud gramo-
phone. Looking back on these incidents Porter has commented: 'The
sound of *Don Giovanni* through the heat of a Brisbane afternoon never
seemed decently classical'.[25]

One way out of the trance would have been to satirize these
suburbs of hell, but in his late teens Porter lacked the confidence to do
much of this. Not that there was any shortage of suburb-bashers in
Australia. Many intellectuals of the inter-war years and beyond felt
bound to fulminate against the neighbourhoods in which they lived.
Walter Murdoch, for instance, mocked the Melbourne environment:

> . . . the awful sameness of Melbourne's suburban streets, with
> their red-tiled houses, neat lawns, gravel paths, *Pittosporum*
> hedges, reflecting a uniformity of spirit, a complacency, a
> positive fear of originality or difference.[26]

Later, in Perth, Murdoch attacked the 'suburban spirit' as 'the everlasting
enemy'.[27] Similarly, A. D. Hope's verse of the 1930s and 1940s assumes a
standardization and repression (especially sexual) in Australian
suburbia, in poems such as 'The Explorers', 'The Brides' and 'The
House of God'. George Johnston's *My Brother Jack* (1964) shows the
persistence of this attitude into the 1960s, although social analysts
such as Hugh Stretton, Craig McGregor and Donald Horne contribute
to a revaluation of this despised milieu of Australian living.[28]

In spite of the torpor which seems to have overtaken Peter Porter
and coloured his view of Brisbane in the late 1940s, he was storing
images of difference which suggest an early appreciation of ethnic
variety in a predominantly Anglo-Saxon Australia, and a special interest
in outsiders. The Chinese market gardeners in 'A Giant Refreshed'
(*OBTB, CP* 3) and the fish and chip shop proprietors from eastern
Europe in 'Requiem for Mrs Hammelswang' (*PAAM, CP* 55−6) are
treated with unsentimental appreciation. The sugar farmer's son and
the girl in the Everest Milk Bar in 'Homage to Gaetano Donizetti'
(*PAAM, CP* 61) metamorphose into Nemerino and bel' Adina from
Donizetti's opera *The Elixir of Love*. Such cosmopolitan variety is not
pressured into conformity but held for consideration and notice in

Porter's suburbs. At the same time, it should be said that the prevalent negative image of the suburbs is reinforced in certain poems, including 'A Giant Refreshed', the first poem in *Once Bitten, Twice Bitten*:

I think now of your prescribed joy
That from a suburban home came —
(The arrogant freckled girl kissed her boy
At the gate by the juiceless garden) —

Although 'the suburbs' might have been a common target for intellectuals' or artists' attacks, they will not suffice as the single reason for the emigration of so many of the young to other countries, especially to England and Europe. The list of Australian expatriate writers before Porter is a long one, and includes Henry Handel Richardson, Barbara Baynton, Christina Stead, Jack Lindsay, Frederic Manning, Miles Franklin, Martin Boyd and Colin McInnes. The Australian critic, A. A. Phillips, has commented that the departure of such writers 'robbed' Australian writing of 'a leaven of venturesome minds',[29] but he does not ask how they would have written had they remained in Australia, or indeed if they would have written at all. For some writers, whether coming to Australia or leaving it, the experience of migration is a major factor in what they write, and how. English-born fiction writer Elizabeth Jolley, for instance, who moved to Western Australia in 1959, is preoccupied in her work with the spirit of migration, the hopes, fears and joys associated with settling in another country. A large and important contemporary literature deals with international migrations and with individuals caught between two worlds: V. S. Naipaul and Salman Rushdie are outstanding exponents of this literature. Understandable as A. A. Phillips's protectionist statement is, counterclaims, in the literary sphere at least, might be made for the value of the international movement of writers, whether as literary travellers or as emigrés. In other areas of the arts, too, it can be claimed that the profession, as well as the individual, are enhanced by expatriation: Nellie Melba, Joan Sutherland, Robert Helpmann and Sidney Nolan all expanded their repertoire, and enhanced their skills, by living for lengthy periods outside their home country.

Alan Moorehead gives a convincing case for expatriation in describing the sense of isolation felt by many Australians in the inter-war period:

As far as possible the local environment was ignored; all things had to be a reflection of life in England ... Everything was imported. And because they believed that the imitation could never be as good as the original, they were afflicted always with a feeling of nostalgia, a yearning to go back to their lost homes on the other side of the world.

To go abroad — that was the thing. That was the way to make your name. To stay at home was to condemn yourself to non-entity. Success depended upon an imprimatur from London, and it did not matter whether you were a surgeon, a writer, a banker, or a politician; to be really someone in Australian eyes you first had to make your mark or win your degree on the other side of the world.[30]

Tasmanian author Christopher Koch, several years younger than Porter, has written of the special power of literary images of London for the aspiring writer; in his case, Dickens, Arthur Conan Doyle and Beatrix Potter were especially potent. London, more than New York or Paris, was the centre. Koch becomes lyrical when he recalls the peculiar historical moment which Australians and New Zealanders shared when they moved to England to live and work in the 1950s:

We sailed, as soon as we reached our twenties, for isles of the Hesperides we never doubted were real. What no native of the 'mother country' could ever understand — what no-one but overseas children of the Empire could ever experience in fact — was the unique emotion summoned up by the first sight of a country known at one remove from birth, and waited for as an adolescent waits for love.[31]

Peter Porter's motives and feelings as he prepared for departure from Brisbane on the *Otranto* on 2 January 1951 were, inevitably, mixed. At twenty-one he was an unsettled autodidact who needed a larger theatre in which to develop his knowledge, ideas and experience. Deeply uncertain of his talent as a writer, he was in need of the company of others who could test and help expand whatever capacity he had. Analysis by hindsight often suggests nothing is an accident. Whether Porter would have actually caught the boat to England if Roger Covell had not already been in London and invited him there is a matter of conjecture. Looking back across twenty-six years in 1977, Porter gave a gloss of inevitability to these events, and placed his own departure for London within a broader notion of collective nostalgia than that proposed by either Moorehead or Koch:

There is, I believe, a huge nostalgia in us which goes much further back than to our immediate antecedents. It is a form of the Jungian anima, and if, in addition, we are temperamentally disappointed, it takes us towards the older societies and away from the more galvanic tribes whose myth is of palpable success here and now.[32]

There are no liberationist expectations in this, no aspirations towards climbing the peaks of civilization. This departure seems necessary, a compulsion. Inwardly, the traveller seeks a society which approximates more closely to his inner sense of himself; departing from Brisbane, where suffering is considered 'vulgar' and only happiness seems allowed,[33] he sets out for a promised land in which it may be possible to partake in 'a permanent sense of being beleaguered, a savouring of European hopelessness'.[34]

Certain leave-takings had to be made. In a sense, the break with the father had already occurred when the ten year old boy was sent off to boarding school. Renewals of contact with memories of the very 'ordinary' William Porter would be made subsequently, in poems ranging from 'Tobias and the Angel' (OBTB, CP 37) to 'Where We Came In' (FF 29), written after his father's death. Just as William Porter later became a link with Australia, so, in the Brisbane years, he had seemed a link with England, despite the fact that his family had been in Australia a hundred years and he had never been to Britain. As well as his love of roses and an attitude of respect for the nation which produced manchester goods, William Porter was one of the last Australians to wear a bowler hat, heavy tweed suits, a waistcoat and chain.[35] He would have understood, in a less intellectual sense, what his son meant when he described his arrival in England as 'an exile's return'.[36]

The son was not leaving behind an authoritarian father but an unassertive one. The son's attempt to 'find himself' in the metropolis of London would involve, in part, a search for alternative authority figures, but not, like T. S. Eliot, a relapse into the institutions of royalty and classicism. Porter's career would show instead an oscillation between the desire for genuine authority figures in the literary world and a recoiling from the influence of such authority — a recognition of the necessity of the father, and of his necessary absence, for genuine creative activity.

Another parting was necessary — during the first port call in Sydney, the young man made a final visit to his grandmother at the house called 'Iona' in Woolwich. With the passing of time this visit has attained a dramatic and musical plangency:

> Inside the house, where the old lady seemed to have faded to a Tchekovian transparency, I fancied I could still hear the boisterous voices of the men and women who had once made the house so lively a place. Ours was what they called a drinking family and the sideboard was still packed with bottles of spirits and cut glass tumblers, unused for years, the same dispensary my mother had officiated from in the thirties.[37]

This house is a 'magic place' of childhood, to which the ruminant self of the author returns in autobiographical commentaries and poems which seem to have little to do with Sydney but everything to do with his concepts of memory and image-making, such as 'Landscape with Orpheus' (*ES*, *CP* 309–10), and 'Essay on Dreams' (*TAO* 38–9). The autobiographical reflections on Porter's last stop before sailing from Australia persist like a refrain:

> My unit of measurement is a child's stature and a child's wonder — even a child's fear. The spiders will go on living in the cellar, the 78 rpm records of Haydn's *Military* Symphony will go on filling the air with the sounds of the Vienna Philharmonic. The snails will still be writhing in the salt and the pepperina tree will wait for the crying child to seek comfort under its branches.[38]

These images will recur in a memory stream in which the place left behind assumes an increasing importance and value.

4

At Home, Away

♦

After breaking free of the paper streamers which linked it with its earth-bound well-wishers, the *Otranto* was fifty days at sea before it debouched its passengers at Tilbury docks on 19 February 1951, three days after Porter's twenty-second birthday. The gradual adjustment to change which ship travel allowed had vastly different socializing effects to plane travel; rather than being thrown, jet-lagged and catatonic, into a new environment, one savoured its slower approach. One had time to entertain illusions of a changed self in new circumstances.

It is tempting to think of travel by ship as an interlude. Australian literature is spiced with such interludes, as in the crossings between Australia and Europe in Henry Handel Richardson's *The Fortunes of Richard Mahony*, and Martin Boyd's novels. In some, the journey spans all or most of the story, as in Angela Thirkell's *Trooper to the Southern Cross* (1934) and Barbara Hanrahan's *Sea-Green* (1974), during which mistakes or discoveries are made and lives changed irrevocably. Notwithstanding the claims for sex-in-the-sky made by Erica Jong in her novel *Fear of Flying*, sea travel provided an opportunity for romance of a more traditional, drawn out kind. Undoubtedly the most significant event in Peter Porter's journey to England was his meeting and romantic friendship, after some four weeks at sea, with nineteen year old Jill Neville from Sydney. Neville's sense of adventure subsequently carried her through a turbulent series of other romantic relationships, and brought her jobs including advertising copywriter, social worker, teacher, journalist, critic and novelist in both Britain and Europe. The connecting link through all these experiences was to be the 'boy from Brisbane'.

When Jill Neville looked down from an upper deck of the *Otranto* and saw the shock of black hair of the awkward young man she had been told was a poet, she felt driven to monopolize his attention. In her first novel, *Fall-Girl*,[1] she used this and other autobiographical material in a series of scenes which show the comedy, romance and pathos of a young Australian woman's transition to, and early years in, London. The chief structural continuity through the novel is provided by the figure of Seth, based on the author's perceptions of Peter Porter in these years. In a novel which teeters uneasily between comic romance and bathos, Seth is presented as an anti-romantic figure, but one who, paradoxically, offers his companion some depth and continuity of affection. The first meeting on board ship is recalled amusingly by the narrator, and perhaps with some self-irony at her naivety:

> I had met Seth on the boat. Someone had said the fatal words:
> 'He's a poet', and click went my brain and took a First Sight
> Snap. He was standing on the deck below me. His cheeks
> were flushed, his hair black, he was cracking his knuckles and
> glancing nervously from left to right ... I had discovered
> someone extraordinary. He was a pearl lying unnoticed, so
> how could it be a real pearl? ... And for the first time I had
> got really close up to a living, breathing poet. It gave me the
> same voluptuous thrill as discovering a mushroom alone on a
> scraggy field.[2]

Anti-romantic details are a necessary counter to the volatile young
woman's high-flown expectations. Instead of champagne, the two
drink Eno's Fruit Salts in each other's cabins. Seth is not a silent and
moody Romantic, rather, he talks volubly, dresses unfashionably, and
works daily on 'an assembly line of poems'. With his 'wide flapping
trousers' and 'gawky gestures' he is unlike 'all the Teddies, Rickies and
Maxs' lined up in the narrator's mother's snapshot album, who had
'spent their youth ... flicking on cigarette lighters with daemonic
timing, jumping elegantly over tennis nets and dancing divinely'.[3]
From being a collector's item, at first, for the bored young woman on
her maiden voyage from Australia (she and he are both virgins in the
fashion of the early 1950s), Seth becomes essential to her repertoire.
Their friendship deepens and persists through several years of an
anguished love affair, followed by various other sexual liaisons on her
part, her pregnancy and desertion by the father of her child, and Seth's
marriage to another woman. The largely autobiographical narrator
perceives her relationship with Seth as an 'uncommon' love, 'the kind
you talk about ten years later',[4] which is roughly the point at which
the author finds herself writing *Fall-Girl*.

When the *Otranto* docked at Tilbury on that grey February morning
in 1951, Porter went on deck with the other passengers to see England,
and it was all exactly as he had imagined. His Australian diet of
English literature, English films and BBC radio programmes contributed
to this feeling; in spite of the almost accidental way in which he had
come to England, it now seemed that he had been drawn to a place he
already knew. It was indeed an 'exile's return'.[5] Porter and Carne took
the train to St. Pancras Station and then to South Kensington, where
they moved into a basement flat with Roger Covell, and registered for
meat and other rations at the corner Oakeshotts store.[6] This was the
first of a number of basement flats rented by Porter with Australian
friends in London between 1951 and 1953. Later, he stayed in Notting
Hill Gate and then Belsize Park Gardens, where he again lived with
Roger Covell. As for most Australians in these years, Porter's London

was the area north of the Thames. Unpublished poems of this time refer to 'basement London' — this is the perspective from which he views the city.

The milieu into which Porter arrived in London in the early 1950s was a poor person's bohemia of casual jobs, pubs and parties. Not surprisingly, many of his companions during this time were other 'colonials', especially fellow Australians. But there was little in Porter's demeanour that would qualify him as a precursor of Barry Humphries's beer-swilling ocker stereotype Barry McKenzie — Australia's picaroon among the Poms — who first appeared in the cartoon *Private Eye* in 1965. The young man from Brisbane whose great ambition was to write good poetry had no desire, either, to act out an alternative stereotype, that of the Poet. In these years, Porter developed a preference for the kind of poet who does not place him or herself above others in a special role, either bohemian or priestly, and this has remained with him. Out of a kind of reticence, and a touch of puritanism, he resisted the excesses of the Dylan Thomas and George Barker neo-Romantic set, who raged up and down Soho through the late 1940s and early 1950s. Yet, often in the company of the lively and sociable Jill Neville, he would attend parties, and drinking, talking, laughing and joking became his pursuits, as he recalled they had been for his mother, with her more ambiguous 'party melancholy'. His ebullient sense of humour and already enormous range of knowledge attracted around him those who were not caught up in dancing or seduction. Comparing his bohemian living of the early 1950s with the lives of commune dwellers in the early 1970s, Porter has commented that there were many similarities, except that his cohorts were characterized by being poorer, more frightened (of sex, especially), and with no supporting drug except alcohol.[7]

Porter soon developed a reputation as being a brilliant talker. Jill Neville has commented that he was 'the greatest talker I had ever met in my life — the funniest, wittiest, most entertaining, most coruscating. His humour was unending'.[8] The two young explorers walked endlessly around London, to and from concerts, plays, pubs, parties. 'I don't know if we were talking about culture or politics or the meaning of life', Jill Neville has remarked, 'but it was such high-level talk — I regret I didn't have a tape recorder in my pocket ... He was full of an incredibly wild, almost Oscar Wildean humour ... an aficionado of the brilliant monologue'.[9] There was a darker, more volatile part of Porter, however, which is evident in some of his unpublished verse of these years. Something of this mood is captured also, in several semi-humorous episodes in *Fall-Girl*. In one scene the first-person narrator and Seth reflect upon a brief conversation with a dejected Australian waiter at a Lyons Corner House, who has been ten years in England:

Seth looked wretched. 'That's what we'll be like after ten

years in this bloody dump of a town. All the life beaten out of,
us. You'll have gone off and married one of these smart young
men in charcoal grey suits, and I'll still be a clerk'.[10]

Feeling herself, by contrast with this self-indulgent pessimism, rather
like a potential film star, the narrator admits that she has already
acceded to 'the English Class Thing'. Her aspirations to gentility are
offended by the cockney accent and by Australian swearing. Not so
Seth, who makes the stabbing comment, 'Orwell's right ... A man is
branded on his tongue in this fucking country'. Seth's comment is not
designed to charm his companion; but it expresses with some relish
the deep gloom and sense of hopelessness to which Porter was especially
prone in his early years in London. Later, in the 1960s, when Porter's
poetry caught the public's attention and he became a voguish figure on
the London literary scene, Marxist critics were attracted to his sometimes scathing satiric attacks on the pretensions of class in Britain. But
although Porter was identifying with the drifters and drop-outs of post-
war London, as in this scene from *Fall-Girl*, and was moving away,
politically, from his father's conservatism to a personal brand of socialism, it would be mistaken to see him as a Marxist. A preference for
socialism had been awakened earlier in his Brisbane days through his
reading of Fabian essays, the complete works of Shaw and the socialist
poets of the thirties. His political education in the 1950s and 1960s
reinforced his inclination toward the left, towards a vision of what
many in the Labour movement perceived in these years as a future for
Britain as a relatively classless society.

But it was thoughts of sex, rather than politics, which dominated
Peter Porter's mind at this time. 'There were/So many ways of losing a
troublesome innocence/But so many ways of keeping it too', he wrote
about his schooldays in 'Eat Early Earthapples' (*1962/63*, *CP* 44),
and the problem persisted. To break with the past, however strong the
desire, was not easy. A tortured sexuality and a strong idealizing
tendency combine in a number of unpublished love poems, addressed
to Jill Neville in the early 1950s, to show the anguish rather than the
pleasure of this first major love affair. An occasional boldness is
regularly followed by retreat. Very little of the ebullience and humour
of Porter's conversation of these years is evident in the sonnets, arias
and cantatas written for the loved one whom he felt destined to lose.
An audacious and playful tone emerges in an unpublished poem called
'A Deliverance for Doctor Donne', but it is Donne's audacity rather
than Porter's which shows through. The lack of confidence is both
personal and poetic. In some respects, Porter's problems as a poet
matched those of Byron and Browning; only after admitting the language
of wit and conversation into their verse would it flourish.

Pessimism and party-going are an explosive mix. A favourite spot
for parties among young Australian expatriates and their friends was

the boat colony on the Thames near the Lots Road Power Station in Chelsea. Jill Neville, who shared a houseboat with a friend, was a focal point at these gatherings. *Fall-Girl* gives a vivid account of a party which became legendary among Australian expatriates, in which the twenty three year old Porter's pent-up frustration, combined with jealousy, suddenly erupted.[11] Australian journalist, Murray Sayle, had invited along a young man called Ralph Howard, who worked in advertising. Jill Neville began to dance with Howard, and to demonstrate her affection for him. In the novel, the narrator's dancing partner, Reg, puts on a Stan Kenton record while Seth lurks behind the *Evening Standard*. She feels as if her joints have been 'oiled by an Esso bottle'. Although she has been having an affair with an older man, she is inclined to follow this 'young and callous man's mating-call'. Seeing this happening, Seth's repressed feelings of 'dismay and wrath and jealousy' burst forth. A fight starts on the houseboat and he throws her in the Thames. Recalling the incident later, in an interview, Jill Neville's account differs little from the episode in her novel:

> I swam out of the Thames and saw Peter and Ralph struggling in the houseboat ... Ralph got away from Peter ... And I ran away from Peter with Ralph running with me up that Chelsea embankment. Peter was running after me with all his friends, who thought I was a bitch. But we caught a taxi, and I escaped to my life of crime.[12]

The obvious relish in the verbal recounting of this comic melodrama is tempered slightly in the novel by an interest in the psychology of Seth, whose complex loyalty to the narrator contrasts with her desire to cut a figure, to shock and thus make a mark in her small corner of London. As for many Australians newly in London, the world seems a stage. Seth, on the other hand, is still withholding himself and when his only serious love relationship is threatened, he reacts with violence, anger and passion. Peter Porter, like Seth, seldom bares his soul and his verse writings of the early 1950s, as of later years, are often self-consciously concerned with questions of 'authentic dress'. But the violent undercurrents also reveal themselves at certain moments and are the more potent for their irregular and infrequent eruptions. When a London reviewer in the 1960s called Porter a 'bruiser from Australia'[13] he was probably unaware of the houseboat incident. If he had known Porter personally, he would have found that he was not at all macho, that he abhorred bullying or violence. But the young man's emotions were vehemently aroused as he watched his lover escaping up the embankment and beyond. Nor did he calm down when he learnt of her subsequent pregnancy (to Ralph) and abortion; the letters they exchanged, and their occasional meetings, were taut with anguish. In retrospect, Porter has called his subsequent nervous collapse 'statutory',[14] understating with irony both the significance of these episodes and the Romantic element in his own life and writings.

Life in London at times seemed not worth living to Porter, although his musical tastes and interests were expanding through attending concerts and discussions long into the night with friends such as Roger Covell and David Lumsdaine. Working as a clerk with International Paints, and elsewhere, gave him no pleasure. The bohemian poets of Soho seemed poseurs, the semi-bohemia of Australian expatriates pointless. On 31 December 1953, in an attempt to retreat from the maelstrom, he sailed from Tilbury towards Sydney, with Brisbane as his destination.

But had these early years in London been so disastrous? Porter had written hundreds of poems, though he had not found a publisher for any of them. More significantly, in terms of his ideas and obsessions, he had written in 1952 a long three-act play in blank verse called *The Losing Chance*. This play may owe something to Porter's early reading of Browning's verse dramas but appears to have been encouraged by contemporary successes in this genre, such as Christopher Fry's *The Lady's Not for Burning*, which had been first produced at the Arts Theatre in London in 1948, and T. S. Eliot's *The Cocktail Party*, which appeared in 1950. In *The Losing Chance*, the action takes place in the present on an island off the North Queensland coast, 'within the smooth waters of the Great Barrier Reef'. The characters include a millionaire American businessman named Tungsten T. Manganese and a Professor William Woebegone. Off-stage is the mysterious figure of the Adversary (via Auden). Introduced into this setting is one of Porter's alter egos, Peter Mainchance, 'a tall slightly stooping young man of about 22: 'poet,/Demon, ... esoteric wanderer, listener-to-music,/ Gatherer of sensations, down-and-out, clerk,/Bourgeois hoper and our coming sacrifice'. It is a wonderfully sardonic self-projection of the 'boy from Brisbane', as Clive James later called Porter.[15] Although set in Queensland, the play explores the protagonist's despair within a framework of Porter's experience in London, and enabled him to locate some of the sources of his personal sense of hopelessness. What he recognizes, and expresses through Woebegone in *The Losing Chance* is a resilience which grows from the capacity to observe despair, both his own and that of others. Thus, when Woebegone says, 'Mr Mainchance, I think you find yourself/Altogether too interesting a misanthrope/ To really despair', the insight is one which gives some direction, if not anything yet as firm as a purpose, to the budding writer. Hope, Woebegone insists, may be found where others might see only despair: 'Let your hurt/Glow with a fire reviewers dream of;/And let your grief grow outwards, filling/Lyrical gardens with the light of Hell'. This might be an agenda for Porter's later work.

While the protagonist in *The Losing Chance* is forced to agree that 'suffering is part of me/And my wholeness', a rival point of view, in the figure of Anson Jones, ridicules this 'world of ego stinging even itself'. A later book, *Narcissus* (1984), produced by Porter in collaboration with Australian artist Arthur Boyd, would explore the twin problems for the writer of paranoia and solipsism. In *The Losing Chance*, the main temptation is suicide. Like a Chekhovian character, however, Peter Mainchance bungles a first attempt and survives to question, in the third Act, the sources of his desperation. This is the liveliest part of the play, in which Porter's invention of Inspector Christopher Smart — the mad poet metamorphosed into a policeman who investigates a suicide attempt — has genuine theatrical flair. (Porter later resurrected this character in the poems 'Inspector Christopher Smart Calls', and 'The Return of Inspector Christopher Smart'.) One way out of the personal morass seems to be an attachment to the desirable Miss Upland, for whom 'movement/Is aphrodisiac'. The Barrier Reef island setting enters, *Tempest*-like, into the imagery of a prothalamion pronounced by the versatile Inspector Smart, whose eccentricity gives a bizarre operatic vitality to the fish of Australia's tropical northeast coast: 'At your wedding the leaping staggering fish/And coral roping out the sea will be/Your breathless partners . . .' The lifting hope of the play's third Act is not quelled by a telegram from the ubiquitous Adversary, who prophesies post-marital boredom: 'Remember the summer nights so long/When tedium turns to suburban song'. A romantic marriage is one route proffered for the imagination then; the other way is to accept and live with the madness the protagonist feels in himself. Inspector Smart advises: 'Wear your madness with authority, Peter'. Professor Woebegone agrees. He foresees the novice moving beyond mere romance or purpose-driven activity to the province of visionaries, 'the green persuasive kingdom' where perception rather than achievement is all-important.

After the play's first, failed suicide attempt, the moral drama infects even Manganese the millionaire, who admits, somewhat unconvincingly, that 'Mainchance has shown us all up/As gropers after certainty./Nothing but the purchase of death/Is certain . . .' This central certainty in Porter's life and writings is too easily transferred to the businessman. In the playscript, however, Mainchance makes his own plunge for certainty by shooting himself, this time fatally, leaving Christopher Smart, the poet-cum-inspector of souls to lament at his failure to lure Peter through into 'tempest-free' times. While the script as a whole does not exhibit great theatrical flair and its tonal variations are sometimes confusing, it does give form to the quiet desperation which informed its author's early years in London, at a time when he felt close to madness and, at times, driven towards suicide.

During 1952, when Porter subsisted in the basement flat at 23A Belsize Park Gardens, working as a clerk by day and developing a satiric attitude towards fashionable Hampstead by night, he was writing poems as well as his long verse drama. In the poetry, a major issue is a deep but inhibiting respect for the major writers of the past and present, which seems to be stifling his own creative energies. In an unpublished poem for Roger Covell called 'Wonder Boy', his persona expresses the burdens of an anxiety of influence: 'Harnessing my envy/ Of the great and classic/(for our own time merely)/I think of my duty . . .' His transplantation 'like a radish' to English soil has given him only a sense of failure. The nervous system of this poem suggests a closeness to autobiography as the fictional persona sallies bravely forth but withers: 'Prometheus from the suburbs/Timidly complaining/By way of insurrection'. The putative culture-hero has a fire in his belly but already senses the limits of its effectiveness — the wonder boy is not a god. In a love poem, 'To Jill Through Cold and Night', the imagery of winter again prevails as the lover implores: 'Unwind me/ Out of the springs that stiffen me,/Out of the love of winter binding'. The inevitable rupture with the first woman he has loved occurs; the hurt and disappointment are patent in the verse of this time, which is often tortured and introspective.

At the end of 1952 Covell left the flat at Belsize Park and returned to Brisbane to marry. Porter spent six months in the flat on his own, and during this time reached his lowest ebb. Thin and distraught, he had quit his job at International Paints and had no work except for a casual job painting and cleaning a boat on the Thames. The breach with Jill Neville had led to what was clearly enough a nervous breakdown, but one which he stoically refused to have treated by psychoanalysis or drugs. During the summer of 1953, following the lead of his fictional Peter Mainchance, Porter twice made attempts to kill himself. Twice he turned on the gas in the kitchen of his flat, hoping to find some peace in death at least. Unlike Mainchance, he failed the first and the second, more serious attempt. This was the nadir of his 'tied up fitful twenties'. His desperation was eased, somewhat, when he was joined in the flat by David Lumsdaine, who shared his love of music. Lumsdaine later became a lecturer in music at King's College, London, before returning to Australia and becoming a successful composer there. In 1953 Porter and Lumsdaine worked together on an opera which was to be a Brechtian version of Voltaire's *Candide*, entitled 'The Best of all Possible Worlds', but this but was never finished. The memory of this aborted project was in Porter's mind when he later called his twelfth volume *Possible Worlds* (1989).

In the latter half of 1953 Porter and Lumsdaine 'dropped out'. It was a time of penury and parties, long days in bed, and sexual experiment with a number of women. The verse of this time shows both an attraction to and repulsion from hedonistic party-going and the reawakening of sexual activity. Arresting phrases or lines of verse occur, some of which reappear in later work, though vitality and control were not yet in kilter. The nervous breakdown did not resolve for Porter the tension between contending puritanical and hedonistic tendencies. Nor did the shores of home suggest to him even in his desperation more than a temporary refuge. 'Home' in one poem of this time is 'a sluggish coast' and the myth of the Australian emptiness which gained such currency in the 1950s in Patrick White's work, amongst others, is rehearsed in lines which suggest no possibility of regeneration there: 'My country's bleeding soul will show/There is nothing in its heart to know/And I am empty like its tongue'.

In another unpublished piece, titled 'Nothing to Declare', the persona of the displaced person arrives in an early formulation when Porter's autobiographical speaker recalls, from an imagined quay, some places in London which have been etched on the writer's personal map: 'When I have gone, Portobello Road and Camden Town,/The hugging boats of Chelsea, the grey disdain/Of Cheyne Walk, hermetic Hampstead,/Various thick Kilburn and those places/Where I have pretended to be a Londoner/Will hardly notice that an integer has moved'. Yet this cold, indifferent city is also personified as 'Stipendiary Judge and endless Artificer', whose judgement upon the empty provincial may contain some wisdom. It is not surprising, then, that when the ship taking Porter back to Australia in early 1954 was only a few days out from port he already knew he had made a mistake in leaving England and not facing up to its judgements and challenges, and wrote accordingly to his continuing correspondent, Jill Neville. Porter stayed with various friends and then with Roger Covell and his wife in Brisbane, but was soon saving for his return journey. A scaffolding company sacked him because of his long hair. (He recalled the boss in a later poem: 'the bristly boss speaking/with a captain's certainty to the clerk,/"we run a neat ship here"' *CP*, 140−1.) He found other work by day and wrote prolifically for English newspapers and periodicals, without success, by night. Mentally, he was back in England. It would be another twenty years before he returned to Australia. The sense of a divided self was projected in 'Sick Man's Jewel' (*OBTB, CP* 17−18) in the image of a person who is paradoxically 'At Home, Away'.

After a ten months absence, Porter was indeed back in England. The feeling that Australia was not his 'home', and so could not provide the security he sought, hardened his resolve to succeed as a writer in England. After spending several desperate months in Guildford, Surrey (the only time he has lived outside London), Porter moved into a room in Artesian Road behind the Westbourne Grove Odeon. Although he had written around fifty poems each year since he was sixteen, it was not until 1955 that poetry became a major focus in his life. His room in Artesian Road suited the gloom which still hung over him, but it offered an element of Orwellian theatre which he seemed to relish:

> [The room] was about as large as a coffin on its side, and looked out over a cat-piss-saturated oblong of green, the chosen tilting ground of a very drunk man and his shrewish wife who

would shout abuse at each other for hours each night beneath my window. There was often blood in the streets, and the noise of bottles breaking and demented people screaming rose on the air regularly after ten o'clock.[16]

Here was a milieu suited to a dispersal of Porter's gloom into the beleaguered souls he saw around him. But the drifting continued. From Westbourne Grove he moved back into the Kensington area, where he slept in the corridor of a friend's basement flat until his flatmate left for Canada. He stayed there until the beginning of 1960, with an apparent indifference to material comforts or status objects, which Jill Neville attempted to ridicule in *Fall-Girl*. Porter has himself referred his 'Oblomov-like indifference' to his surroundings in this flat which was known as 'the tomb'.[17] What was particularly important during this time was that Porter's sense of place was becoming localized in a moral map of the Kensington and Chelsea area, as well as other parts of London, which would inform his poems of the late 1950s and 1960s:

> I got to like this part of London, but my prejudices about Kensington and Chelsea are as precise as a police map — street by street and even one side of a street after another, I approve or disapprove of the whole area. In my early days in London, I had spent a lot of time at the boat colony near Lots Road Power Station, but even then, long before boutiques and The Drug Store, I took no pleasure in the King's Road itself ... Each part of London has a likeable or dislikeable quality, which makes sense of Dr Johnson's famous observation about never becoming tired of the city. It is a world within a world.[18]

This was a London which would inscribe itself in a variety of ways in Porter's work. It was not the picturesque London of the tourist brochures: Porter has never visited the Tower of London and has been to Westminster Abbey only a few times, and then always on literary or official occasions, such as the memorial services for Auden and Larkin. The old heart of Empire provided for Porter a 'map of ordinariness', within which the temperamentally disappointed Australian could think, feel and act. Yet it was a map within which points of incandescence might occasionally flare.

The major spur to advancement came with Porter's introduction into an informal association of writers which became known as 'the Group'.[19] Soon after his return to England from Australia Porter had met and discussed his poetry with Julian Cooper, who had known Philip Hobsbaum while at Cambridge University. Hobsbaum had convened regular meetings since 1952 with a group of young poets connected with the magazine *Delta* at Cambridge, including Peter Redgrove and Christopher Levenson, in which the rigorous New Critical procedures of F. R. Leavis were applied to contemporary poetry. The slightly older Ted Hughes also had connections with this Cambridge group.[20] When Hobsbaum moved to London in 1955 to teach, he transferred similar weekly meetings to his flat in Kendal Street, off the Edgware Road, and Porter and a number of other aspiring younger poets were invited to attend. The basis upon which such writers were recruited appears to have been 'hunches' by leading figures as to what they might later write. There was no formal membership and no subscription.[21]

The significance of this association to Porter was greater than has sometimes been recognized. In response to Blake Morrison's assertion that Porter's 'brief allegiance to the Group ... never counted for much even in the 1960s',[22] Philip Hobsbaum, after checking his records, retorted that from late 1955 to 1965, when the Group adapted itself into the Poetry Society, Porter 'seldom missed a meeting and in fact read at the Group himself on eighteen occasions, more than any other participant except George MacBeth and Peter Redgrove'.[23] For a person not generally a joiner of clubs or societies, this is indeed a record of sustained interest, loyalty and involvement. Over a period of nine years Porter presented 106 poems for discussion at meetings of the Group, many of which appeared later, often considerably revised in the light of these discussions, in his published books of the early 1960s. At another level Morrison's comment is perhaps accurate: Porter has always been an eccentric writer and a thinker in his own right who has resisted easy classification within schools, camps or groups; and the Group, while enhancing his sense of audience, did not deter him from exploring in verse his own particular obsessions and fears.

The obvious attraction the Group held for Porter was that it allowed him to meet men and women to whom poetry mattered. Their range of nationalities, allegiances and outlooks would also have been appealing. Hobsbaum, Redgrove, Anthony Smith and David Wevill had attended Cambridge and Edward Lucie-Smith, Alan Brownjohn, George MacBeth and Adrian Mitchell were from Oxford. But the internationalism of the Group was also apparent, with Taner Baybars from Turkish Cyprus, Zulfikar Ghose from Pakistan, and Margaret Owen from Belfast, rubbing shoulders with writers from the English provinces, including Martin Bell from Southampton, and Rosemary Joseph from Northumberland. Given such a variety of backgrounds, Roger Garfitt's comment about the Group that 'divided they stand, united they fall' is easily enough understood.[24] Porter was bemused but pleased at his acceptance: 'I was very grateful to these people, who ... accepted me as one of them overnight, like that: no question of you're a colonial, or un-educated, or who are you, or anything like that'.[25] Such a group seemed to emphasize the possibilities of a literary capital, where a wide variety of talents could co-exist and learn from each other without stamping on each other's individuality.

With hindsight, the Group may be seen as a dual reaction, to both the neo-Romantics of the 1940s — Dylan Thomas, George Barker, Lawrence Durrell, W. S. Graham and others, and the Movement poets of the early 1950s, such as Robert Conquest, D. J. Enright, Elizabeth Jennings, Kingsley Amis, Thom Gunn, Philip Larkin, John Wain and Donald Davie. Porter saw the Movement as a strong reaction, perhaps an over-reaction, to forties Romanticism, 'a time-bomb planted by the Labour landslide of 1945, a bomb which went off only after the restoration of Tory rule in the 1950s'.[26] Its major poet was Philip Larkin, whom Porter admired as the most accomplished poet of the post-war years, as an unmistakeable voice. Porter later dedicated the poem 'Going to Parties' (*FF* 55–6) to Larkin, a piece which affirms both the understated power of Larkin's poetic voice and its author's recognition of limits in both art and living.

The Movement, to which the Group has been likened, was not in Porter's view motivated primarily by politics:

> It stood for Little Englandism, anti-phonyism, anti-pretension, realism of an austere sort, hostility to Modernism and dislike of obscurity or panjandrum-attitudes in poetry. In practice, it often produced a low-key naturalistic, disenchanted verse, preoccupied by moral discrimination rather than by aesthetic uplift. One thing Movement poets were united on — verse must be formal, shapely and unclouded by turgid language. Purple passages and flights of fancy were out.[27]

For all its surface control, a wilder element was obvious in certain of the Group poets. Some of them, including Porter and George MacBeth, seemed well attuned to the sentiments of the Angry Young Men of the London theatre scene in the mid to late 1950s, John Osborne and Arnold Wesker. Porter and George MacBeth in particular became adept at the verse monologue which rages and rails at the injustices and absurdities of society, and Alan Brownjohn, a teacher and later a Labour Party candidate for Parliament, wrote verse potent for its low-key social criticism. The range of attitudes and styles of these and other Group poets renders them unclassifiable in the way the Movement poets' manifestos and practice linked its members.

Other forces were at work in British society, and beyond. The British Empire was dispersing into a decentralized Commonwealth, with a resultant loss of purpose among those who had seen their principal role as colonizers. Colonial attitudes were, however, slow to die. The Suez Crisis of 1956 threw Britain's 'moral leadership' into further doubt, just as Australian Prime Minister Menzies's hasty support of Britain in this misguided intervention led to deep distrust in sections of the Australian populace. The Soviet invasion of Hungary and the Suez fiasco created a climate of political debate hotter than at any time since the end of the Second World War. Even the political neutrality of the Movement poets was shaken by these events, and a number of them moved to the right politically.[28] The generally younger Group poets, who largely shared the anti-romantic and anti-sentimental preferences of the Movement writers, nevertheless appeared to be more attracted to the political left. Comparing the Movement and Group writers, G. S. Fraser commented: 'There is more political radicalism [among the Group poets], a sympathy with the New Left'.[29] Porter shared this preference, and developed a close friendship with the somewhat older Martin Bell, a member of the Communist Party who, in Porter's words 'took Communism seriously, both for its doctrine and for its strategic directives', but left the Party after learning 'the full horror of Stalinism' in the immediate post-war period.[30]

Members of the Group were not greatly concerned with manifestos or political posturing however. Central to the Group's activities were the weekly poetry workshop meetings, which some members attended regularly, and others quite irregularly.[31]. A mailing list and cyclostyled sheets provided scope for continuity and close reading of the texts. Frank discussion was encouraged and the only rule was that the writer could not intervene to explain his or her intentions. An important principle was enacted: that, 'the process by which words work in poetry is something open to rational examination'.[32] In keeping with this view, no alcohol was served at these meetings though they would sometimes continue at nearby pubs. Hobsbaum commented that, 'The pressures against anyone writing at his best are very strong. The writer needs a community to keep him in touch with his audience'.[33] This was certainly true of Porter, whose preparation for and involvement in meetings, as well as the friendships he developed, offered him his first real hope of reaching a wider reading public.

In any communal activity, certain individuals stand out. Philip Hobsbaum's role as revivalist of the Group in London was of particular importance, though Porter appears to have been more sceptical from the start than those who had read English at Cambridge about the Leavis-inspired patterns of tutelage; furthermore, in Porter's view the restrictions of the convener's personality meant that none of the poets did his or her best work while Hobsbaum was in charge.[34] Nevertheless, Hobsbaum's particular moral and analytical sense helped them in technical ways, especially in their use of traditional verse forms. As time went on, Porter increasingly recognized the negative impact of Leavis's attitudes and values on contemporary literature and criticism, and wrote a stinging rebuttal of these influences in an article in *The New Review* in 1974.[35] But the disciplined approach to discussion of verse forms, meaning and morality were clearly of great significance to an unpublished writer who, for all his ambition and flashes of talent, had not been trained in the rigours (nor rigor mortises, for that matter) of a university education. When the venue for meetings changed to Ted Lucie-Smith's home in Chelsea in 1958 so too did the atmosphere, and individuals developed more in their own ways. Roger Garfitt has discerned in Lucie-Smith's verse a 'master of the modest, perfectly-articulated statement', and that his 'courteous, fluent irony can be a surprising guide'.[36] It seems likely that these qualities were evident in the person of Lucie-Smith in his role as chairman, as well as in his verse.

For Peter Porter, two individuals stood out in the Group — Martin Bell and Peter Redgrove. Bell was the eccentric radical from Southampton who, like Porter, hailed from outside the Oxbridge élite, an idealist from a lower middle-class provincial background, where the reigning values of respectability and sexual repression had delayed but not deformed his poetic ambitions. Porter clearly admired his fighting spirit: 'There was never anything provincial about Bell's attitude to literature, and however hard he fought the class war he was not intimidated in the slightest by literati from the public schools and ancient universities. He knew he had read more than they had, and he knew he could write better'.[37] The fellow-feeling in this statement is patent both in its over-compensating exaggeration and in its assertion of the energies generated by deprivation. Porter's observation that Bell's wartime experience had 'freed him from English provincialism' is also revealing.[38] For Porter, metropolitanism implied the goal of a comprehensive cultural reference which Bell's wide reading in European (especially French) as well as English literature and history satisfied. For Porter, as for other members of the Group, Bell was the key figure in meetings, their 'epicentre'. But it was outside the arranged meetings that Porter most fully appreciated Martin Bell's brilliant talk, which ranged 'in all directions and into hundreds of special regions'.[39] Moreover, the two friends shared certain aberrant prejudices in terms of Group orthodoxy: they admired Auden and the 1930s, and disapproved of D. H. Lawrence, whom Leavis had consecrated. Opera was a shared love, especially Stravinsky and Mozart, with Verdi as Bell's favourite. After the breakdown of his marriage in the mid-1960s, his departure for Leeds in 1967 and a long period of alcoholism, Bell died in 1978. From his self-appointed exile, he had made Porter his literary executor.

The other major figure for Porter in the Group, Peter Redgrove, had invited Bell to meetings after encountering him in a pub at Chiswick.[40] Redgrove was born in London, but from the early 1960s chose to live in England's West Country. He had studied natural sciences at Cambridge before becoming more interested in poetry and psychoanalysis. A giant of a man, Redgrove struck Porter immediately as 'an absolutely original person':

From the very beginning I knew that I could never be so absolute in my pursuit of poetic reality as he was, and that I would never be able to believe that in language and its sacred ritualisation as art man had a tool which could make him a god, which has always been Redgrove's prime canon.[41]

There is no question here of Porter's looking for a model on whom he could base his own work. The practice of writing poetry out of his own needs and desires was already well established. Instead of increasing the burdensome anxiety of influence, Group contacts reduced it. Redgrove, as person and poet, was entirely different from Porter in his background and thinking, and this is what appealed to the Australian. As a contemporary of Ted Hughes at Cambridge, Redgrove was somewhat overshadowed by the Laureate-to-be, but Porter has remained a strong supporter of Redgrove, believing him to be the most gifted poet working in Britain, the poet who, although his writing may be 'overstrenuous, hypertrophied or obsessional', is the one with 'the largest vision and the most spectacular trust in language'.[42] Although he does not deny Hughes's power as a poet, the preference of the present age for Hughes over Redgrove signifies, for Porter, a preference for 'despairing chic' over optimism. This preference for Redgrove seems to be born not merely of perverseness or loyalty to an old friend; it is typical of Porter's continuing interest in rescuing worthy artists from undeserved neglect, and in so doing broadening the critical response to their work. (Among past poets whom Porter has set out to 'rescue' for himself — and having no likeness to Redgrove — are the Earl of Rochester, Christopher Smart and Christina Rossetti.) A further aspect of Redgrove's work which appeals to Porter, and which differs from his own, is what the Australian poet Les A. Murray has called 'the quality of sprawl' — the long lines and rhetorical orchestration giving a sense of magic to place, in Redgrove's case, of Cornwall and the West Country.[43] While Porter remains centrally a poet of the built rather than the natural environment, who places a high value on rational discourse, even as he disrupts it, the lyrical and liturgical element in Redgrove's work, and its capacity to hint at a mystery behind ordinary objects and events, are recurrent elements in a quite different way in Porter's writings.

The Group meetings were complemented in the late 1950s by a monthly 'open house' for poets at the home of G. S. Fraser and his wife in Chelsea. One of the participants, Anthony Thwaite, has described these occasions:

Poets came, with manuscripts, and read, drank, stubbed out their cigarettes on George's first editions of Yeats, Eliot, Pound, etc., vomited on the Fraser carpets, quarrelled on the stairs, sometimes bashed one another with fists or bottles, and in general behaved as poets are (or were) supposed to behave.[44]

Thwaite recalls his first impression of Porter at these meetings, as part of an initially indistinguishable 'troika or trinity or three-headed monster — quietly snarling, vaguely truculent and conspiratorial. It took a while longer to establish that this was — or they were — three distinct persons — Martin Bell, Peter Redgrove, and Peter Porter'.[45] The interpretation of Porter's Australian background, which has been a recurrent attribution by reviewers and critics throughout his already extensive career as a writer in London, is quite revealing. Thwaite observed that, in 1957, in England, the term 'Australian' carried 'such a load of associated boring provincial peripheralism that it must have been hard to bear such a burden'.[46] Thwaite recognized, as did Margaret Owen, that a clue to Porter's background in those years was a certain vulgarity in his use of language, a vulgarity which Porter himself may have perceived more specifically as an inheritance from his mother's chamberpot humour rather than from his native land. But the brittle gentility of English society which Porter was offending should not be underestimated. Two instances illustrate this. In the first, Porter's use of the word 'piss' twice in a poem about cats was questioned by the controllers of the BBC's Third Programme; in the second, the British Museum withdrew from their exhibition of manuscripts by contemporary poets a poster showing one of Porter's poems, 'An Anthropologist's Confession' (*PAAM, CP* 59–60), in his own handwriting, which included words such as 'vagina' and 'semen'.[47] Although Porter was affected by a sense of the enormous cultural riches in Europe, which he strove in an over-reactive way to know and understand as rapidly as he could, he seems not to have made any conscious efforts to curb his Queenslander's accent. His acceptance by the 'finishing school' of the Group in the late 1950s and early 1960s meant that originality and difference were prized rather than rejected, in spite of the persistence in British society at large of those hegemonies of which Patrick White had been a victim at Cheltenham School some thirty years earlier. White's Australian origins seemed a 'deformity' and effectively shut him up: 'I hardly dared open my mouth for fear of the toads which might tumble out, and the curled lips, cold eyes waiting to receive renewed evidence of what made me unacceptable to the British ruling class'.[48]

But within the Group Porter's Australian background provided much more than an exotic diversion; indeed, Margaret Owen perceived him as an original mind, whose identity gained additional force by what she assumed derived from his background:

No-one else had Porter's note of pain and indignation. But he also had a kind of gracelessness which was potent and surely

Australian. I can't think of anyone else who would have had a
first line like 'The rain is sick upon the ground outside' and,
later on in the same poem 'by dawn when the waxy words
have died/In the entering ear, comes the wet/Responsible
kiss ...' I use the word graceless with caution but some of the
writing in the group seemed pursed up and gracile by comparison and what I want to convey is the urgent, unabashed
lack of charm.[49]

Porter was to Margaret Owen an 'acid, gloomy young man' whose
glasses could give him a 'Mr Magoo Jun. aspect' in the street or in
pubs, but like others who knew Porter in his late twenties, she was
deeply impressed with him as a conversationalist, both as speaker and
listener: 'He was always a good listener, perhaps one of the best
conversationalists I had met because he always heard what you said
and not just what was going on in his own head'.[50] Her chief criticism
of his poetry was its ellipses and ambiguities which, with other members
of the Group, she would question, but concluded that these resulted
less from compression than from 'downright individually perverse
connections, perhaps because his correlatives were deep in his
Australian background'.[51] Clearly, Porter was already to a large extent
his own man and would take only so much notice of others' views.
Owen suspects that 'a slight impatience with Leavis dogmatism helped
push him along towards his own kind of particularity'.[52] From these
and other comments, it seems that Porter's own curious amalgam of
originality was strengthened rather than dissipated by his first major
socializing experience as a poet. In the early 1960s he would be
launched on the British poetry scene as a 'social' poet, but the deep
personal compulsions which had made him already an inveterate writer
of verse would drive him continually to explore the boundaries of his
talent, in his own particular ways.

5

CARNIVAL WINTER

◆

Seldom have a city and decade come together so neatly to fit the slogans as did 'swinging London' in the 'swinging sixties'. 'Swinging London' appears to have been an American invention, coined by the London-based American editor of *Encounter*, Melvin Lasky.[1] The expression was popularized by a *Time* magazine cover story, in which Lasky was quoted as saying:

> London is the only European metropolis that has managed to maintain a combination of greenness and greyness, vitality and yet a certain gentleness. Paris hasn't got it. Rome is oppressive. Berlin is a special case. All the others are just villages.[2]

Vigour, colour, excitement, experimentation were all to be found in London. In retrospect, the decade would assume a warm afterglow of nostalgia, as of cheerfully misspent youth, and a London *Times* editorialist was not untypical in presenting it as a time when Britain 'truly emerged from its postwar depression and became a country of joyful and envied achievement'.[3] This mythical reconstruction is typical in its personification of a Britain which was seen to be breaking through rather than breaking down. The liberationist ethic which has informed many accounts of the decade is based on greater sexual freedoms, the end of Tory rule, the beginning of Harold Wilson's Labour Government in 1964, vigorous attacks on censorship and privilege, and an increasing energy and invention in popular music, drama and the visual and verbal arts. This is the decade of the Beatles, the Rolling Stones, Radio Caroline and the celebrated attacks on censorship in the *Lady Chatterly's Lover*, *Fanny Hill* and *The Naked Lunch* trials.

Of course a darker side to the media-inspired euphoria also existed. Scandal (as in the Profumo affair) and satire (as in *Private Eye*) marched side by side with the 'ban the Bomb' demonstrators. The sardonic, rebellious figure which Peter Porter thrust upon the poetry stage in London early in the decade contributed to a prevailing mood of irony and excitement. This figure was not simply conjured up for the occasion: it had long been brewing inside him, since his personal crises of the 1950s. Although Porter's malcontent figures in celebrated poems such as 'John Marston Advises Anger' (*OBTB*, *CP* 17), 'Metamorphosis' (*OBTB*, *CP* 10−11) or 'Beast and the Beauty' (*OBTB*, *CP* 12), grew in part from the theatrical iconoclasm of the 'angry young men' school of playwrights, their moods, emphases and verbal constructions were also original to Porter and arose from his own experience, in particular a growing awareness of the experience of exile. The 'social' poet who was introduced to the London public as a swashbuckling satirist of contemporary mores and manners was a young man deeply divided, and involved in the resolution of personal crises.

Yet an account of the public events in Peter Porter's life in the 1960s sounds like a record of increasing stabilization and success. In 1959 he moved from Bumpus Bookshop in Oxford Street when he obtained a job as a copywriter at Notley's advertising agency. In 1960, he married Jannice Henry, a nurse from Marlow in Surrey, whom he had met in 1958, soon after she returned from six months teaching English in Italy. Together they attended the Amsterdam Arts Festival in Porter's brief first encounter with Europe. Their first daughter, Katherine, was born in 1962, and a second daughter, Jane, in 1965. Porter's first book of poems, *Once Bitten, Twice Bitten*, was published by Scorpion Press in 1961, and was rapidly followed by a selection of twenty-six poems in the popular *Penguin Modern Poets* series, with Kingsley Amis and Dom Moraes, in April 1962. Porter's work received critical acclaim from many quarters. Selections of his poems were published in *New Poems 1962*, Lucie-Smith and Hobsbaum's *A Group Anthology* (1963) and Alvarez's revised edition of *The New Poetry* (1966). One reviewer bravely albeit prematurely called Porter an 'established' poet.[4] A second collection, *Poems Ancient and Modern*, was published by Scorpion Press in 1964, and in 1969 the same small press produced Porter's third volume, *A Porter Folio*.

In a decade when public readings were becoming common, Porter found himself in considerable demand. In 1962 he addressed a conference on 'Poetry and the Universities' at University College in London, and was later quoted in *The Times* as saying that he had been merely 'shadow boxing' with his writing in Australia, with only a few friends as listeners, but that now he perceived the beginnings of connection with an audience. (Such comments, however apposite, would earn him no friends in Australia.) In 1963 he was described by a reviewer as an 'extrovert' when he read at the Royal Court Theatre in Sloane Square with Paul Roche, Alan Brownjohn and Christopher Logue. His was the kind of poetry, the reviewer noted, which communicates well and 'registers without the need for special advocacy'.[5] At another festival of poetry at the Royal Court Theatre in 1965, Porter met the Australian artist Arthur Boyd, who was arranging the stage design. The two began what has proved one of the most successful artistic partnerships in modern art and poetry. Their work together on books of art and verse which skilfully counterpoint each other first bore fruit in 1973, with the publication of *Jonah*. External markers such as Porter's published books and public appearances suggest success: the apparently stable home life of the family man in his thirties seems reinforced by a rising gradient of professional achievement and recognition. There is a pulling in a different direction from the hedonistic sixties. This pattern of progress reached a high point in 1970 with the publication of Porter's fourth volume, *The Last of England*, by Oxford University Press.

But the realities of this decade for Porter were more fraught and uncertain than his outward achievements suggest. On the practical level, how, among the babble of contending voices, did Peter Porter make his voice heard in London? When three poems by Porter were published in the Cambridge magazine *Delta*, in 1958, they were his first to appear in print since the *Toowoomba Grammar School Magazine* contributions in 1946. Yet by the end of the 1950s Porter had composed some five hundred pieces. The image of a fresh and fully formed talent, in 1961, which press reports encouraged, was illusory. Porter was already an inveterate writer with a long backlist of manuscripts and rejection slips from magazines and newspapers when he was first noticed as a writer of talent. It might seem from the outside that the 1960s provided a moment, which the talented young poet waiting in the wings turned to his advantage. However Porter's experience of 'making a name' as a writer in London shows a higher degree of accident than such retrospective inevitability allows.

◆

The publication of Porter's first book, *Once Bitten, Twice Bitten*, occurred as a result of an interest taken in his writing not by a member of the Group, but by Edwin Brock, a poet who also worked in advertising. The two often read together at meetings and Brock was later unwittingly helpful in getting Porter out of advertising — he invited Porter to work with him at his agency, S. H. Benson, which Porter hated so much he left after only five months, in September 1968. Brock suggested that Porter show his poems to John Rolph, director of Scorpion Press, which had published as its first book a selection of Brock's own work. Porter was introduced to Rolph and his partner over dinner at an Italian restaurant at Cambridge Circus, and presented the publishers with a thick file of his verse. Impressed with the 'polished professionalism' of Porter's work, Rolph offered to publish a selection chosen by the author.[6] A happy accident occurred in the publication of the book. Of the 500 copies published in 1961, 309 copies, including the twenty-five 'specials' for signing, were delivered from the bindery not in the grey-green covers as ordered but in purple cloth, which *Printing News* later described as 'striking'. In the days of sedate book designs, at least in poetry, it seemed a bold statement for a first book. Unfortunately the second issue of 191 copies were bound in the dark greyish green cloth as ordered.[7] The story behind the title of this volume is also entertaining. Porter originally offered 'The Cage of Discontent', from the poem 'John Marston Advises Anger', as its name. The publishers, however, preferred *Once Bitten, Twice Bitten* from the poem of the same title, evidently because of its apparent relevance to their name, Scorpion Press. Porter's first book contained forty-two poems. Before its publication in 1961 only eight of his poems had previously appeared in print in England.[8]

In spite of its small print run, its considerable expense at 15 shillings a copy, and the virtually unknown press from which it sprang, *Once Bitten, Twice Bitten* won responses from a wide spectrum of reviewers and critics. Not surprisingly, in view of the poems of social critique and satire, reviewers for the left-wing British press were generally more appreciative than those for the right-wing papers. Richard Kell, the left-liberal *Guardian's* reviewer, after noting Porter's 'concentration on death and physical and spiritual disease', found the book 'energetic, witty, skilful and serious'.[9] While noting the book's occasional reminders of Auden in its compressed satirical summaries and psychological diagnoses, Kell also recognized Porter's original ways of thinking in verse:

> Under the clever satire there is sometimes a deep pathos, between the lines a reticent compassion. Technically the verse is excellent of its kind — hard, brisk, strongly shaped with a deliberate irregularity in its stresses. Occasionally the wit results in an obliqueness which, where the matter is trivial, simply calls attention to its own silliness ... But the striking, rough-edged verse easily accommodates such imperfections.[10]

Where Kell recognized the compassion, cleverness and poetic force of the book, the right-wing *Daily Telegraph's* reviewer, Anthony Cronin, could see only surfaces: 'King's Road character sketches, coffee bars, drunken parties, jeans and sweaters'. Rather than giving due credit to the new poet's irony, Cronin called him 'confused' in his values (that is, critical of upper middle-class philistinism) and damned with faint praise the Australian's 'genial enough talent'.[11] As was the case with most newspaper reviews, there was no attempt (or space) to substantiate such criticism. Another kind of conservative response came from poet and critic Donald Davie, in the *Spectator*,[12] with whom Porter was to have continuing critical disagreements over Davie's preference for 'smooth' verse and right-wing ideology. While acknowledging that Porter was a 'promising poet', Davie failed to recognize the distinctiveness of Porter's preoccupations and styles from 'the better British verse of the Fifties'. Davie also criticized the metrical irregularities in the poem 'A Christmas Recalled' — in which the child's point of view would surely be less convincingly rendered in smooth verse. By calling the Australian newcomer 'a much shaggier, a much rougher workman' than Thom Gunn, Davie implied a lack of discipline and control.

While class-based ideologies and images informed some British reviewers' responses to *Once Bitten, Twice Bitten*, another typology based on national differences also operated. The adjective 'Australian', which at this time became a regular descriptor for Porter, usually connoted the stereotype of a brash, untutored masculinity. As Margaret Owen had noticed in readings of the Group, some of Porter's images were vulgar by 'well bred' English standards. But a tendency among some of London's intelligentsia to typecast all Australians as wild denizens of Kangaroo Valley was clearly misguided in Porter's case. What it primarily left out of account was 'that soft thing, the self', the vulnerable self which inhabits corners of Porter's first book of poems. It also underplayed his intelligence and learning.

Al Alvarez, who was collecting exempla of a new poetry which was brash, aggressive and an affront to British gentility, set a fashion among critics in his praise of the 'public' or 'social' Porter, which he linked with his Australian background:

> Peter Porter, an Australian, writes the kind of tough, aggressive, prize-fighter verse that has become fashionable in the last year or two. *Once Bitten, Twice Bitten* (Scorpion Press, 15s.) has echoes of Gunn, Hughes, Redgrove, and even Amis. Yet Porter is not simply, or even very derivative. He seems genuinely to be trying to use these influences to establish an Australian poetic voice which does not seem simply a comic dialect: he wants a language which can move easily from knowledge to power.[13]

The assumption here that Australian 'dialect' would immediately be thought of as comic in the literary capital has a history going back to Oscar Wilde and beyond. What interested Alvarez was its new capacity, in a poet such as Porter, to offer a counteractive power to the effeteness of British poetry, and in this he was closer to the mark. Like regional and working-class dialects, which were admitted into London theatre in the 1960s through actors such as Albert Finney, Tom Courtenay and Michael Caine, the Australian voice in poetry could challenge the traditional keepers of the keys. While the language which Alvarez perceived in Porter's first book placed him as a fighter and challenger, he seemed slightly bemused by an apparent love-hate relationship with the polite upper middle-classes:

> Porter is much possessed by Chelsea. He can't stand the place, with its Oxford-tweedy admen, debutante intellectuals, be-jeaned teasers, smart marriages, artistic pretension and ghastly parties. Yet he can't stop talking about it. It seems an unneces-

sary form of masochism. After all, London is mercifully large; no-one has to be victimised to the point of mania by one boring part of it. I feel that the quickest way for Porter to fulfil his distinctive role as a poet is simply to change his address.[14]

As it happened, Porter did leave his South Kensington flat — he did not live in Chelsea — after he married Jannice Henry; they moved to the Paddington area, which would in its turn infiltrate his later poetry as another 'moral map'. But Porter's cognitive and emotional mapping of London has never been as changeable as Alvarez's flippant ejection advice suggested it should be. During the 1950s Porter had become familiar enough with Kensington and Chelsea to sense both the attractions and traps. This was a necessary precondition for his brand of satire and social critique.

The ascription of the 'Australian' identity to Porter in England was carried further by the anonymous *Times Literary Supplement* reviewer (perhaps G. S. Fraser), who linked the outlook in Porter's first book with a 'democratic' tradition in Australia:

Mr Porter is a young Australian, fiercely democratic like so many of his countrymen, and yet, settled in England, angrily fascinated, like an Elizabethan malcontent, with what he sees as a society of expense-account men putting on the airs of an aristocracy, or of aristocrats surrendering to expense-account standards.[15]

In this perceptive and sympathetic review, the *Times* critic noted the poet's invitation to read *Once Bitten, Twice Bitten* with a 'sociological eye'; he also noted with approval the similarity of some poems to brilliantly condensed short stories. Like Alvarez, this reviewer saw strategic advantages in the point of view of the outsider. It is worth noting here that Colin Wilson's book, *The Outsider*, appeared in 1956, and in paperback in 1963, and was achieving cult following during this period. A vantage point such as Porter was establishing, although a basement room outlook, was held by some British reviewers in the 1960s to be a literary advantage, however uncomfortable it might be for the poet.

The publication of Porter's first book was not widely noticed in Australia. One early response, Ray Mathew's review in the *Bulletin*,[16] prefigured many later discussions of Porter's work in its ambivalent desire to claim Porter as an Australian, but to chastise him for leaving his native land. The British critical response to Porter's mentor Auden, after he left England for America in 1939, had provided similar difficulties for British critics, who tended to divide him into the 'English' Auden and the 'American' Auden, with a strong preference for the former and a castigation of the latter. Porter himself has been a strong supporter of the later, expatriated Auden's work. Questions of patriotism and national pride were of course greater in Auden's case, given the timing of his departure on the eve of war, but Australians in the post-war years were also sensitive about the slur which might be implied by the departure of the talented from their shores. Mathew's review of *Once Bitten, Twice Bitten* headlined the issue of expatriation: 'From Brisbane to Kensington'. The reviewer asserted:

> Mr Porter writes like an Australian — not only because of his expatriate intolerance of London highlights but because he has the Australian faith in fair-dinkumness as more important than fine artifice.

This observation has the makings of an interesting argument, but his subsequent assertion that Porter fails on both grounds — in Australian 'honesty' and British 'artifice' — is too glib. It presumes, before it argues, that the no-man's land of the expatriate is uncreative territory, where the expatriate is doomed to artistic failure. A more careful reading of Porter's work and career dispels this myth, though it remains a shibboleth of certain critics. One of the present book's central arguments is that Porter's 'spirit' — his spirited response to life — is stimulated by expatriation and a sense of exile.

Far less prescriptive than Mathew's review were Sylvia Lawson's comments in the Sydney-based magazine *Nation*.[17] Like some of the British reviewers, Lawson found parallels between Porter's poetry and Osborne's play *Look Back In Anger*:

> Like Osborne's Jimmy, this Porter finds himself out of gear, and doesn't want to be in. He suspects cultural veneers, despises middle-class goals and upper-class politics. The public wrong and the private grievance are all mixed up in his chagrin, but he leaves his inner confusions largely unexamined, and uses his anger as a driving force.[18]

Lawson engages with the manoeuvres of the early Porter with some of the finesse and sympathy accorded to his later work by another critic in Australia, Dorothy Green.[19] While acknowledging the influence of Auden in Porter's 'plain-spoken habits of style and grim tones of voice', Lawson points out how social criticism is present in Porter's book:

> ... not, as in the minor work of the thirties, in the form of forced conclusions; it is ingrained in the view point, and helps to give the work its quarrelsome vitality. Peter Porter is obsessed by the horrors of admass, all that is oversized, over-coloured, neon-lit and screaming; his language, plain and

cryptic by turns, slangy and littered with trade-names, contains the admass, jungle-world and implies his own rage at the penetration of jargon into thought. London life in these poems is a boiling and jangling surface ... [20]

Lawson acknowledges the pity extended to individuals in these poems, especially the poor, the old and the dispossessed, but discerns anger against 'the Haves' as the dominant emotion. Her perception of the poet's 'absolute rejection' and 'contempt' of Australia is less sustainable, as earlier discussion of the problematics of Porter's developing mythology of Australia, in poems such as 'A Christmas Recalled' and 'Forefathers' View of Failure', has shown. Moreover, 'Phar Lap in the Melbourne Museum' exhibits an admiration, tinged with irony, for 'Australian innocence', and for the 'naturally excessive'.

The desire for relevance to the contemporary mood in Britain at the beginning of the 1960s no doubt accounted for a selection of poems in *Once Bitten, Twice Bitten* biassed towards satire and social criticism, which critics and reviewers subsequently took up as the keynote of this book. The three 'new' Australian poems which appeared in *Penguin Modern Poets 2* — 'Somme and Flanders', 'Ghosts', 'Eat Early Earthapples', together with 'Reading MND in Form 4B' — somewhat redressed this imbalance, and provide a partial basis for Michael Hulse's observing a fundamental affection for Australia in Porter's writings.[21] As is often the case with migrants, however, Porter's feelings and responses could swing wildly from idealized affection to angry rejection. At times the chief target, for one who sought a niche in the metropolis, was Australian 'provincialism', which connoted ignorance of the sophisticated arts along with judgemental puritan values.

Porter's overriding reputation as a 'social' poet at this time received reinforcement from a variety of sources. The first and most decisive was the reading of 'Your Attention Please' (*1962/3, CP* 38−9) on BBC radio on 15 October 1961. The actor who read the free verse composition, Denys Hawthorne, managed to neutralize his native Irish accent sufficiently to persuade many listeners that he was, in fact, a BBC announcer telling them of an impending nuclear attack.[22] The carefully orchestrated rhetoric of the poem begins with the following lines:

The Polar DEW has just warned that
A nuclear rocket strike of
At least one thousand megatons
Has been launched by the enemy
Directly at our major cities.

So effective was the pronouncement that listeners were shaken from their seats and flooded the radio station with alarmed calls. On 23 November *The Daily Mail* carried a headline beyond parody: 'Voice of Doom on the Third: H-Rocket Poem Panics Listeners Expecting an Opera'. The newspaper report quoted extensively from the poem, but ran it on as if it were in prose form, without its expressive line-end pauses. The paper further reported a BBC directive to department heads warning them not to cause their listeners alarm or despondency with programmes about nuclear attacks or outer-space invasions. This allusion to Orson Welles's infamous Martian invasion broadcast in 1938, which caused panic in the streets and a number of suicides, was a backhanded tribute to Porter's skilful adaptation of bureaucratese, as well as to the publicity sense of BBC producer and poet, Anthony Thwaite, who subsequently became one of Porter's close friends. Porter's own view of 'Your Attention Please', which has been a very popular poem in anthologies and at public readings, is rather harsh. He has called it a 'botched effort' in metrical terms, but successful in its introduction of an element of ambiguity and in its purpose of satirizing officialdom.[23] This underestimates the writer's obvious skill in establishing a voice in the mind of reader or listener which is the bearer of apparently authoritative cultural meanings. The improvised poem works by rhetorically sustaining its fiction while parodically undermining the social foundations. Like the later 'Mort aux Chats' (*PTTC, CP* 184), a poem which satirizes the language and attitudes of inter-communal prejudice, 'Your Attention Please' succeeds in spite of its author's disclaimers about the value of such directly confrontational verse.

In such ways, Porter's work was at last attracting attention. His mode of protest was typically ambitious, being against both the society in which he lived and the more general conditions of living. Moreover, rather than seeking a popular success through counter-culture satiric magazines such as *Private Eye* (launched in 1961) or London *Oz* (1966), Porter aspired to 'high' culture outlets such as the *Listener*, the *New Statesman* and the *Observer*; by 1962 his poems or reviews had appeared in each of these publications. The one poem Porter did publish in *Oz* was the verse sequence 'Metamorphoses', which appeared in the first issue in 1966. The editor of *Oz* was Jill Neville's younger brother Richard, who remembers Porter as modest and sensitive ('a perfect gentle knight') beside his image of himself and fellow *Oz* revolutionaries as 'wild men from Borneo'.[24] Other Australian expatriates or travellers who appeared in early issues of *Oz* included Germaine Greer, Clive James ('our token right winger', according to Neville), Bruce Beresford and Bob Ellis. Like Greer and James, Porter sought to make what mark he could at the heart of a dying empire, while Beresford and Ellis returned to Australia to make their careers in the more youthful arts scene of their home country, in film and journalism respectively.

The designation of Porter as a 'social' poet was never more than a half truth. Of the forty-three poems in *Once Bitten, Twice Bitten*, only seven locate themselves centrally in the milieu of contemporary London. These poems, admittedly, are among the most strikingly successful in the volume: 'Conventions of Death', 'Metamorphosis', 'Beast and the Beauty', 'John Marston Advises Anger', 'Made in Heaven', 'Death in the Pergola Tea-Rooms' and 'The Smell on the Landing'. Another poem, 'For John Clare from London' (*OBTB, CP* 27−8), encapsulates the growing attraction of the city for Porter's speaker, not for its parks and gardens or cultural attractions, but for its access to a madness which correlates with his own. The explicit parallel drawn between the deeply disturbed regionalist John Clare, writing with detailed detachment about the Northamptonshire countryside, and the quietly desperate Australian of the late 1950s, giving a similar attention to the streets of London, is a serious one: Porter is often at his most earnest when jesting with what seem unlikely comparisons. In these London poems Porter was writing what Ian Hamilton has called 'poetry in its street clothes, poetry which could ingest and include the world'.[25] Whereas Hamilton and other Romantics who favoured a stylistic minimalism preferred poetry to be hard, gem-like and lyrical, Porter was voraciously eclectic. A great metropolis had opened before him and, in his desire to take it all in, his Australian temerity and torpor were transformed. Here was a scene big, decrepit and various enough for him to explore within it his doubts, fears and, on occasion, his aspirations.

Drawing on his experience of the city around which he walked or took the underground trains (he has never had a driving licence or owned a car), Porter observed, recorded and reinvented people and places with the imaginative gusto of a participant in a carnival, a carnival of mask-switching, as in Mozart's and Da Ponte's *Don Giovanni*, where the new identities of one's wildest dreams might be revealed:

P eter is now an irritant for pearls.
O f all the things I'd like to be, a Casanova,
R andy, vulgar, would be best. A lover
T raining the South Ken. girls in his two-seater,
E gged on by Hickey, Tanfield, sexy in a sweater,
R aping the honest, loveless, childhood Peter.

'Reflection on my own Name' (*OBTB, CP* 14) projects an autobiographical amalgam of the fears, shames and envies of what is felt as a continuing adolescence, the 'late-developer' syndrome to which Australians have often felt themselves consigned in the face of metropolitan sophistication. 'Reflection on my own Name' counterpoints 'Eat Early Earthapples', (*1962/3, CP* 44), transposing the Toowoomba ethos to South Kensington. The self-mockery implied in the opening line's echo of the sentimental musical *The Sound of Music* ('Peter — a name I share with many others —') gives the represented experience a certain ironic distance, but the self-pity hinted at in the poem's concluding line indicates an indulgence in unexplored emotions. The poet and critic John Fuller noticed this tendency in Porter's first book: 'One eye is ... praiseworthily trained upon the minutiae of contemporary life, parties, lusts, vivid metropolitan personae, and so on; the other, unfortunately, is swivelling about wildly in some private darkness and finds it difficult to transmit the vision successfully'.[26] Like other British critics, Fuller was more interested in the externals, the sociological dimension of Porter's early work. The 'rape' of the childhood self promised to be another painful and difficult process, which nevertheless seemed necessary for artistic and personal development.

In 'John Marston Advises Anger' (*OBTB, CP* 17) this process is under way. The poem was written in 1959 after Porter had seen Marston's play, *The Dutch Courtesan*, with Diana Watson-Taylor. Porter had met Watson-Taylor at Bumpus Bookshop and they had become lovers. Although the affair was over by now, Porter still felt deeply upset at having been 'ditched' by her at the beginning of 1959. Almost all the poems in *Once Bitten, Twice Bitten* concerning her were written in the years 1959−60. Thus, although the relationship with Jill Neville had been formative, Diana Watson-Taylor was in a sense the unwitting 'onelie begetter' of Porter's entry into the public arena, with his satires and bitter, morbid tones. At the time, he was sharing a flat with an old friend and also 'consorting with' Jannice Henry, whom he would marry the following year. In Porter's view, he had been abandoned because Diana was 'upper class and had been trained to believe that money married money for its own protection and that social miscegenation didn't work'.[27] For her part, Diana Phillips (formerly Watson-Taylor) has said that, although she found Porter intellectually and physically attractive and enjoyed his company very much, she never intended to marry him.[28] The feelings of the two theatre-goers can only be guessed at as they watched the dramatized anger of Marston the satirist, and the betrayals and jealousies of his characters. Porter has said that his poem is a locus for a 'rhetorical outburst', giving vent to his many confusions — sexual, social, financial and professional — at the time.[29]

The speaker in 'John Marston Advises Anger' is a young man from north London who situates himself somewhere between the fashionable young 'Chelsea set' and those middle-aged moral arbiters who dismiss this sub-culture as 'seedy'. In an urgent, rough-edged monologue which has its roots in Browning as well as Marston, Porter's speaker encourages the identification of the corruption and moral confusion of late Elizabethan court life with that of affluent, fashionable London:

> The same thin richness of these worlds remains —
> The flesh-packed jeans, the car-stung appetite
> Volley on his stage, the cage of discontent.

Porter's initial choice of 'The Cage of Discontent' as the title of his first volume indicates the extent to which it implies the tone and outlook of the book as a whole. As the concluding phrase of 'John Marston Advises Anger', it fixes the mood with the finality of a falling curtain. Although the concluding lines may suffer from adjectival overload — a characteristic of some other early poems — they keep pace with the vehemently expressed catalogue of instances which precede them, and the tempo and rhythms have a strong dramatic sense.

Yet certain lines in the poem draw attention to a more personal, underlying hurt, reminding us of T. S. Eliot's complaint about the lack of an adequate 'objective correlative' for Hamlet's confused feelings. In Porter's poem the lines which draw attention to some such problem are perhaps the most memorable: 'What hurts dies on paper,/Fades to classic pain'. The sudden shift here to an interpolation which questions the relationship between life and art is postmodernist before its time. In view of the biographical situation of the author and his former lover, however, it suggests that writing is an anodyne for the pain of separation. The neatness of the aphorism mocks this process while appreciating it. Yet what gives the poem as a whole its force is that correlatives to the emotion are strongly located in a surrounding society:

> ... Love goes as the MG goes.
> The colonel's daughter in black stockings, hair
> Like sash cords, face iced white, studies art,
> Goes home once a month. She won't marry the men
> She sleeps with, she'll revert to type — it's part
> Of the side-show: Mummy and Daddy in the wings,
> The bongos fading on the road to Haslemere
> Where the inheritors are inheriting still.

This is a virtuoso speech by a modern Malheureux. It has the explosive vitality of a young man who is carried away by the rhetoric of his anger and cynicism. It is both cleverer and more succinct than Jimmy Porter's tirades in *Look Back in Anger*, while it shares their angry disaffection. It is also effective as a piece of social criticism, naming just enough status details (the MG, black stockings, white make-up, bongo drums) to locate the young woman in her time and milieu. If the reader does not know Haslemere as gentrified Surrey — a town whose most famous literary associations are with Enid Blyton and Lord Tennyson — he or she can nevertheless guess that it is a place of settled, conservative values beyond the metropolis. Porter's chief concern is with the metonymic value of Haslemere within the social hierarchies which the poem delineates. In the concluding lines of this passage the tempo is of rhythmical recession as the bongo drums from London parties are replaced by the deeper and more ominous drum-beat of a way of life which is settled for all time at birth: 'Where the inheritors are inheriting still'.

For Porter's London contemporaries, the social details in 'John Marston Advises Anger' gave a certain frisson to the poem — the names of coffee houses (l 9), the 'Poetry and Jazz' evenings pioneered by Christopher Logue (ll 25−6), the vision of a world promoted by Condé Nast, owner of glossy magazines of which *Vogue* was the flagship (ll 27−8). The past is not invoked in the neoclassical pattern as an ideal to which the fallen present should do homage; both worlds, that perceived by the late Elizabethan malcontent and the affluent, fashionable milieu of contemporary London's West End, have a similar status. The difference is that 'our death's out of sight', concealed from all eyes but those of the jaundiced, nay-saying, truth-telling poet. So there *is* a persuasive purpose in this poem, as in other 'London' poems by Porter, which lends weight to the views of liberal thinkers and critics such as Frederick Grubb, who saw Porter as 'a deadly analyst of metropolitan manners, and also an inverted, cantankerous, yet, in the end edifying moralist'.[30] He seemed on the right (that is, the left) side, even if his outlook was more Fabian and idealist than practical and reformist. His 'under the house' Brisbane perspective had transferred to 'basement' London and produced in Porter a modern malcontent whom Grubb portrayed as a Don Quixote of London's West End, with Martin Bell as an 'outrageous Sancho Panza' by his side.[31] For this critic, Porter had come from Australia into 'the golden heart of commercialism — a heart which beats in the breasts of all classes' and there created 'a horrifying sense of the power of money to actually determine questions of value and the fate of individuals'.[32] Grubb made the acute observation that Porter avoided pompous moral opinion and used a '*complicity* with his material' to invite his readers to do the judging.[33]

The degree of Porter's 'complicity' with his material may be considered biographically. It is clear, for instance, that what Porter experienced as his 'ditching' by Diana Watson-Taylor in 1958 focussed an indignation which he felt about upper middle-class British 'inheritors' of wealth and advantage. A pattern of rejection and loss which began with his mother's death was repeated in his other major relationships with women before 1960, especially Jill Neville and Diana Watson-Taylor. The unpublished verse which the end of the affair with Jill Neville had occasioned was predominently introspective and tortured; the poems which related to Diana Watson-Taylor were more precisely located in a social milieu, which she herself has observed were partly accurate, partly an outcome of Porter's volatile feelings at the time.[34] The upper middle-class social setting which both attracted and repelled Porter was based to some extent on the life of Diana Watson-Taylor's parents who, in 1959, had just moved from Hampstead to a large house in Canterbury (not Haslemere, as in the poem). Their conspicuous wealth, philistine values and apparent abandonment of responsibility for their daughter when she dropped out of university provoked anger in Porter, which spilled over into a more general diagnosis of a society which they were held to represent. Other poems in *Once Bitten, Twice Bitten* which Diana Watson-Taylor has recognized as concerning her, or based on times which she shared with Porter, include 'Metamorphosis', 'Beast and the Beauty', 'Made in Heaven' and 'Party Line'. Another poem, 'Lament for a Proprietor' (*OBTB, CP* 15−16), is partly based on the man from Cambridge whom she was seeing while the relationship with Porter was developing in London, and whom she subsequently married:

> This was the end of a man but also died
> Ten suits, twenty shirts, Clare College ties
> And scarves, a radiogram, one hundred dance discs
> And Vivaldi's Seasons, shells picked up
> On Sark and Ibiza, Phaidon and Skira books
> Coverless and crooked . . .

Diana Watson-Taylor had just returned from a holiday on Sark when she first met Porter and some but not all of the other details relate to the persons concerned. It should be stressed that in 'Lament for a Proprietor', as in other poems, Porter freely introduces fictional attributes alongside actual ones even where the inspiration is obviously derived from particular experience. His drive is towards the gestalt rather than the case history, towards general observation rather than self-diagnosis, and in this sense, he avoids being a 'confessional' poet. The mock lament of this poem might be read as a curse or 'hex' upon an advantaged opponent, but Porter's control of tone and capacity to generalize through the accretion of significant detail give it a wider social and philosophical reference. 'Lament for a Proprietor' shares some of the asperity of Pope's satiric verse portraits. Above all, it scorns the achievement of identity through material possessions.

   Porter's focus on the clothes people wear in 'Lament for a Proprietor' is paralleled in a more self-satirizing way in 'Metamorphosis' (*OBTB, CP* 10−11). Given the title of the poem, which summons up literary parallels such as Ovid's stories of transmutation or Kafka's narrative of a young man who finds himself changed into a monstrous beetle, it would be unwise to assume anything directly autobiographical in the first-person speaker in Porter's dramatic monologue. Yet the fascinated attention to clothes and brand names in the London poems relates both to his early experience in advertising and to a consuming interest in the disguises offered by the metropolis. Lying behind both, we can now see, was his doomed affair with a woman who seemed 'out of his class'. In an article on clothing fashions in *The Evening Standard* in

1961[35] the journalist quoted the opening lines of 'Metamorphosis':

> This new Daks suit, greeny-brown,
> Oyster-coloured buttons, single vent, tapered
> Trousers, no waistcoat, hairy tweed — my own

The article also quoted a response to the poem from a spokesman for Simpsons of Piccadilly, who manufactured Daks: 'Ah yes ... That sounds like the Rutland new shade. Very nice suit. But hairy — never!' Porter revelled in such evocations in his first book, which both complied with, and defied, the advertisers, and these captured the attention of newspapers. Few were interested in the more complex drama involved in tracing the dramatic poetry of a young man's dissolution, despair and search for disguises; the clothes in which the drama was dressed had a more immediate appeal. In 'Metamorphosis', though, the proverbial element — clothes make the man — collides with a crisis of identity. The predicament is characteristic: a faltering, and ultimately futile, attempt by a young man to re-create himself in an image of the commercial world, in the wake of a broken love affair. The poem, no less than the persona, asks what is 'real'. The speaker's concluding 'recognition' is grimly humorous, melodramatic and, by comparison with other images of the self offered up by the commercial city, it seems real:

> As in a werewolf film I'm horrible, far
> Below the collar — my fingers crack, my tyrant suit
> Chokes me as it hugs me in its fire.

The monster self will out; and however exaggerated this may seem (recalling Nessus's shirt), it is of a piece with the bizarre metamorphoses of an unreal city.

Whereas in 'Metamorphosis' the werewolf self emerges cinematically as the first-person narrator sips his Worthington with an ex-lover in a London pub, the monster in the companion piece, 'Beast and the Beauty' (*OBTB, CP* 12), emerges as the poem's third-person protagonist 'sits alone in Libraries, hideous and hairy of soul'. The juxtaposition of figure and setting give a similarly hyperbolic sense of the bizarre, verging on the humorous. In each poem, though, the story of a deeply felt love affair which has been of no real account to the young woman lends pathos to the male protagonist's predicament. Yet in these poems the sociological dimension does seem more significant than the psychological; we are presented with a case of provincial innocence encountering entrenched prejudice. In the second stanza of 'Beast and the Beauty' we are shown the young man's initial delight in his girlfriend's apparently sophisticated home life:

Her sophistication was his great delight:
Her mother and father drinking, throwing things,
The unhappy marriage, the tradespeople on Christian
Name terms — all the democratic sexiness — mornings
With the Pick of the Pops and the Daily Express

In the continuing allegory of Porter's self-realizations throughout his poetry, this is an early encounter with false sophistication, which not only rapes but murders the innocent:

But the sophistication chose to kill — the itch
Was on the inside of the skin. Her family of drunks
Were shrewd, wine-wise young barristers and gentlemen —
Farmers fought for her hand. In the loft there waited trunks
Of heirlooms to be taken seriously...

The young man is rejected. His world shrinks to eating alone in Lyons' corner houses or sitting in those libraries, 'waiting for a lustful kiss to bring/Back his human smell, the taste of woman on his tongue'. The theatricality of this short verse narrative is a way of projecting powerful feelings of anger and bitterness in the face of rejection. No anodyne — least of all work — seems sufficient to repress the dominating sense of despair, which encompasses rejection not only by a single woman but by the whole apparatus of England's inherited wealth and privilege. Here, as elsewhere, Porter determined to extend his meanings well beyond autobiographical experience.

Porter's diagnosis of western metropolitan society, epitomized by London, is ominous, a deadly illness hangs over it. In 'Conventions of Death' (*OBTB, CP* 10) the illness is perceived to be in ourselves. The narrator is obsessively convinced that he is mad — with desire, lust, need:

What I want is a particular body,
The further particulars being obscene
By definition. The obscenity is really me,
Mad, wanting mad possession: what else can mad mean?

This vision is transferred to the streets of London. In a tradition which includes Baudelaire's *Les Fleurs du Mal* and T. S. Eliot's *The Waste Land*, the city becomes a metaphor for a prevalent state of disorientation and despair. The paradox of such reconstructions is that while signifying a death-in-life, the city is shown to quicken the pulse and heighten subjective perception. Thus the opening stanza of 'Conventions of Death', while proclaiming doom, is informed by an almost jauntily surrealist act of perception:

We live under the stately mushroom shadow —
A cliché to walk with going up Bond Street.
The dead parked in the Triumph dealer's window
Are this year's models, require no upkeep.

The observer here is not out of control: madness is an experience
known to him but not surrendered to. Total disorientation is held at
bay with an act of will and an accommodation of the bizarre. As in
some of A. D. Hope's early satires, Porter mocks the conventional as a
'lying philosophy', offering a heavily ironic response to the problems
of urban dislocation in a poem first published in 1960, the year before
his marriage:

So give up thinking, work hard, buy a car,
Get married, keep a garden, bring up kids —
Answers to all the problems that there are,
Except the love that kills, the death that lives.

The obvious scorn here for marriage as a solution to life's problems
recalls Chekhov's comment that nobody who is frightened of loneliness
should ever get married. In subsequent poems of the 1960s very
few bear upon the routines of a family life and children; that part of
the 'ordinary' events of the poet's life goes largely unrepresented.
Domesticity recurs in poems of the late 1970s and 1980s, however, as
one of the lost places to recall and rediscover, for its unhappiness as
well as its pleasures.

Like other poets and artists of the modern city, Porter presents his
images from *within* its confines. The harmonious iconography of the
Renaissance city viewed from *outside* as a congruent whole is perhaps
impossible to the modern city dweller. As Edward Timms has
commented, the 'amorphousness' of the modern metropolis precludes
such a perspective.[36] Modernists in literature and painting, influenced
by Bergson, Freud and William James, built their art upon assumptions
that fragmentation was a necessary correlative to the experience of city
living, that the arts should consist of 'fleeting impressions registered
by fluid states of consciousness'.[37] Porter's running quarrel with
Modernism was not a total rejection of its premises (into which well
he would drop his own bucket from time to time) or its images of the
city. Rather, his work asserted the importance for humanity and the
artist of a strong sense of the past; and of the value for the poet of
knowing and understanding a literary tradition. In a way, this was a
rejection of the popular idea that immediate experience is the first
cause of poetry. Music, painting and books could, in Porter's view,
assume such importance in one's life as to be considered primary
experience.

'Walking Home on St Cecilia's Day' (*OBTB, CP* 13) shows how, against the odds, harmony could sometimes prevail in Porter's 'unreal city'. For Porter, a non-believer in orthodox Christianity, the patron saint of music is the most worthy of worship; St Cecilia's Day (22 November) is the most significant date on the calendar. And what does music chiefly accomplish? Nothing, suggests Porter through his auto-biographical persona, and that is precisely its beauty. Assailed by a sense of his own mortality he walks through the streets towards an unspecified 'home', 'Rehearsing wrongly Mozart's own congruity'. Even at moments of personal desperation, as here, a self-deprecating humour and irony remain: the speaker hums Mozart to himself and fails, like most others who try, to get it right; the great composer's 'congruity' lapses through its human agent into incongruity. Such music remains however as a counterpoint to the ordinary worries and injustices of living, whereby 'Nodules on noses grow, pet cats get killed,/The lush and smooth upstage the scrag and thin'. Music is both a gift of the maker and mechanically produced, a 'miracle on the ground'. The sense of personal unhappiness which has dogged the speaker for thirty years may be relieved, he concludes, through music:

> There is a practice of music which befriends
> The ear — useless, impartial as rain on desert —
>
> And conjures the listener for a time to be happy,
> Making from this love of limits what he can,
> Saddled with Eden's gift, living in the reins
> Of music's huge light irresponsibility.

This poem announces a different agenda from the critical realism and sociological dimensions of the more specifically 'London' poems in *Once Bitten, Twice Bitten*. Its deftly counterpointed recognition of both 'Eden' and 'limits' presents the polarities of Porter's vision, and leaves no doubt that music is considered its finest vehicle.

The poem 'Annotations of Auschwitz' (*OBTB, CP* 30–2) presents another type of music. This piece was quoted often by early reviewers and was written in the first instance for the Australian composer David Lumsdaine, who stayed with Porter in Belsize Park Gardens in 1953. Parts of the verse composition were set to music by Lumsdaine as a cantata. The published poem is a series of nightmare images presented flatly in seven sections as if in note form, or annotations. In them, the horrors of Auschwitz enter the psyche of the Londoner:

> On Piccadilly underground I fall asleep —
> I shuffle with the naked to the steel door,
> Now I am only ten from the front — I wake up —
> We are past Gloucester Rd, I am not a Jew,
> But scratches web the ceiling of the train.

The visual, almost cinematic immediacy of this section, and others, seemed necessary to Porter in dealing with a subject so freighted with culturally-acquired feeling. Porter has commented that this piece was directly inspired by Alain Resnais's *Nuit et Brouillard*, a documentary which used footage shot by the Germans themselves in Auschwitz. In the film this was intercut with images of what the camps looked like subsequently — shrines with grass growing over the gas chambers and ovens. But a problem of technique remained for the poet: cinema does not transfer directly to the more compressed form of verse.

With an eclectic resourcefulness which was becoming characteristic of Porter's attempts to discover the right form for his observations and feelings, he found in Wallace Stevens's 'Thirteen Ways of Looking at a Blackbird' a model of perceptual variation which suited his own purposes. Porter's historical subject, Auschwitz, being outside his direct experience, had to be derived from films, photographs, books and other 'secondary' sources. This was not a major deterrent. Porter's growing sense of the importance of cultivating such 'vicariousness' as a defence against 'viciousness' had been explored in the past of his own family in 'Forefathers' View of Failure' (*OBTB*, *CP* 3–4). By choosing Stevens's 'Thirteen Ways of Looking at a Blackbird' as a formal model for 'Annotations of Auschwitz' he was implicitly rejecting another avenue, that of William Carlos Williams and the Imagists. Rather than following Williams's procedure of deriving a self from the perceived world, Porter chose Stevens's approach which involved identifying the self as a state of soul which informs the perception.[38] The perceiving self in Porter's poem has links with the self of 'Metamorphosis', 'Beast and the Beauty' and other London poems in *Once Bitten, Twice Bitten*. An obsession with 'hairy' clothing and a child self which salts the 'puny snail' are hints at an autobiographically-based continuity. The poem's fractures are of a kind which link this personal self with the wider historical horror.

Porter himself felt that 'Annotations of Auschwitz' was 'inadequate for its subject': the sixth section, in particular, was in his view 'heavy, weary and sententious'. More than this, he considered the basic fault of the piece was that it 'tries to handle a subject in a very short span — the unique awfulness of which can't be coped with in language'.[39] The ambition of this enterprize was itself remarkable for a young Australian in London. It indicated that he was ready to take on not only the fraught present of western metropolitan life, as he saw it, but also, in his idiosyncratic way, the burden of suffering which recent European history had imposed. G. S. Fraser considered this an outstanding early achievement by the virtually unknown poet, and, focussing on the poem's concluding, highly cinematic image, placed his achievement in an historical perspective:

Peter Porter has a very powerful poem about Auschwitz in

which he sees broiler-chickens turning on their spits in a restaurant window, as expressive of the same ruthless human efficiency as the German gas-chambers. It is almost as if the horrors of the Second World War, from Auschwitz to Hiroshima, having failed to penetrate deeply into the imaginations of men in their forties who served through that war (we anaesthetised that, and we were on the right side) had pierced through, twenty years later, to the imaginations of a younger generation.[40]

But Fraser, like others, worried about whether or not the tendency of Porter and other young poets who were 'anti-genteel', in Alvarez's sense, would enhance an assumed shared interest in inculcating humane values. Fraser was raising questions here about the moral purposes of art, from a perspective which Porter would vigorously question, and reject, in the 1970s. The truthteller, he would conclude, should not also be bound by the requirement to 'improve' humankind, though he or she might incidentally increase its imaginative capacity to improve itself.

Perhaps the most successful poem in Porter's first volume is 'Death in the Pergola Tea-Rooms' (*OBTB*, *CP* 21−2), in which questions of belief are confronted by the reality of death in a London winter. In forty-five lines of predominantly blank verse, Porter sets the scene, focusses on his central drama on the banks of the Thames in Marlow ('The old rationalist is dying in the Pergola'), then reflects upon the meaning, or lack of it, in this man's impending death. The narrative pattern is unremarkable, seeming to follow the lead of realist fiction writers of the time such as Alan Sillitoe and John Braine. But the sound-patterning of the poem, its evocative detail, and the all-pervading atmosphere of winter are finely introduced and controlled. No other early poem illustrates better the 'realism' for which Porter has been justly praised. Yet realism in the sense of accommodating accurate details of milieu is only part of the story, as the opening lines indicate:

Snakes are hissing behind the misted glass.
Inside, there are tea urns of rubicund copper, chromium pipes
Pissing steam, a hot rattle of cups, British
Institutional Thickness. Under a covering of yellowing glass
Or old celluloid, cress-and-tomato, tongue-and-ham
Sandwiches shine complacently, skewered
By 1/6 a round. The wind spitefully lays the door shut
On a slow customer — ten pairs of eyes track
To his fairisle jersey; for a few seconds voices drop
Lower than the skirmishing of steam.

Precision of detail is important in anchoring this situation in a place

and time, as if the unidentified observer/speaker were testing
the strengths and limits of a materialist world view. An inspired
combination of clipped 'i' and hissing 's' sounds contributes to the
construction of the observer's voice, its thin-lipped, wintry rejection of
pleasure or hope. Typically, Porter's poem begins 'inside' with
manufactured objects, rather than 'outside' with nature, and throughout
the piece this oppositional metaphor suggests the strategies which
humans employ to maintain a rational hold on life in defiance of
winter's warnings: the poem offers no metaphor of 'above' or 'beyond'
which could transcend these earth-bound conditions. Winter is all.

By locating his realist drama in a public eating place, among the
recognizable objects of such institutions, the writer implies that the
disease of spirit which he diagnoses here is as widespread as the
society which engenders it. This is an example of what Jonathan Raban
meant when he wrote of the decisive entry of 'society into the poem' in
the 1960s.[41] Porter's poem offers no special personage, no prophecy;
instead, it offers a view of a society which is blind to its own mortality.

The narrative movement of 'Death in the Pergola Tea-Rooms' is
minimal. A man is dying. Nothing to get heated about. A doctor
arrives:

> Gets out of his '47 Vauxhall, sucking today's
> Twentieth cigarette. He stops and throws it
> Down in the mud of the howling orchard.

No surge of emotional energy is proffered in these flatly recounted
events; no romantic, nostalgic oratory as in Dylan Thomas's 'Do not go
gentle into that good night'. Porter's aim is not to infuse the event with
an inspirational outburst, but to enter the spirit of hopelessness — to
convey the feel of its presence on the pulse. To the rationalist, this
death is ironically 'a business of wills'; the compassionate onlooker,
with a strong cast of rationalism in his own outlook, observes the
unequal battle of a single human will against the 'carcinoma commune';
and a dark irony pervades the subsequent presentation of a rearguard
action which the mind mounts on the body's behalf against the man's
imminent decease. Porter skilfully uses the imagery of meetings to
indicate ways in which 'reasoning' and 'keeping busy' may be strategies
of evasion. While the dying man is initially associated with a Labour
Party tradition of humanism and rationalism (as in Fabian socialism,
for instance) the imagery broadens, so that he becomes not a restricted
case study, but one which dramatizes a general condition. In keeping
with the naturalistic approach, Porter is concerned to show the dogged
will of the man of reason as a pitiful counter to his body's dictates.
Cups of tea and the 'warm patch where the cat has been' give only
temporary solace against the encroaching winter.

Whose voice in the poem suddenly announces 'There is no God'?

Is it Porter, his quiet observer, or a summative statement of the whole scene (the scene, as it were, and the whole society lying behind it, making their own statement)? Whatever the answer to this, Porter here breaks from his carefully knitted surface realism in an oracular pronouncement which prefigures others later in his work. The 'death of God' controversy, popularized by Bishop Robinson in the 1960s (via Dietrich Bonhoeffer and others) is given dramatic immediacy and choral authority in a statement which seems to leap from the logic of the poem's narrative.

The conclusion of 'Death in the Pergola Tea-Rooms' is a stripping away of illusions:

> . . . It is winter, the windows sing
> And stealthy sippers linger with their tea.
> Now rushing a bare branch, the wind tips up
> The baleful embroidery of cold drops
> On a spider's web. Inside the old man's body
> The draught is from an open furnace door — outside the
>   room,
> Ignoring the doctor's mild professional face,
> The carnival winter like the careful God
> Lays on sap-cold rose trees and sour flower beds
> The cruel confusion of its disregard.

In these lines Porter anticipates the reader's wish for a comforting coda, and rejects it. The 'careful God' seems less a contradiction of the sudden atheistic intervention than an elaboration: God exists in the poem as the nature of the 'carnival winter', which freezes animal and vegetable life in a show of supreme indifference to their fate. The poem is thus deeply discomforting and its melancholy is entirely lacking in sentimentality, as it contemplates the existence of a hard 'God' and the fact of human impotence before the forces of natural destruction. For Margaret Owen, it was this poem above all which came to represent Porter's vision:

> The 'real world' of Wallace Stevens, 'where ordinariness, poverty and sadness are inseparable from human existence' — which world had somehow never been acknowledged in my scene — came before me with unforgettable force in this poem. The cancer-yellow hangs over all, from the caf, through the winter-balding river bank to the sap-cold rose trees, the sour flower beds — and there is such plangent melancholy about its reality that it came to represent Peter's vision, then and still.[42]

Few of the other London poems in *Once Bitten, Twice Bitten* have the assured tone of 'Death in the Pergola Tea-Rooms'. A Byronic flourish and self-regard characterizes pieces such as 'Away, Musgrave, Away' (*OBTB*, *CP* 23), 'South of the Duodenum' (*OBTB*, *CP* 26) and 'Euphoria Dies' (*OBTB*, *CP* 26−7). In these poems a robust, game-playing irony seems intent on tickling and teasing death into life. Porter's other major theme, love, is still, poetically, at a social rather than an existential level; the idea of death had preoccupied him more persistently. In the poem 'Made in Heaven' (*OBTB*, *CP* 18−19), however, the underlying subject is love. In this poem he exposes an ironic discrepancy between the advertised world of marriage and its reality from the woman's point of view. Blake Morrison considers this the best of Porter's early satires[43] and Michael Hulse has called it 'fully effective satire'.[44] The advertising copywriter's ear is attuned to his profession's clichés, but is not buying them:

> As things were ticked off the Harrods list, there grew
> A middle-class maze to pick your way through —
>
> The labour-saving kitchen to match the labour-saving thing
> She'd fitted before marriage (O Love, with this ring
>
> I thee wed) — lastly the stereophonic radiogram
> And her Aunt's sly letter promising a pram.

For a poet who seems to have been a generally unsuccessful copywriter, in spite of working in the 1960s alongside other literary advertisers such as Fay Weldon, Trevor Cox (William Trevor) and Gavin Ewart, Porter is adept at undermining the premises of advertising. Beside other organs of satire in the 1960s such as the television programme 'That Was The Week That Was', and magazines *Private Eye* and *Oz*, Porter's satire is, as Michael Hulse has pointed out, 'an altogether more stable, adult thing'.[45] Measured against major literary satirists of previous ages, (Juvenal, Pope or Swift, for instance) poems such as 'Made in Heaven', however, lack a certain ruthlessness. Porter often betrays a concern for victims, as he does for the woman in 'Made in Heaven'. In addition his rage is not maintained. As Blake Morrison puts it, 'he loves the thing he kills'.[46]

Porter himself believes that he has never been a very satirical poet because he lacks a strong, central viewpoint. This makes for an ironist rather than a satirist, informed by a sense of general human failings, including one's own. As he remarked in an interview:

> The irony always seems to me to be built into life that the attainments of human beings are in fact magnificent — if you think of Scarlatti and Piero and also individual people's lives. And yet all around you is compromise, shiftiness, egotism,

vileness. It's the contrast between what man can be and what man customarily is; and then the recognition that the person who can see this is himself like that.[47]

With this outlook in mind, it is not surprising that while Porter's writings have satiric elements they are seldom sustained formal verse satires. Significantly, he is closest to an out-and-out satirist when translating another writer. In translations of Martial's satiric and often obscene epigrams, a project carried on intermittently through the 1960s, culminating in the publication of *After Martial* (1972),[48] Porter assumed the voice and attitudes of another spiritual exile, one who had felt similarly drawn to forbidden excitements at the centre of a deliquescent empire. Martial was a provincial Spaniard from Bilbilis, a town not far from where Barcelona now stands. He moved to Rome when he was about twenty in 64 A.D., lived there for thirty-five years and then returned to Spain. His 1500 or so short poems or epigrams are valued for the light they shed on Roman life and manners. Porter has commented that Martial's 'unique achievement' was to have 'almost vanished as a personality, substituting for himself a detailed picture of the civilization he lived in'.[49] Certain elements of Porter's poetic achievement in the 1960s have a similar sociological value, as has been pointed out, but the persistence of an autobiographical self and an associated and accumulating mythology of self provide a more problematical and interesting guide through the mazes of Porter's metropolis. His free renderings of a selection of Martial's epigrams can also be read in this light, as a composite mask figure of Martial/Porter who relishes the almost bewildering variety of human vanity and folly which the metropolis throws up. There are occasional parallels, too, with other writers whom Porter admires, including Jonson, Molière and Pope. A comic buoyancy and delight in the absurdities of urban fashion, ranging from clothes to sexual behaviour, wells forth. In his free translations of Martial, which are intended to capture the spirit rather than the letter of the Spaniard's encounters with Rome, the Australian in London gleefully indulges in experiments with anachronisms which elide the two metropolises and periods. Later, in a poem called 'The Prince of Anachronisms' (*TAO* 5), the practice of placing details out of their historical time would become an article of poetic faith. In his renderings of Martial, it is an experiment in freedom within limits, the only limits being imaginative audacity and a perceived relevance to contemporary metropolitan life. The poems are anything but staid: they show an ebullient imagination flouting taboos, picking up a variety of verse forms and adapting them to individual purposes, mixing past and present in a spirit of carnival, enjoying the freedom and scope of the Martial 'mask'.

In the free translations from Martial as in much else he wrote in the 1960s, Porter steadfastly resisted the disease which Al Alvarez had found to be so lethal in contemporary English culture: gentility. But such resistance can provide only a rough measure of a poet's worth. Alvarez unwittingly demonstrated its inadequacy as a criterion of value in his introduction to the revised edition of *The New Poetry* (1966).[50] The poems discussed were Larkin's 'At Grass' and Hughes's 'A Dream of Horses'. Larkin's poem, in Alvarez's view, was 'elegant and unpretentious and rather beautiful in its gentle way' but 'a nostalgic re-creation of the English scene, part pastoral, part sporting'; his horses were '*social* creatures of fashionable race meetings and high style'. Hughes's poem, by contrast, was less skilful but more urgent. Hughes's horses have 'a violent, impending presence' and they 'reach back, as in a dream, into a nexus of fear and sensation'. These comments leave no doubt as to which of the two poets Alvarez prefers, having stated his preference for flouters of the 'gentility principle'; but the 'principle' itself offers little more than a flimsy prejudice in favour of vitalism over conservative social values. Interestingly, Alvarez omitted Porter's popular poem 'Phar Lap in the Melbourne Museum' (*OBTB, CP* 36) from his selection in *The New Poetry*, obviously preferring more experimental and psychologically disturbing pieces such as 'John Marston Advises Anger', 'Beast and the Beauty' and 'Annotations of Auschwitz'.

According to Alvarez's criteria of poetic worth, 'Phar Lap in the Melbourne Museum' falls somewhere between Hughes and Larkin. The choice of a museum specimen as subject might seem to relegate it to Sunday-viewing status only — at the genteel edge of the spectrum — but Porter's project here, as in other poems, is to retrieve the dead for our common discourse: contemporaneity alone is insufficient. His concern is not didactic in the Gradgrind sense; Porter's horse is not a mere 'graminivorous quadruped', nor is it mocked. On the contrary, the racehorse Phar Lap is a talismanic hero of the Australian people, and Porter's object is evidently to explore the nature of this hero worship and its significance. The historical Phar Lap was actually born in New Zealand, a fact conveniently forgotten by most Australians and Peter Porter in this poem.[51] Australia's most famous horse won thirty-six races in New South Wales and Victoria between 1929 and 1931, including the Melbourne Cup in 1930. The Australian film *Phar Lap: Hero to a Nation* (1983), reflecting the legend, subsequently made much of the horse's mysterious death in California in 1932. (Did the Americans poison him?) Porter briefly alludes to these aspects of the legend. Interestingly, he has chosen to revise the summative description of the horse from 'thoroughbred bay gelding' to the more historically accurate 'handsome chestnut gelding' in *A Porter Selected* (1989). But throughout the poem historical accuracy seems less relevant than Phar Lap's status, along with other figures such as the cricketer Don Bradman and outlaw Ned Kelly, as an object of hero worship for Australians. In these ways, Porter's poem has more historical range and interrogative force than Larkin's, but is less concerned with vitalist impulses than Hughes's. However, the stuffed horse in the glass case at the museum has more *human* application than do Hughes's creatures, and sums up the aspirations and fears of a people — 'A horse with a nation's soul upon his back'.

Typically, Porter does not gloss over the fears nor the limited aspirations which he inherited as an Australian and which, as an expatriate, he has distanced himself from. Chief among these suppressed fears, which the Phar Lap legend exposes, is the possibility of betrayal by one's friends. A second fear is of a more general conspiracy: 'To live in strength, to excel and die too soon'. The response to this fear seems incongruous: Australians give their heroic animal an immortality behind glass. In spite of this and the 'dirty jokes' (presumably concerning Phar Lap's gelding), Australians, like Ted Hughes and D. H. Lawrence, seem to prefer the idea of a virile stallion. This symbol of a people remains:

> It is Australian innocence to love
> The naturally excessive and be proud
> Of a handsome chestnut gelding who ran fast.

These concluding lines imply an ambivalence about 'Australian innocence'. On the one hand, an uncomfortable resemblance is suggested to the ignorant and narrow imaginations of Porter's forebears as they are represented in 'Forefathers' View of Failure'. On the other, the tone here is basically one of acceptance and even of sneaking admiration. The baroque imagination which seems most in agreement with Porter's contortions of conscience and desire in the 1950s and 1960s can be a burden, and a return to innocence seems attractive. But it is an expatriate's perspective which is offered upon that innocence. While physically distanced from the country of birth Porter remains involved through myths and a shared national memory, which he now regards with irony, but half-affectionately.

In another historical context, 'Phar Lap in the Melbourne Museum' lies somewhere between A. D. Hope's poem 'Australia' (1939) and Bruce Dawe's 'Life-Cycle' (1967). Hope's 'Australia', as Leonie Kramer has pointed out,[52] is a response to its author's return to Australia from Oxford in the Depression years of the early 1930s. After its condemnation of the torpid cultural life of Australia's cities, 'Where second-hand Europeans pullulate/Timidly on the edge of alien shores', Hope's poem turns dramatically away from its alienated European perspective, the 'lush jungle of modern thought', with its 'learned doubt' and mimicry of fashionable terminology, towards the hope of visionary and prophetic thought in Australia itself: 'Hoping, if still from the deserts the prophets come,/Such savage and scarlet as no green hills dare/Springs in that waste ...' By 1967, Bruce Dawe recognized himself unselfconsciously as a citizen of one of Australia's cities (Melbourne), and in 'Life-Cycle' presents an image of himself as one of the crowd at an Australian Rules football match. There is no yearning here for alternative prophetic voices — creating religion out of a game is a tribal ritual which we may mock even as we join in. Moral superiority should not be assigned to either Hope or Dawe on the grounds of their critical positioning or allegiances. It is worth noting, however, that Porter avoids both the intellectual isolation felt by Hope, on the one hand, and the enclosing tribal warmth expressed by Dawe on the other, and in 'Phar Lap in the Melbourne Museum' this contributes to a certain clarity of thought and vision. Conversely, Porter's poem lacks the idiomatic vigour and directness of Dawe's homespun affections. Michael Hulse has pointed out that 'Exiles often affect coldness and condescension in talking of their homelands, but of Porter the opposite is true: of course he is clear-headed and unsentimental, but his poetic response to Australia is close and companionable'.[53] Although this statement must be adjusted to account for variations in Porter's feelings about Australia, as we shall see, over the total period of his voluntary exile, Hulse's view is more accurate than that of some Australian critics who, in their nationalistic zeal, have lashed out as at a renegade.

◆

Porter's reputation as a skilled poet grew quickly after the publication of *Once Bitten, Twice Bitten*; and the favourable critical response to this first volume was followed rapidly by his inclusion in *Penguin Modern Poets 2*. In addition, most reviewers found his verse more engaging and interesting than that of his fellow poets in the Penguin volume, Kingsley Amis and Dom Moraes. The contemporaneity of Porter's work, its up-to-the-minute engagement with styles and status details clearly held wide appeal. For some reviewers, however, a certain 'knowing' cleverness seemed disabling and was felt to be a direct outcome of the poet's work as an advertising copywriter. Such charges also surfaced after the publication of Porter's second and third volumes, *Poems Ancient and Modern* (1964) and *A Porter Folio* (1969). Reviewing *Poems Ancient and Modern*, for instance, Al Alvarez suggested that certain copywriting habits had entered Porter's style, especially 'a false concentration of the surface only'.[54] 'The World of Simon Raven' (*PAAM*, *CP* 53), which influenced Gavin Ewart's 'Fiction: The House Party', and 'Shopping Scenes' (*APF*, *CP* 74) *are* two poems which stay pretty much at the level of verse journalism, without offering any real intoxication with words or ideas, and might support these contentions. But Alvarez had the wrong poet if he considered Porter about to be 'bought' by a copywriter's modes of thinking. As a *Times Literary Supplement* reviewer pointed out, Porter has a 'contemplative mind' not a 'Condé Nast' one.[55] This is a fundamentally important distinction which one superfical reviewer of the early volumes, Martin Seymour-Smith,[56] failed to notice.

An occasional brittleness of tone and an over-reliance on headlining, phrase-making and resounding last lines does however reveal some element of transfer from advertising to poetry in the early books. Porter was sensitive to these charges and perhaps protested a little too much in an interview in 1976:

> [Working in advertising] had no direct bearing at all upon my poetry and if in fact some people see signs of brashness, journalism, jumping to conclusions, smartness, cynicism, etc. in my writing I can assure them it was there long before I had anything to do with copywriting. At the same time I ought to stress that I was a very poor copywriter ... I was able to stay in advertising because I was prepared to do the boring work which the geniuses didn't want to do, which meant that I did all the small print and cut-along-the-dotted line and the showcards and the sixpence off and the trade ads and the nuts-and-bolts and the Women's Royal Army Corps and all this sort of thing that the more so-called creative copywriters wouldn't do.[57]

A fellow employee at Notley's from 1960, the Irish-born novelist and short story writer Trevor Cox (better known by his pen name William Trevor) has confirmed Porter's impression that the best artists were generally not the best copywriters.[58] Advertising provided paid employment in Australia also for young fiction writers such as Peter Carey, Barry Oakley and Morris Lurie and, as Carey has pointed out: 'We were not really part of the advertising world. We weren't even very good at it; we sat in a strange little room and had our copy rejected by someone who was'.[59] Porter's experience was similar, except that he left advertising whereas Carey stayed and became successful, until leaving himself to write full-time. For a time, poets were favoured by Notley's in London in the 1960s, a firm which later became the corporate advertising giant Saatchi and Saatchi. The copy chief, Ernest Marchant Smith, surrounded himself with copywriters who were in the arts. Peter Redgrove arrived to work there soon after 1960. Oliver Bernard and Ted Lucie-Smith were also employed at Notley's and, from the mid-1960s, Gavin Ewart. From the start, Porter treated advertising as a job rather than a profession; his real dedication was always to poetry. However, together with meetings of the Group, his work in advertising sharpened a sense of audience as well as informing him more closely about the features and motives of commercialism which he satirized in poems of the 1960s, such as 'Made in Heaven' (*OBTB, CP* 18–19), 'Farewell to Theophrastus' (*APF, CP* 116–18) and 'A Consumer's Report' (*TLOE, CP* 131–2).

The commercial world, its motives and drives, did not remain 'outside' for Porter, however, as a mere object of ridicule. The poem 'Tobias and the Angel' (*OBTB, CP* 37) illustrates this. The first-person speaker in the poem appears to be Tobias, son of Tobit in the apocryphal *Book of Tobit*. Its other, and principal, objective source is one of the most popular paintings at the National Gallery in London. Attributed to the Florentine master Verrochio, this painting shows Tobias, accompanied by the angel Raphael and a woolly dog, returning home with the fish whose gall will cure the blindness of his father Tobit. As Michael Levey has said, the painting suggests that 'on life's journey men should have faith ... The painting speaks of an artistic climate of absolute certitude and optimism'.[60] In the personal allegory of Porter's work, this poem indicates the return of the poet's imagination and feelings to his father and the commercial world which he inhabited in the textile business in Brisbane. Rather than chicanery and corruption, he finds there goodwill and habitual decency, with none of the rank commercialism with which his London poems are so preoccupied.

Sharing the mood of Verrocchio's painting, 'Tobias and the Angel' evokes an idealized image of a previous, less confusing era of provincial capitalism, in which regular habits, decency and order prevailed, and days could end simply, 'Like my father closing his Day Book on his trade'. The poem is a tribute not only to the father, then, but to the whole ethos of the mercantile world to which he belonged. The closing lines, while resisting religious authority, are close to a benediction:

> ... Our house
> Is not a tabernacle, miracles are forgotten
> In usefulness, the weight and irony of love.

This is a 'house' shared by father and son. These lines indicate the way in which even an allusive and apparently difficult poem such as this may still be informed with a quiet strength of feeling. The poem shows that the 'bruiser' from Australia with socialist leanings and a well exercised capacity to satirize 'the inheritors' of wealth in Britain also recognized the presence, and necessity, of commerce and trade. This recognition, which extends from the idealized images in 'Tobias and the Angel' to more complex images of contemporary commercialism, lends Porter's verse an often 'impure' quality, in which trade-offs between the high ideals of art and its everyday requirements are negotiated. He has none of the aristocratic disdain for the business world which afflicts some poets, nor is his socialism that of the narrow ideologue.

While acknowledging the necessity of business for survival, however, Porter has never been seduced by its rewards. His own explanation would be that he has simply been unsuccessful in making money, but the truth is that earning a living by writing, especially poetry, is precarious, and yet this is all he has cared to do. Of his work in advertising from 1960 until he took the plunge into full-time writing as a free-lancer in 1968, Porter has said:

> I used to take [my job] seriously to this extent, that it paid for my bills, and that I also believed that I ought to do some work for my boss since he was paying me, but I didn't take it seriously as anything other than another one of the strange sort of runes you have to go through to survive in our society . . .[61]

Between 1960 and 1968, however, Porter gradually began to build up ancillary work as a literary journalist. While still working in advertising, he found time to write and have published some thirty-one reviews or commentaries. Of these, twenty-one pieces appeared in *The New Statesman* and *The Listener*, the others being published in *London Magazine*, *Writing Today*, *Ambit*, *The Jewish Quarterly* and *Encounter*.

A number of the early reviews were of Australian books, and encouraged London's 'highbrow' newspapers to identify Porter as someone with knowledge of that country. But these books were often set in the Australian outback, about which Porter admitted he knew almost nothing, and had little interest.

The myth of the Australian as 'bushman' persisted through the 1960s, in spite of complaints by Australian fiction writers such as Hal Porter and Michael Wilding that its associated style of 'bush realism' was crippling.[62] Furthermore, the outback, or the bush, also persisted as literary London's preferred myth for Australia. In a review of writer Jock Marshall and artist Russell Drysdale's *Journey Among Men*,[63] which appeared in *The Listener* in 1962, Porter may be found testing the book against a sense of himself as expatriate Australian and Londoner. In this review Porter rejected the implied exoticism of the Australian outback and, with it, the tough masculine image which he had found so frightening through a schoolboy's eyes in the poem 'Eat Early Earthapples':

> The men in the title (*Journey Among Men*) include bore-sinkers, stockmen, pearlers, main roads workers, dingo hunters and above all pub-keepers. They remain mostly shadowy outback caricatures, exhibiting a narrow range of bush humour centred on drunkenness, on complaining drudgery, and tall tales.

Porter's estrangement from this Australia is obvious enough, but he ends on an up-beat note as he considers the land itself, and its creatures:

> Fortunately, the book is not primarily about men at all. It has
> a subject [the land] which survives all obtrusive vulgarities of
> emphasis and style, and the birds, animals, insects and plants
> which live in it.

When a personal reconciliation with Australia occurred later, in the
mid-1970s, Porter would indeed find this Australia more fascinating.
But in 1962, as he considered his joint inheritances as an Australian
and a Londoner with an English wife and baby daughter he was
irritated, not only by the quality of the writing, but also by the book's
'truculent nationalism and jejune anti-British cracks'. In this statement,
as in others later, he expressed a division of loyalty between his
adopted country and an Australia of the future, which would be more
mature than the defensive anti-Englishness some of its current spokespeople practised.

More revealing than the comments on Jock Marshall's travel writings
are Porter's comments on Russell Drysdale's drawings, in *Journey
Among Men*, which were found to be 'rather monotonous, though
skilfully done', and having 'none of Nolan's iconographic power'. The
preference here for the expatriate Nolan is instructive. Porter was
praising an 'iconographic' power above illustrative art, with an implied
assumption that the development of powerful symbols such as Nolan
produced could best be done at some distance, psychological and
physical, from their source. Here Porter was developing, perhaps
unconsciously, a rationale for the artist as exile. Nolan, who had
arrived in England in 1950, was, after a decade based mainly there,
facing identity problems of his own. These were exacerbated by critical
responses from Australia that he had 'sold out' on his home country.
When his *Leda and The Swan* series was shown in Sydney in 1961,
John Olsen commented that the paintings were 'contaminated with the
over-ripe atmosphere of Bond Street'.[64] The implication behind this
criticism was that an artist who loses his roots in the country which
gives his work its primary inspiration is lost. Beyond any personal
sniping, this is a version of a serious argument to which Porter would
return later, pre-eminently in his lively quarrels in verse and prose
with Les A. Murray in the 1980s. For the moment, he was staking out a
territory of iconography for the expatriate Nolan over the Australia-
based Drysdale's illustrative powers. Porter's later work with another
expatriate painter with prodigious iconographic powers, Arthur Boyd,
would lead to a growing recognition of the need for both distance from
and involvement with the people and landscapes of his birth.

London reviewers of Australian books had a more powerful
influence on Australian opinion in the 1960s than in the 1980s and
1990s, when a diverse and relatively independent professional literary
journalism began to emerge. In this context, Porter's jousting with
Australian legends and myths of identity was potentially inflammatory.

In reviewing a symposium of papers titled *Australian Civilization* in 1963[65] he commented: 'The New Right in Australia is now fully in the open'. Given his strong association with the left-wing weekly *The New Statesman* and the espousal of his work in these years by socialist commentators and critics, Porter's identification of 'the New Right' here would seem an immediate condemnation. But as reviewer he avoided making facile judgements of right or left. What was at issue was the way something was done, the quality of thought and presentation. It was not good enough, he contended, for Australian or any other myths to be 'served up to the world in the deepest dredgings of sociological jargon'. Here an impatience with such jargon commenced in public, and Porter was not cowed by academic learning. Though he genuinely respected perceptive and illuminating scholarly work, he became annoyed when non-fiction writers turned their backs on commonsense and communication with a reading public in favour of impressing each other. (Poets were given more licence in this respect, but not much.) It was necessary, he believed, for thinking people to be able to entertain alternative hypotheses: the rewriting of Australian myths from a middle-class perspective, for instance, might have some value as 'a corrective to the lazy commonplaces of old' and ought to be considered. In such critiques and commentaries, the swingeing malcontent of Porter's early poetry contended with another persona: the seeker after cultural sophistication and wisdom; a depth in cultural and social commentary which some intellectuals in the 1960s seemed intent on denying in favour of professional advancement, and which, Porter came to believe, high-quality metropolitan literary journalism could redeem. This belief would enable him to survive as a writer in the competitive literary jungles of London.

6

THE STRETCH MARKS ON HISTORY

◆

Personal and public voices are closely intertwined in Porter's development of an historical perspective in his poetry during the 1960s. In 'The Battle of Cannae' (*APF, CP* 81), for example, a younger Porter is recalled choosing as his hero Hannibal, the Carthaginian general who won a famous victory against Rome in 216 B.C., and about whom Porter had read when he studied Ancient History at school.

After thrilling to his imagined war games, alone in happy exile in his yellow Brisbane bedroom, the boy emerges from his 'dove-clapped house and sandy street' to an Australian Saturday afternoon which offers him the more frightening physical war games of sport. In one of those urgent reversions to interior monologue which were becoming a signature in his work, Porter realizes, in an epiphanic moment, that 'No right but clemency can save a life' and he 'travels back to find the stretch marks/on history', arriving at 'the same eclipse', the deep trauma which readers of his work will identify as the early death of his mother. The poem reveals how personal need and history were closely allied in his imagination at this time; the fearful certainty of death and the search for his origins. 'The stretch marks/on history' therefore provides a metaphor for journeys of exploration into the genesis of self, nation and species, the individual and society, beginnings and ends.

The search for national origins became an important poetic quest for the voluntary exile: it helped establish a mythology of the past and a rationale for the present. 'Sydney Cove, 1788' (*PAAM, CP* 50–1) is a major poem of this time in which Porter adapts Auden's style of verse historicism in order to trace his own historical origins as a European Australian. The idea for this piece came from John Cobley's collection of documents and diaries, *Sydney Cove, 1788*, which Porter had reviewed for the *Listener*.[1] Summarizing his reading of the book, Porter stated in this review:

> Once the female convicts were landed there was trouble continually: concubinage, drunkenness, hard work, misery and death ... Backs were flogged raw and took months to heal on a diet of salt meat and flour. Over the whole scene presided the formidable presence of the Governor himself. He reduced the ferocious sentences to flogging, forbade trickery or cruelty to the Aborigines, organized exploration, and encouraged agriculture. He was never in any doubt that this country would prove to be one of Britain's most valuable possessions.

Porter was intrigued by what lay behind the recorded words of these 'unwilling pioneers'. 'Each one', he mused, 'must have been a remarkable person, pushed by circumstances for a short time to epic size'. In particular, his attention was arrested by the anonymously written words of a diarist recording Christmas Eve, 1788: 'Amelia Levy and Elizabeth Fowles spent the night with Corporal Plowman and Corporal Winxstead in return for a shirt apiece'. This sentence is reproduced exactly towards the end of 'Sydney Cove, 1788', providing a dramatic instance of the way in which the lives of ordinary sufferers may 'flare up briefly and light the imagination', securing themselves an epic niche in history.

Porter's transmutation of these documents on European settlement at Botany Bay is worthy of consideration. Using a verse form of three-line stanzas with rhyming first and third lines, Porter built his poem through a series of apparently flat declarative statements which nevertheless suggest by changes in linguistic register a number of emotional shifts. An interesting feature of the poem is the use of a first-person narrator. The presentation of this figure as a participant-observer of the 'founding of Australia', creates a certain immediacy. The speaker, however, is not a fixed, recognizable 'character' of the Theophrastian or even Browningesque kind. Rather, he is a composite figure, mobile in a way approved of by postmodernist critics, combining historical recuperation with contemporary critique.

'High' and 'low' moments in 'Sydney Cove, 1788' combine in a complex melodic key within the frame of an apparently flat and regular format. An uncanny knack of running certain stanzas on (such as the fifth and sixth) prevents the format acting as a straitjacket. In later stanzas a mixing of the portentous and the pertinent lends a strangely surrealist aura to these reimagined events:

We wake in the oven of its cloudless sky,
Already the blood-encircled sun is up.
Mad sharks swim in the convenient sea.

The Governor says we mustn't land a man
Or woman with gonorrhoea. Sound felons only
May leave their bodies in a hangman's land.

Heat, blood, violence and disease are the ingredients of Australia's European beginnings. But the tone shifts to an unusually lyrical conclusion, the closest approach in Porter's verse to pantheism:

The cantor curlew sings the surf asleep.
The moon inducts the lovers in the ferns.

Whereas Martin Seymour-Smith[2] failed to see beyond the 'coarse' and 'vulgar' in such poems, Christopher Ricks commented more clear-sightedly on the 'various kinds of loving' in this piece, leading to its 'precarious conclusion' in which 'romantic aspiration is buffeted but not belittled'.[3]

This poem accords with Porter's more general belief that such apparently inauspicious beginnings for a nation, in which pleasure or happiness can be snatched only temporarily from the jaws of extinction, may paradoxically provide the best kind of myth for succeeding generations — better, indeed, than the myth of hope:

> I have always believed (though it is a broad myth) that Australia differs from the United States in being born in horror, not in hope, so that the country's progress could only be an improvement, while America has watched its dream of hope disintegrate into corruption and inanition.[4]

Here, as elsewhere, the notion of a fallen world seems easier to cope with imaginatively than utopia. In some ways, Porter is 'at home' with the first settlers and convicts. Indeed, the poem can be taken to represent a position of temperate patriotism, an emotion of which Porter is normally highly sceptical.

'Sydney Cove, 1788' is a key poem in Porter's second collection, *Poems Ancient and Modern* (1964). Dedicated to his father, William, and his daughter, Katherine, this volume follows the lead of its title (from the Anglican Hymnal) in mixing past and present themes and styles, and giving scope to lyrical or existential treatments; the lyric would be extended more definitively in the third volume, *A Porter Folio* (1969), in 'Three Poems for Music' and 'The Porter Song Book'. The most noticeable development in *Poems Ancient and Modern*, however, is its concern with historicism. Having gained the attention of readers and critics, Porter did not rely on the formula of the popular 'London' poems but went on to extend his range of poetic forms and subject matter. Almost inevitably, there was a slump in public attention, as if, having 'found' the new poet, the journalists and critics did not know how to describe his 'development'. 'Oh where are you going? said Peter to Porter' as Christopher Ricks put it.[5] Unlike some other reviewers, Ricks did not regret the loss in 'availability' of certain poems which bring figures of the ancient past such as Septimius Severus or Hulagu Khan into the present by a species of witty anachronism. What was lost in availability, for Ricks, was 'more than made up in compact suggestion'. Nor was Cyril Connolly deterred by the excursions into history. Having praised Porter's first volume, the author of *The Enemies of Promise* saw signs of improvement in *Poems Ancient and Modern*, due largely to its author's 'careful reading of Martial and a marination in the last days of the Roman Empire'.[6] But, like many older critics, he was also keen to give advice. The wit is wonderful, he says, but Porter should write 'a few (other) serious poems which are complete in themselves (like 'The Sins of the Fathers') and not just a fragment of the bullet-proof waistcoat which his personality seems to be constructing out of his power of language'.

Connolly's difficulty in locating a vulnerable self in Porter's second book indicates, not so much a disappearance of this figure as its acquisition of heavier disguises, of masks which change more rapidly. In 'The Sins of the Fathers' (*PAAM, CP* 54−5), which Connolly admired, this persona's nerves seem close to the surface in his projection of images of a 'pelican' daughter, who, as in *King Lear*, reverses the usual structures of patriarchal tyranny and repays the father, the poem's speaker:

> I pay out my ungrudged money
> And long attritions of my love
> To raise a daughter. In her pram, she
>
> Looks up at the plane trees. Daddy owes
> Her his flesh like a fattened steer.

This unsentimental evocation of a father's relationship with a child lurches at the poem's end towards a mood of guilt and self-pity in the bizarre image of the father metamorphosed into a trapped bear, dancing. Other poems reach further back into the poet's personal history of feelings: 'Two Merits of Sunshine' (*PAAM, CP* 56−7) is an exploration of the boy's fixation on his mother, and the significance of her loss; and 'The View From Misfortune's Back' (*PAAM, CP* 57−9) recalls a boarding school/orphanage. 'Homage to Gaetano Donizetti' (*PAAM, CP* 61) is a more extrovert poem, celebrating adolescent lust and the validity of vulgarity in appreciating music and life.

A number of other poems in *Poems Ancient and Modern* are closer to the moods of late adolescence than they are to a settled or quiescent parenthood. 'An Anthropologist's Confession' (*PAAM, CP* 59−60) is a powerful short story as poem, told in twelve five-line stanzas, alternately rhymed and syllabically in the style of George MacBeth. Its theme is violence and lust. Blake Morrison described it as 'a remarkable poem about envy of the sexually successful ... an anthropologist watches with vicarious pleasure the rape of a beautiful young girl by a goatherd, then revenges himself by having the goatherd killed by the village headman'.[7] On the other hand Thomas Dilworth observed that 'the poet-moralist is clearly and clumsily visible'[8] in this piece. While a scorn for the arrogance of such interventionist anthropologist figures may be easily enough deduced from the poem, there is however also a powerful directness in the observation of the goatherd's lust. In 'Nine Points of Law' (*PAAM, CP* 67−71), 'Happening at Sordid Creek' (*PAAM, CP* 61−2) and 'How to Get a Girl Friend' (*PAAM, CP* 63−5) the twin problems of lust and guilt are vigorously disclosed.

Typically, while the title of a poem hints at sexual excitement, the contents offer more ambiguous pleasures. 'Happening at Sordid Creek', for instance, structurally mimics Ambrose Bierce's story 'An Occurrence at Owl Creek Bridge', and the poem's chief insight seems close to that of the Jacobean playwrights, especially Webster: 'The world's/A hospital, we won't get well'.

For many writers, history is sharpened by a sense of place. For Porter, the commitments of work and the needs of two young daughters meant that, apart from the two-week holiday in Vienna, with Jannice, in 1963, he did not visit Europe again until the early 1970s. 'Vienna' (*PAAM*, *CP* 49−50) and 'Homily in the English Reading Rooms' (*APF*, *CP* 99−100) derive from Porter's experiences in the Austrian capital. These poems are preludes to later 'European' poems and indicate ways in which Porter began to incorporate historical material, and the idea of history, into his work. It is significant that Porter's first European holiday was to the world capital of music, the city with which the names of Haydn, Mozart, Beethoven, Schubert, Brahms and, more recently, Alban Berg and Schoenberg, have been associated. But Vienna was also a city of faded grandeur, a former centre of the Austro-Hungarian empire, now stripped of its possessions. 'Vienna', which first appeared in *The New Statesman* in November 1963, begins with the profits and losses of empire:

> This Imperial city
> Needs no Empire: turks, saints, huge nineteenth-
> Century geniuses,
>
> Poets with the tic, the spade
> Bearded patriarch, who raised to the nth.
> The power of love, these came
>
> Here like spokes to their axle.

A far cry from a ghost town in the genuinely empty Australian sense, Vienna lives on through the ghosts of its famous artists and thinkers: haunted, neurotic poets such as the expressionist Georg Trakl, perhaps, but above all by Sigmund Freud, 'the spade/Bearded patriarch'. This is the only direct reference to Freud in Porter's poems of the 1960s, but Freud's *Civilization and its Discontents*, first published in 1929, is at least as important to the archeology of knowledge in Porter's writings of this period as another 'spade bearded patriarch', Karl Marx. Freud's theory of instinctual repression as a trade-off for civilization continued to haunt the poet's thinking as a valid formulation; and it remains in his writings beyond the socio-political notions of class conspiracy which some of the 'London' poems seem to propose. When, in a later poem ('The Story of my Conversion', *LIACC, CP*

204−6), Porter entertains the idea of Freud applying for a job in an Australian hospital, he points up the incongruity of the world's spokesman for neuroticism living in a country where so many apparently deny the experience of pain, suffering and repression, and where happiness is insisted on as a public and private reality in spite of evidence to the contrary. This point of view is expanded in the poem 'In the New World Happiness is Allowed' (*TCOS*, *CP* 254−5).

Vienna is Porter's symbolic centre of neuroticism, but has its compensations for the angst-ridden tourist. Attics and rooms tell of Haydn's suffering, and Schubert's, but always, behind such scenes of desolation, is the music they created. Political and economic power has flown from the centre, but, in a wonderfully surreal image, 'The statued squares/Don't know that wet Hungary/Isn't theirs'. The quality appreciated here is not the clean slate of the New World innocent (the 'tall Minnesotan' is ridiculed for his hygienic shallowness), but the beautiful artefacts of the Old World, which give scope to the dreamer: 'God can fly/In this architecture'. Nature, in which Porter at this stage and through most of his life is far less interested, 'is one of the boasts/ Of lost prestige'. His attention is caught not by the woods around Vienna, but by the baroque splendour of its buildings. As a New World arriviste with an Old World sensibility, Porter was perhaps less afflicted at this time with an anxiety of influence in relation to Europe than those who considered the traditions automatically theirs: his appropriation of the city is bold and original. 'Vienna', then, is an important poem in Porter's canon. It reveals an early concern with emblematizing certain values of European civilization which defy the forces of political and economic deterioration. For Porter, the greatest artistic glories are born when political and economic power is in decline. It is not surprising that he compared living and working in Vienna and London, two former centres of empire which were both congenial to him. In 'Homily in the English Reading Rooms', which arose in part from the visit to Vienna, the place itself is not named but appears as a city of fable akin to one of Italo Calvino's in *Invisible Cities*. In Porter's version, however, the city is haunted by images of violence, dissension, loneliness, despair and suicide. Freud's city here has turned to nightmare, which for Porter, as for Freud, provided a major source of artistic inspiration.

◆

In the five-year gap between the publication of *Poems Ancient and Modern* (1964) and *A Porter Folio* (1969), Porter was working full-time, helping to raise two children, writing literary journalism, attending meetings of the Group, and giving public readings. In his home life he appeared as the image of the integrated family man, and professionally his reputation as a poet of note was growing. Whatever appearances might have suggested, however, Porter was not in a state of calm stability during this period. *A Porter Folio* (1969) shows that he was far from comfortable with his achievements, and had no intention of resting on his laurels. Though the longest of the individual volumes, reviewers noted that it seemed an 'interim' book, containing five-finger exercises in a variety of poetic forms ranging from tight, elliptical utterances to more expansive dramatic lyrics. It also shows its author's concern with the stretch marks on history, especially as he contemplated the evolution of the species, and himself as a part of it.

Dedicated to the virtuoso poet of the 1960s, George MacBeth, whose work in this decade welcomes the avant-garde and shows many shifts of style and interest, *A Porter Folio* is in some respects a poet's workbook. The volume contains more gnomic and involuted poems than Porter had written since the 1950s, and its relative inaccessibility to a reading public, whose attention Porter seemed not primarily concerned to hold, was confirmed by the inclusion of only five poems from this volume in *A Porter Selected* (1989).

Connoisseurs among the critics nevertheless enthused. Cyril Connolly, for instance, stated: 'Mr Porter is an Australian whom we are lucky to harbour, for he has a genuine talent "smelling of love, like a piece/of oiled wood"'[9] (a quote from 'The Recipe', *APF*, *CP* 105−6). While praising the talent, however, Connolly pointed to a lack of warmth and self-exposure in the book: 'If we are not going to be deeply moved (as happens so seldom in contemporary poetry), let us enjoy the intelligence, the perception, the sheen of words at which [Peter Porter] is adept'. Brian Jones observed the 'mastery of new skills, including an Audenesque ability to shift rapidly from a relaxed, talking voice to aphoristic statement, and added that Porter was using this technique in the dramatizing of 'an eclectic wandering around European culture'.[10] An anonymous *Times Literary Supplement* reviewer compared Porter's role in this book to that of T. S. Eliot:

> Mr Porter parades his culture in every sense, so that England may take a last look. The role (Eliot's role) of the visitor who knows the place better than the inhabitants retains its power . . . The resulting poetry, even at its most clumsily rushed, pulls off the trick of staying in contact while conveying no sense whatsoever of belonging.[11]

Porter was chided nevertheless for failing to acknowledge that he had now gained an audience in England and had responsibilities to it. The *Times* reviewer asserted that notes explaining his allusions and epigrams should have been provided, as T. S. Eliot had done in *The Waste Land*. The only extended review of *A Porter Folio* in Australia, by Carl Harrison-Ford,[12] iterated some of the points made by English critics and noted that the book was 'at once more relaxed and more detached' than previous volumes and that it lacked 'overall cohesion'.

Yet there are paths to be found through *A Porter Folio* other than one come upon by attending to technical exercises and the passing parade of culture. An important other dimension was observed by two poet-critics, Ian Hamilton and Anthony Thwaite. Hamilton acknowledged Porter's 'lethal satiric gifts' but also pointed out that a reader 'never loses sight of the protagonist':

> The hint of self-mythologising in the title (we are also offered 'The Porter Song Book') is no accident of vanity: Porter has always been the chief character in Peter Porter's poems — a cultivated colonial bogged down by the metropolitan grind, a music lover fascinated and repelled by plastic culture, an aesthete and a moralist who's fairly sure he can't be both, and so on.[13]

The protagonist-self which could be seen emerging, however, was at least as much the one acted upon as the one imposing itself on circumstances.

The self presented in this volume had changed from the summer hermit of the 1940s in Brisbane to what Anthony Thwaite called an 'ingenious chameleon':

> It is a body of work that is highly personalised without being personal, from the early 'Reflections on my own Name' to 'Porter's Metamorphoses', 'The Porter Song Book' and the very title of this new volume.[14]

Thwaite contended that when Porter was less oblique in his presentation of self, as in 'Five Generations' (the autobiographical poem of his Australian lineage omitted from his *Collected Poems*[15]) his 'characteristic sardonic voice seems muffled'. Summing up, Thwaite referred to a 'deliberate irony' in Porter's self-reference, since:

> What it adds up to is evasion and sleight of hand, the ingenious chameleon taking his colour from his context. Porter is a set-faced illusionist, creating scenes, monologues, charades, all of them dense with proper names of people, places, titles, identifiable things: and yet the creator, like Joyce's artist, is withdrawn.

Thwaite's is a sympathetic reading, and is no doubt close to the author's intentions when he refers to the pleasures of inventive fictionalizing.

The opening poem in *A Porter Folio* is 'The Last of the Dinosaurs' (*APF, CP* 77–8), the title suggesting a humorous jibe at the poet and his prehistoric pursuits. Typically, however, the joke is turned on its head and becomes a justification for poetic activity. As a whole, the

piece is a jaunty elegy on the death of the last-surviving dinosaur, in
the mode of later elegies on the death of cats (e.g. 'Dis Manibus', *FF*
51−2), though with more ironic distance from the subject. The lightness
of tone is maintained, suggesting a parent thinking about his children's
illustrated books of dinosaurs, personalizing them as children do. The
levity is evident in the joking with names: 'Chalky' of the Cretaceous
or chalk age might suggest an old teacher, a Mr Chips figure, and
'Tyranno − sore arse − Rex' rehearses children's interest in anal
matters. Porter gets his ages and stages of evolution right; within the
Mesozoic era the stegosaurus is the last major species before the
Jurassic period, for instance. But the poem is not freighted with such
learning, rather, it is carried along by the joking, wistful, self-deprecating
tones of a speaker who evokes vast periods of evolution:

> So much time and blue.
> That great arc telling
> the centuries with its pivotless
> movement, tick, tock, tick.

In Porter's development as a writer, this indicates a stepping back from
the metropolitan bustle. At the same time, as so often, he jests in
earnest:

> Some day a mind is going to come
> and question all this dance −
> I've left footprints in the sand.

Like the last dinosaur, the poet may be at the tail-end of an evolutionary
phase, but has trumped extinction by leaving a mark in the sands of
time.

A children's world is again a source of contemplative pleasure in
'Seahorses' (*APF*, *CP* 89−90). Most of the Porter family's summer
holidays in the 1960s were taken on the coast of Dorset at Charmouth.
While 'Seahorses' may have arisen in part from Porter watching his
daughters at play on the beach (he recalls the poem being triggered by
buying some dried seahorses in a packet in Dorset), the poem also
reaches back into his own childhood memories of the coast near
Brisbane, thus mingling past and present preoccupations. Porter is
never simply a descriptive poet and in 'Seahorses' notions of love,
beauty and death combine in a succinct evocation of completeness and
harmony. The tiny dead creatures suggest heroic journeys:

> Seahorses were vikings;
> Somewhere they impassively
> Launched on garrulous currents
> Seeking a far grave . . .

Their beauty is in their death: 'They were the only creatures which had to die/Before we could see them'. A gentleness and reverence of tone is apparent in the speaker's contemplation of 'the unbending/Seriousness of small creatures' which have been brought to the attention of humans by 'the sea's rare love'. Although the piece ends on a note of realism and bathos with the contrasting image of 'the vast and pitiable whale/Which must be quickly buried for its smell', the talisman of the seahorse representing beauty in death and the idea of death as our principal link with the non-human world are held tight at the poem's core.

A 'long view' of history and evolutionary processes is evident also in 'Fantasia on a Line of Stefan George' (*APF*, *CP* 81) and 'My Late T'ang Phase' (*APF*, *CP* 80). In the first of these, the German lyric poet Stefan George is invoked, presumably for the model he provides of a compact classicism in form. Because Porter's imagination always urges him to break boundaries, he allows himself the licence of a musical fantasia while keeping his poem short and compact — he will have his classicism and devour it too. Although the poem begins as a seemingly very personal lyric ('I shall die if I do not touch your body'), it proceeds to reflections on the processes of evolution as the speaker stands in the National History Museum:

> ... The blue whale
> And the passenger pigeon worried for me,
> Surely my fingers were losing their prints,
> I had no history in my bones, I was
> Transparent as a finished gesture.

Like 'The Last of the Dinosaurs' this poem might be read biographically, as a stage in the poet's personal evolution as he worries about the prints of his Australian past wearing away. (He does this explicitly, later, in 'Evolution', *PTTC*, *CP* 159–60.) 'Where you are is life', he muses, existentially, but the obliteration of memory would make one sub-human. The 'rain forest' in the 'inverse paradise' of his personal and Australian past is the place to which he can trace back the 'thread of fear' which runs from the evolutionary mother to her offspring; her mortality haunts him still.

'My Late T'ang Phase' is more firmly located in the poet's domestic life, in which he was bored and restless. The last line of the poem, 'They dug up the dead who were smiling' refers back to the title: the poet might himself be dug up one day, smiling the smile of a rotten skull on a dead body, like T'ang Dynasty figurines. And what would that show of his inner life? And would it even matter? With a technique of rapid cross-cutting between the Paddington flat, its little girls and cats, and the imagined artefacts of the Chinese T'ang dynasty, the poet reveals present experience as a tiny gargoyle on the great wall of history. In this context, his fellow-poets' attempts to 'find new forms which make kites/Of confessions' may be important to them, but are doomed to insignificance in the total scheme of things. To argue, as Porter did later, that poetry is a 'modest art' was, against this kind of panoramic perspective, a calculated defence.

The 'quarrelsome vitality' which Sylvia Lawson had noticed in *Once Bitten, Twice Bitten* resurfaces in *A Porter Folio* in the gallimaufry entitled 'Fair Go for Anglo-Saxons' (*APF*, *CP* 107—10). The first of the six sections in the poem 'Provincial Messiah' refers to those British writers in the mid—1960s who, in Porter's words, were 'busy being self-proclaimed geniuses, especially in the British provinces, or as we referred to them, Provincial Messiahs'.[16] In lampooning those whom he believed to be ignorant poseurs, Porter was stepping into a literary minefield, where he was bound to be attacked as an anti-populist. (His real concern, however, was that poetry should not be side-lined any more than it already was.) Furthermore, in spite of his own 'outsider' status as an Australian in England, he risked being branded as an advantaged metropolitan in a decade when the regions in Britain were reasserting themselves culturally. In some quarters this charge *was* laid against Porter, but his admiration of certain writers who worked in the provinces (Peter Redgrove and Martin Bell, for instance) should have dispelled this view. It has been pointed out by Blake Morrison, among others, that as a reviewer Porter has displayed a very catholic range of appreciation, including support for many writers outside metropolitan centres:

> Peter has a tremendous antagonism towards favour; and some-thing larger which is, I suppose, Oxbridge and English upper middle-class poets, and literary values which come out of that. He shares with people in Newcastle and Scotland the view that this is bad for English culture, and Philistine. So, he has this very interesting balancing act where he is identified by some as metropolitan and by others as an outsider. It makes him interesting and unpredictable as a reviewer and receptive to surprising things.[17]

'Provincial Messiah' took aim, principally, not at individuals or non-metropolitans but at two insidious enemies in Porter's canon, ignorance and hubris:

> to think no one is any good
> except a poet from Alberta,
> Catullus and some friends
> who run a mimeographed magazine,
> to misquote Ovid, misrepresent Pound,
> misunderstand Olson and never
> have heard of Edwin Arlington Robinson . . .

to be all this is better
than to have talent and have to write well.

To attack an avant-garde in this way was inevitably to sound to
some like a spokesman for the establishment. Porter's strength, however,
was that he had never perceived himself as a member of any London
establishment, nor was he perceived as such by most London critics
and commentators. Moreover, his target was ineptness. In later years
he would criticize bigger literary fish, such as Ted Hughes, Craig Raine
and the major poetry publisher, Faber and Faber, on the grounds that
they were similarly arrogating a 'special status' role. Envy, and anger
that inferior poetry to his own was being listened to, are insufficient
motives for such lampooning. What lies behind it is a psychological set
and related ideology which Porter has himself described as Australian:

> I'm an old-fashioned Australian. I believe that a writer should
> be a real democrat, that he should behave in the same way as
> other men in most public places and you certainly shouldn't
> know he's a writer from his demeanour.[18]

What is identified here is not so much an Australian tendency to
decapitate the tall poppy as to suspect the motives and ability of those
who put on special airs and graces.[19] In one of the 'Sanitized Sonnets',
(*TLOE, CP* 143) Porter's persona comments: 'I have never said sir/to
anyone since I was seventeen years old'. The other side of this
democratic and egalitarian outlook is that Porter considers the greatest
works of art open to all who can learn to appreciate them.

Like most of the Movement poets before him, Porter distrusted the
'special person' status which Dylan Thomas, George Barker and other
1940s neo-Romantics seemed to affect, both in their lives and
their poetry. He later extended this criticism to some Australian neo-
Romantics in the 1970s who appeared to consider themselves above
others, in spite of showing an ignorance of tradition and being of
dubious talent. Art as therapy or self-promotion was not enough — a
longer, historical view was necessary. Cultural commentators and critics
had a duty to encourage talent, not faddishness. Associated with this
was a belief that the 'great' poets of the past (Shakespeare, Donne, Pope,
Hardy and others) survive in the imagination beyond their personal
idiosyncrasies, their biographical excesses or limitations. In the context
of Porter's own work, this view takes insufficiently into account the
construction or 'invention' by readers of personae for authors; thus
Porter's own 'quiet eccentricity'[20] in the pursuit of an unshowy
ordinariness is a representation of self which contains contradictory
and problematical elements for the reader. The reader of Porter's work
is faced with continual collisions between ordinariness and grandeur,

simplicity and sophistication, a prosaic present and the caves of imaginative riches which lie behind.

A Porter Folio invites the reader to indulge some of these riches in music, whose 'huge light irresponsibility', as we have seen, could provide relief from both past and present problems. Allusions to music abound in this book. The poem 'Competition is Healthy' (*APF*, *CP* 90−1), for example, has as an epigraph a line from a Bach cantata which is based in turn on the Sermon on the Mount from the Gospel according to St Matthew. For Porter, the biblical quotations in this and other poems were memorable first and foremost by their associations with music. By contrast, 'St Cecilia's Day Epigrams' (*APF*, *CP* 103), is a jokey poem which, in its second section, ridicules the simplistic equation of life and art made by music critics in the cases of Beethoven, Palestrina and Wagner. A line from Auden and Kallman's opera text, *The Bassarids*, provides the basis, and last line, of another piece (*APF*, *CP* 85). In-jokes for aficionados of classical music recur. In the third of 'Three Poems for Music' (*APF*, *CP* 101) allusions to European classical legends, the Christian Bible, and echoes of Browning, Marvell, Shaw and Shakespeare contribute, in semi-subterranean fashion, to a macaronic hymn to music. Two of these pieces, together with 'Poetry' from 'The Porter Song Book', are all that survived into the *Collected Poems* from a work written for the English composer Nicholas Maw, for whom Porter also wrote an eight-poem cycle based on the letters of Dorothy Osborne and William Temple, entitled 'The Voice of Love' (1966). Whether the third of the 'Three Poems for Music' (sometimes read separately by Porter as 'The History of Music') is entirely successful outside its original context is a matter of opinion. The pastiche quality of the couplets and their many allusions would almost certainly reduce its appeal outside the small band of musical initiates for whom it was originally composed. But the poem does contain some of Porter's deepest convictions and some memorable lines, such as the following:

> When Orpheus plays we meet Apollo,
> When there's theology to swallow
>
> We set it to music, our greatest art,
> One that's both intellect *and* heart

Religion, for Porter, can be assented to only through art, and in particular through music: Orpheus, the earthly musician, introduces us to Apollo, the god. This is a similar outlook to that of Neville Cardus who, as an atheist, felt that the line 'I know that my redeemer liveth' was true only when Handel made it so. But the depth of Porter's feeling for music is reserved for the concluding couplet:

And Paradise till we are there
Is in these measured lengths of air.

Porter's deep love of music is no trembling obeisance before
the great names: Mozart, Haydn, Bach, Beethoven and others. His
engagement with the acknowledged musical greats is as robustly inter-
rogatory as his approach to Shakespeare, Donne, Milton, Pope and
Hardy. However, he feels personally drawn to those artists, in music,
painting and literature, who fall short of perfection; indeed, he is more
fascinated by some of the shortfalls in art than by its full-blown
success stories. This tendency is particularly evident in 'St Cecilia's
Day, 1710' (*APF, CP* 102), which is dedicated to W. F. Bach, the firstborn of J. S. Bach's twenty children, whose life was steeped in trouble,
disappointment and alcohol. Porter's relatively deprived boyhood and
youth, with a father who apparently provided no artistic guidance or
example, contrasts vividly with that of Wilhelm Friedemann Bach,
whose famous father doted on his eldest son and gave him his main
musical instruction. In this poem, then, Porter seems to suggest that
such a start in life is a recipe for disaster; it is better for the artist,
perhaps, to begin in philistinism and with no parental expectations.
Porter's sympathy extends across the centuries to this figure who
represents for him both a guard against his own disappointment and
an example. W. F. Bach was not a great originator but had some talent;
only against the high standards expected of him was he a failure. Here,
Porter's 'anxiety of influence' was being worked through in his direct
address to the eighteenth-century musician who did not succeed,
either by his own internalized standards or by those which society
imposed on him:

Having nothing new to say but born
  in the middle of a warm skill,
  . . .

              you are the star
  to hail across centuries
  of competitive rubbish

Porter apotheosizes here a forgotten figure, and arouses our sympathies;
in doing so, he conveys an almost sentimental attraction, not unlike
Graham Greene's, to the idea of failure. But two other ideas come
through in the poem: first, that like polishing tables, art is 'natural', in
that it has to live in, and use the materials of, the ordinary world; and
second, that art is mistaken if it is too optimistic, for its chief function
is the redemption of unhappiness. At the end of the poem Porter
reverts to an image from his own childhood:

>   ... We are all
> children lying awake
> after the light is put out.

The manoeuvre is typical in its linkage of personal experience (Porter's childhood fear of the dark and insistence that his light be left on) with that of the historical subject and a more general human condition. Like Thwaite's image of the chameleon, the self here takes some of its colour from its subject but is not lost from sight: however camouflaged, it reveals itself in certain kinds of identifications and occasional tones which suggest urgent personal experience.

The compositional problems alter when Porter turns away from exemplary figures drawn from musical history to find his own verbal equivalences for music. The fifteen short lyrics in 'The Porter Song Book' (*APF, CP* 110–15), for instance, are an early flirtation with poetry which lightly resists the referent: there is no direct subject in any of these pieces, which work instead through rhythm and image to set a mood. As Cyril Connolly noted,[21] the chief influence here is Hugo Wolf, the nineteenth-century composer who studied at the Vienna Conservatory (perhaps another of Porter's 'spade bearded patriarchs') and who brought the German *lied* to its highest point of development. Porter seems to deliberately withdraw his autobiographical persona from these lyrics and, instead, provides hints of scenes, monologues and situations like those we may experience when listening to music. Deliberate anachronisms abound, for past, present and future are legitimate elements in the suspended states of mind and feeling which music — and by analogy poetry — may induce. In the third lyric, 'The Sanatorium', the predominant mood is implied by an allusion to Thomas Mann's *The Magic Mountain* with its isolated sanatorium, its coughing and deaths. A free-ranging impressionist technique allows rapid switches of association to the 'long verandahs' of an Australian childhood and to the brilliantly apposite image of 'hangers in hotel wardrobes' which 'rock/Occasionally at doors opening'. The poem ends with a vision of dying, and perhaps illumination:

> Turning his eyes to her, he sees
> A wall without bricks lit
> By endless sunshine brightening.

These lyrics are not confined within a narrow range of 'romantic' treatments; Porter is too interested in bizarre realism to allow his verse to wallow in romantic cliché. Thus the twelfth poem, 'White Wedding', which derives in some respects from Mörike's 'Bei Eine Trauung', ends in a wittily ominous lament: 'Alas, alas, God's rosters are declared,/ Tears dry quickly fanned by Prayer Book prose, /If love can't be, then hatred may be shared'. Two others, 'The Visitation' and 'Lyric found in a bottle', show the lover as monster and worm, respectively. Against such aberrant images, the occasional moment of lyric intensity is redoubled, as in 'Clairvoyant':

> But still you gaze through lines and dream
> Of living: here is peace, the reach
> Of an old river with lawns to its edge
> And an opposite bank nobody has seen.

What informs this lyric, like others in 'The Porter Song Book', is a haunting sense of the uncertainty of living, and the mysterious certainty of death.

If music offers a relief from contemplation of the personal and social history, the images of the father which re-emerge in *A Porter Folio* require some reappraisal. On a personal level, the relationship of the poet with his father appears as a significant element in the tribulations of exile. It is there hypothetically in the consideration of the figures of Johann Sebastian Bach and his son Wilhelm Friedemann, in 'St Cecilia's Day, 1710' and again, more directly, in a scene recollected from the poet's own childhood in 'The Recipe' (*APF*, *CP* 105−6), a poem which suggests a certain volatility of attitude and feeling about the Australia Porter had left behind some fifteen years earlier. The speaker tries different ingredients in the expatriate's recipe, then alights as if by chance but with deadly emotional relevance on the following:

> Take instead a veined and freckled man
> with his ten-year-old son
> treading among the uguary shells
> talking about the world's drawn blinds,
> the boarding house of hell where meals are prompt.

This half-repressed image of childhood rejection is an important scene in the continuing drama of the expatriate son's relations with a father whom he in his turn has left behind; a drama which is resolved only with the father's death and the realization, expressed in 'Where We Came In' (*FF* 29), that 'The son was now the father'.

In the mid−1960s, after twelve years in England and in his mid-thirties, Porter evoked images of his father in another poem of deep irony and pathos, titled 'Competition is Healthy' (*APF*, *CP* 90−1):

> . . .Out of reach of the philharmonic
> That old man is planting his garden.
> He nestles each seedling in the soil,
> Contrives a cotton grid to keep the sparrows off,
> Sweats in conscience of his easy goal.
> Unknown to him his son has scattered

>Radish seed in the bed and the red clumsy
>Tubers shall inherit the earth.
>Take no thought saying 'What shall we eat?
>What shall we drink? or Wherewithal shall we be clothed?'
>We shall eat the people we love,
>We shall drink their fluids unslaked,
>We shall dress in the flannel of their blood,
>But we shall not go hungry or thirsty
>Or cold. The old man writes with a post office nib
>To his son. 'The Government has cut the quotas,
>Here the bougainvillea is out,
>The imported rose is sinking in the heat.'

In an emotionally charged image, the father is presented here in a post-Edenic garden of childhood which the son, in silent rebellion, has spoiled, as the boy Peter Porter had himself done by scattering radish seeds among his father's roses when he learnt he was being sent to boarding school. This is a dramatic instance of the competition motif which counterpoints an alternative ethic of love and generosity in this poem. 'We shall eat the people we love' is a travesty of Christ's Sermon on the Mount, just as earlier examples of the competitive ethos of politics and economics are travesties of human justice and tolerance.

'Competition is Healthy' was written in 1964, when the right-wing Republican candidate Barry Goldwater was challenging Democrat Lyndon Johnson for Presidency of the United States of America. Porter's 'underprivileged rich' who pray for a Goldwater victory are carriers of the disease of competition as a justification for existence; as such, they are precursors of Prime Minister Margaret Thatcher in Britain in the 1980s, whose assumptions and values are satirized in later work. Typically, Porter's persona is also implicated: he too is marked with the sign of Cain; he too is a competitor and a moral cannibal. Deeply discomforted, he knows himself caught in the flux of historical forces, a consumer like the rest. Only too aware of the mixed economies of living for simplistic romantic rebellion, he sees his best chance (if not his main chance) as an ironic exposure of such distorted values, in the hope that a degree of enlightenment and change may follow. This is a meliorist's view of gradual reform through a process of increased understanding, in which the reformer's motives and involvement are themselves under ruthless scrutiny.

'Competition is Healthy' is not a fully resolved poem, and here that is perhaps a strength. Its collage technique reflects something of the dispersed nature of *A Porter Folio* as a whole. The book is not marked by a central or controlling tone; rather, it gathers up a range of approaches, styles, ideas and images in the latter half of a decade — the 'swinging Sixties' — when, as Porter himself subsequently realized, he was undergoing a metamorphosis without knowing it.[22]

◆

The publication of *The Last of England* (1970), by Oxford University Press, marked the collection and consolidation of work of the late 1960s, following Porter's sacking ('the asbestos handshake') from the advertising firm Denhard and Stewart in 1968, and his decision to try and make a living as a free-lance writer. (Porter had left Notley's in April 1967 to go to S. H. Benson, and thence to Denhard and Stewart.) A notice on the cover of *The Last of England* reflects the kind of reputation its author was achieving in British literary circles:

> Widely regarded as one of the best 'social' poets now writing in England, [Porter] writes of the English, and of what it means to be living in England now, with a sharpness of perception and expression derived in part, perhaps, from the fact that he is Australian.

The cover shows Ford Madox Brown's famous picture 'The Last of England', with a Victorian couple on the deck of a boat, rugged up, hand-in-hand. Their destination, like Porter's, is unclear, though their dress certainly identifies them as English. Having lived almost twenty years in England by 1970, married to an Englishwoman and with two children there, Porter was now ready to reconsider his origins and identity. He announced:

> I have made the decision to change myself from an Australian into a modern Englishman (I cannot change my accent and I have yet to change my passport). So, in a sense I am an immigrant saying farewell to my past and the country my family went to in the middle of the last century.[23]

This was a metamorphosis which would never occur. A running battle with origins, loyalties, purposes and directions continued and the Australian passport remained in his hands. An unsettled persona persists in the poetry, much of which is migratory in subject and spirit. Questions of history — of the self, nations and species — remain an urgent topic, as a recurrent drama of identity is played out.

Characteristically, the title piece of *The Last of England* projects mixed messages. A doomed, prophetic note is called up by the title, and this is reinforced in the first of the poem's two stanzas by images of poison, chemical rain and other horrors of modern England; even the seas around it seem 'exiled'. But a lyrical calm takes over in the second stanza:

> Sailing away from ourselves, we feel
> The gentle tug of water at the quay —

Language of the liberal dead speaks
From the soil of Highgate, tears
Show a great water table is intact.
You cannot leave England, it turns
A planet majestically in the mind.

The 'we' here speaks with authorial force. Its only caveat — and it is a significant one — is that the majestic vision of a liberal England speaks to the writer from a home of the dead, Highgate cemetery. The idea of England for Porter, as for T. S. Eliot before him, principally comprises the dead with whom he may speak through the incomparable medium of the English language. The chief difference is that those with whom Porter wishes to converse are 'liberal' (a term he uses in a wider context than social doctrine here) and humane, while Eliot's loyalties were to a classical and royalist tradition. At this stage, Porter's principal concern was for a 'mainstream' literary tradition with which he could engage. It is not surprising therefore, that contemporary American poets such as Robert Duncan, Charles Olson and Charles Bukowski, who were actively dismantling an English poetic tradition to which Porter was expressing adherence, should receive short shrift from him in reviews at this time.[24] The immigrant had not come to the literary capital to be a mental fringe-dweller. He had not studied traditional English literature at university level, and was unconstrained in his desire to ingest the best he could find from all ages.

Australian reviews of *The Last of England* were equivocal. R. A. Simpson referred to the fact that Porter was only just beginning to receive wider recognition in his native country, and that much of his writing was the response of 'an intelligent well-read man of his times — a man confronted with the circus that is the second half of this century'.[25] While stressing Porter's racy contemporaneity which could depict death in the twentieth century 'as natty and nasty as Enoch Powell', and could mention 'Paul VI's more dreadful quotes', Simpson warned against seeing him as a social reformer who was trying to push a single political idea or solution. Carl Harrison-Ford noticed the humour in the verse and commented that the volume demonstrated Porter's ability 'to take up serious issues without falling into that area of portentous high seriousness that tends so often to be breathtakingly forgettable'.[26] Peter Joseph, in *New Poetry*, was impressed with Porter's experiments with a variety of verse forms but bemused at whether the slash (in 'The Sanitized Sonnets') 'is employed seriously or as a joke'![27] More discerningly, K. L. Goodwin noticed Porter's gift for parody, sometimes overdone, but observed that he was 'basically writing about his own attitudes, his own thoughts, moods and postures'.[28] After identifying at least two distinct poetic voices, 'one declamatory and high-minded, the other intimate and wry' and preferring the latter, Philip Roberts suggested that, although 'Australia may not be able, by now, to claim [Porter] as her *very* own, ... there would be no harm in trying'.[29] This view was not shared by Edgar Dent (Roger Milliss), who, after criticizing Porter for an alleged lack of passion and 'the big statement of a Brecht or a Pablo Neruda', recommended that 'a spell in the antipodean sun would bring a little life back into his bland and lustreless eyes'.[30] Diagnoses like Dent's, which blamed the English climate, physical or moral, for the misery and pain expressed in some of Porter's work, recur.[31]

British reviewers, being more used to cultural ownership, appeared less concerned with the question of who 'owned' Peter Porter at this stage than the Australians. Nonetheless, some responded positively to his resonant expression of emotional identification with his adopted country in the concluding lines of 'The Last of England'. For the most part British reviewers concerned themselves with changes in technique or subject from earlier books. Continuity with previous work was found in satires of contemporary urban life such as 'Short Story' (*TLOE, CP* 122−4), 'A Consumer's Report' (*TLOE, CP* 131−2), 'Real People', (*TLOE, CP* 147−50), 'Applause for Death' (*TLOE, CP* 152−4), and another group of Martial epigrams. The *Times Literary Supplement* and *New Statesman* reviewers both observed a greater assurance in Porter's use of 'the grand manner' in this volume, and a capacity to move more confidently between the intimate, personal voice and a more detached, satirical manner.[32]

Questions of Porter's stature in relation to other contemporary poets were becoming relevant. Michael Schmidt considered Porter the most impressive poet writing in London in the 1960s:

> His world includes allusion to present urban realities, the polluted pastoral, to music in all its styles, to literatures from many languages and ages; to various religions and to many histories, myths and legends: a formidable store of information deployed with an almost unerring intelligence. The speed and suppleness, and yet the complexity of allusion, and the consistent structure of implication ... the truth to voice as well as truth to form, make this poetry not far short of major.[33]

This enthusiastic review formed part of Schmidt's argument that British poetry of the late 1960s was not inferior to American poetry of the same period. It was John Fuller, however, who noted most precisely the particular strength which Porter was developing: 'a conceptual scope, an ability to turn example and observation into fluent statement'.[34] It was the development of this voice which could range from everyday colloquialism to oracular utterance, that enabled Porter to examine the stretch marks on history in his idiosyncratic and highly evocative way.

Porter was no orthodox historian in *The Last of England*. Although a relative selflessness is evident (the book contains no tracing of Australian roots, for instance, and a diminution of the first-person pronoun), the power of subjectivity remains everywhere. Historicism, the belief that all social and cultural facts are socially determined, is one of the thoughts entertained briefly, then dismissed, in 'Stroking the Chin' (*TLOE*, *CP* 132−4). The poem puns on the stroking, or playing of the Ch'in, a Chinese lute:

> Things don't happen this way,
> I write them down this way
> and I pull from my chin
> two hairs I've allowed to grow −
> so much for historicism, I say.

The ode 'Europe' (*TLOE*, *CP* 127−9) reinforces a notion of the historian, like the fiction writer in 'Short Story' (*TLOE*, *CP* 122−4), as an almost arbitrary composer of signs. Jonathan Raban has described 'Europe' as 'a patchwork quilt of European history, from the sack of Carthage to the present time ... a marvellously rich transmutation of an entire cultural history into a sort of comic strip novel, a fiction of fierce jumps in time, of chance resemblances, of the tumbling rag-bag of imagination'.[35] In technique and tendency, 'Europe' stands at the opposite end of the spectrum from the naturalistic masterpiece of the 'London' poems, 'Death in the Pergola Tea-Rooms' (*OBTB*, *CP* 21−2). It is as if the techniques and outlook of mimetic realization have given way, under the impact of an idea of Europe (which Porter had hardly set foot in during the 1960s), to dream as the only valid form of realization. If the Movement and certain Group poets were major influences on the earlier critical realism, one can detect American voices behind the controlled chaos of Porter's 'Europe': Wallace Stevens, Olson, and even Berryman. As in many other aspects of Porter's work, Auden also provided a model, in his historical poems, and the flexuous accommodation of thought, image and idea sometimes recalls Empson.

Like stream of consciousness fiction, 'Europe' resists categorization and conclusiveness. Yet its arrangement into quatrains with rhyming couplets concluding each stanza suggests the clinching of ideas. Porter uses the couplet as a tease, seeming about to offer authoritative comment and then withholding it. Only the final two stanzas, with their air of omniscience and calm penetration, offer a point of balance after the bewildering onslaught of names, details, ideas and images. Returning to the long view of history which had been playfully contemplated in 'The Last of the Dinosaurs' (*APF*, *CP* 77−8), Porter introduces Oviraptor, the human-sized biped of the late Cretaceous age which ate other dinosaurs' eggs, as progenitor of rapacious Europe, where 'we' (the poet and his cohorts) are:

Launched in the wake of our stormy mother
  To end up on a tideless shore
    Which this is the dream of, a place
    Of skulls, looking history in the face.

The concluding effect, bold and direct though it is, fails to show a path through the labyrinth. To face Europe and its history, it seems at this point, is to face the chaos of nightmare and rapacity, and this is the only legitimate way to render it. Later, on closer acquaintance with certain parts of Europe, in particular Italy, Porter's vision would clarify and become more focussed around specific places and their associations, which could even, on occasion, transport him towards states of rapture.

For the time being, however, the idea of Europe remained somewhat theoretical. From one point of view Europe provided an image of the imperial exporter of people and goods to places like Australia: 'Europe has tipped these workings/on a spastic shore' ('The Recipe', *APF*, *CP* 105−6). From another, it was a source of intellectual and emotional stimulation through reading, music and art. The 800 page histories of Florence to which Porter's persona referred in 'My Late T'ang Phase' (*APF*, *CP* 80) have been complemented by further voracious reading at local libraries in London, so that he can now write in the fourth of 'The Sanitized Sonnets' (*TLOE*, *CP* 143), echoing Keats, 'Much have I travelled in the realms of gold/for which I thank the Paddington and Westminster/Public Libraries ...'. It would be mistaken, however, to suggest that such excursions into the problems of historical represent-ation indicate a complete retreat from contemporaneity at this time. In 'The Workers' (*TLOE*, *CP* 125−6) Porter presents a satiric portrait of a modern urban woman of the comfortable upper middle-classes, whose novelettish sensibility transforms the actual miners who dig up the ground into minions serving her desires. The poem progresses from previous satires in its linking of a certain sensibility and style with the material conditions of power, and its deft epigrammatic insight, that:

On the broad back of money
are the fine moles of sensibility −

A sense of one's place in time is also sought in more autobiographical pieces in *The Last of England*, such as 'The Sadness of the Creatures' (*TLOE*, *CP* 136−7), 'At Whitechurch Canonicorum' (*TLOE*, *CP* 138−9) and 'On This Day I Complete My Fortieth Year' (*TLOE*, *CP* 139−41). Porter has himself described 'The Sadness of the Creatures' as 'a bleak reflection on the themes of predatoriness, carnivorousness, marriage and death'.[36] The short unrhymed lines and laconic tones begin some-what playfully with domestic details of cats and children and arguments interspersed with a spoof on the then voguish William Carlos Williams's style of presenting such details: '*The water I boiled the lobster in/is cool enough to top/up the chrysanthemums*'. But the poem moves towards an understated eloquence in the speaker's representation of scenes from a marriage, which now frightens him more than annihilation by a stroke or the bomb. Here is its focus:

> ... the picture of a lit room
> where two people not disposed
> to quarrel have met so
> oblique a slant of the dark
> they can find no words for
> their appalled hurt but only
> ride the rearing greyness

The setting is specifically a third-floor flat in London, like Peter and Jannice Porter's in Cleveland Square. The represented experience goes beyond surface facts, however — it reveals the 'appalled hurt' of two individuals who find themselves, against their better instincts, predators. There is respite from time to time, 'jokes and love and reassurance', but not enough, and a Salvador Dali-like horror suddenly emerges:

> it seems a trio sonata
> is playing from a bullock's
> skull and the God of Man
> is born in a tub of entrails

This grotesque image is not diminished when one contemplates its artful arrangement. Its recognition both of an informing sensibility and of a shared experience give it an audacity and emotional range well beyond the flat ironies of much imagist verse, which is both mimicked and leant upon here. The emergence of a voice in this poem whose representativeness derives from a weight of felt experience is a significant development.

History arrives at its most autobiographical expression in this volume with the poem 'On This Day I Complete My Fortieth Year' (*TLOE, CP* 139–41). The attentive reader will be wary of Porter's confessional mode, however, which achieves its authenticity less by baring the heart than by placing the individual experience in a context which lends itself to comparison and generalization. The allusion to Byron's last poem, 'On This Day I Complete My Thirty-Sixth Year' is more sustained than many other allusions in his work; it operates structurally to highlight the felt ordinariness of Porter's life beside that of the acknowledged epitome of European Romanticism. Yet, by a curious fusion of wit and irony, Porter's 'ordinariness' achieves the force of a resonant honesty, anti-Romantic in mien but hinting at a grandeur behind appearances. The poetic manoeuvres are deftly achieved. Byron's short fourth line in each quatrain of his poem, with its two strong beats, ends each of his stanzas with a flourish; in the first stanza, for instance, his line is 'Still let me love!' Porter's poem, while retaining the short line, eliminates the heroics in a smothering of syllables: 'someone has to suffer'.

Porter's Brisbane background, when set against Byron's Newstead Abbey, is both diminished and enhanced, making his beginnings not an exalted 'pilgrimage' but a more workmanlike 'apprenticeship':

> to have a weatherboard house and a white
> paling fence and poinsettias and palm nuts
> instead of Newstead Abbey and owls and graves
>   and not even a club foot;
>
> above all to miss the European gloom
> in the endless eleven o'clock heat among
> the lightweight suits and warped verandahs,
>   an apprenticeship, not a pilgrimage —

Porter's persona does not *really* want to substitute his shabby genteel origins for Byron's aristocratic Abbey with its gothic trappings. The important question for Porter is what contributes to imaginative vitality. For him, it will not be an heroic end liberating the Greeks; after all, he is already forty and Byron died (feverishly, not in battle) at thirty-six. Porter knows he has already lived past the age for heroic, Romantic death. Nor should his art conceal the poverty of its origins. He is consoled by the fact that it is from 'detritus' that art may emerge:

>   ... a pile
> of moon-ore, the workings of the astonished
> mole who breathes through your journalism
>   'the air of another planet'

The 'astonished mole' is a surprisingly apt image for the author who, burrowing in the dark, surprises himself from time to time with imaginative discoveries. In his 'unchosen way', Porter's autobiographical speaker has marked out poetry as the 'gift' he must follow. From the early 'hot promises' to the later 'mid-channel' period which he was entering at forty, he may have contributed, in his mentor's (Auden's) words, 'epiphanies of a poor light'.

Porter's self-deprecation and guarded understatements should be recognized here for what they protect: the promise of a purpose, a working life for which an apprenticeship has been served. What is necessary now is a 'piling on' of 'fuel for the dark'. In reviewing his personal history in the light of Byron's poem, Porter was conducting an experiment which had unexpected results. On the occasion of his own fortieth birthday, Sydney poet John Tranter wrote a poem which alludes structurally and in ideas to Porter's.[37] In such ways are traditions in poetry refreshed and remade; and history takes another turn.

7

FUEL FOR THE DARK

•

The early 1970s was a *Sturm und Drang* period for Porter, a phase of intense activity in his life and writing career in which reason and sanity were severely tested by the buffeting of fate and circumstance. Much of his writing in these years has the pressure of a coiled spring. In 'On This Day I Complete My Fortieth Year' (*TLOE, CP* 139–41), an inverted tribute to Byronic Romanticism, he presented his persona 'piling on fuel for the dark'. The imaginative energy and prodigious productivity of his early forties seems, in retrospect, almost manic — a race against time and imminent extinction.

Such frenetic activity seems an outcome of both personal and social factors. Porter's domestic arrangements in London had not, to the eye of an outside observer, changed. With his wife and daughters, he continued to live in Cleveland Square near Paddington Station. He and Jannice had first moved to a flat in the square in April 1960, when they shared the ground-floor at No. 43 with Jill Neville. At the beginning of the following year they moved to the basement of the same flat, and were married there on 24 March 1961. Jill Neville continued to live above them. In early 1963 the Porters moved to a ground and first floor flat at 27 Cleveland Square, where they remained until March 1968, at which point they moved to the third floor flat at 42 Cleveland Square, where Porter has remained. The square, which was less fashionable in the 1960s than it became in succeeding decades, is approached through streets of small hotels, the signs everywhere of international transients. The square itself, with its Victorian porticoes and a central garden for residents with a locked gate, suggests stability and order within the wider restlessness of an international city. This was the oasis of order which Jill Neville later projected the Porter family as inhabiting in *Fall-Girl*.

In their private lives, however, Peter and Jannice Porter were deeply disturbed and restless through the late 1960s and early 1970s. Porter has warned against too literal or autobiographical an interpretation of 'The Sadness of the Creatures' (*TLOE, CP* 136–7), which was suggested by the marriages of friends as well as by his own, and, as has been shown, touches on wider themes and correspondences. No external measure of happiness can be applied to a marriage and much of Porter's discontent was long-term, stemming back at least as far as the death of his mother. The fact that so much of his poetry of this period was directed outwards, in a public voice and manner, may itself suggest a turning from intimate sources of pain and unhappiness. That there was pain, sadness and recrimination in the marriage as well as comfort, support and pleasure seems certain. Both partners were plagued by the Furies of previous relationships in which they had suffered rejection. The lives of the Porters seem to have diverged in the early 1970s, in a not unfamiliar pattern for that time, the wife turning inward to children, family and home, and the husband outward to more public activities and roles before, in December 1974, in an act with shattering consequences for others, Jannice Porter took her own life. Thereafter, ironically, 'she' (or some version of the former woman) was projected as the major character in some of the finest and most moving poetry of the twentieth century, as Peter Porter's concerns turned again to the personal and domestic. But that is a story for the next chapter.

Porter's public engagements and achievements in the early 1970s were in keeping with his growing identification as an important writer in Britain and internationally. Earning a living and keeping a family required short-term contracts on a variety of fronts. Thus Porter took up invitations as visiting lecturer at the University of Hull in 1970−71, and at Reading University, in 1972. In January 1973 he took over from Ian Hamilton as fiction and poetry editor of the *Times Literary Supplement*, a position he held until May 1974. As poetry reviewer for the *Observer*, a position he commenced in 1973, he began to build an influential role with a wider audience. As a book editor, he produced *A Choice of Pope's Verse* (1971), a PEN anthology of contemporary poets, *New Poems 1971−1972* (1972) and, with poet Anthony Thwaite, *The English Poets: From Chaucer to Edward Thomas* (1974). His love of Italy led to his collaboration with photographer Roloff Beny and Anthony Thwaite to produce the text for *Roloff Beny in Italy*. Porter was also in demand as a reviewer and literary commentator for BBC radio, and wrote two plays which were produced by that station, *The Siege of Munster* (1971) and *The Children's Crusade* (1974). But the most important publications in this period were five books of poetry: *After Martial* (1971), *Preaching to the Converted* (1972), *Living in a Calm Country* (1975) and, with Arthur Boyd, two critically neglected but outstanding works, *Jonah* (1973) and *The Lady and the Unicorn* (1975). In recognition of Porter's growing international reputation, a record, *Peter Porter Reads from His Own Work*, was produced by the University of Queensland Press in 1974. The record was released to coincide with the poet's return to his home country for the Adelaide Festival of Arts, after an absence of twenty years. This visit proved to be a crucial turning-point in his life and later work.

The literary journalism which had become Porter's stock-in-trade requires some comment here. Without doubt Porter's principal commitment was to poetry; the rest he was inclined to refer to as 'hack work', in its pejorative eighteenth-century sense. But a debate was opening up in the early 1970s regarding the function and value of metropolitan literary journalism, and in this debate Porter attempted to carve out a rationale for this side of his developing career. In a major article in an early issue of *The New Review*, entitled 'Grub Street *versus* Academe',[1] Porter analyzed and deplored the gap which had developed between the universities and a general intelligent public, represented by readers of 'high-brow journalism'. The gap was caused principally by a resurgent 'monastic' view of the universities led by a second-generation of Leavisites who considered contemporary literature beneath serious attention. Those few younger critics or reviewers from the universities who made the cross-over were subject to 'the Klemperer principle':

> Renowned in his younger days as a champion of modern music in Berlin, Otto Klemperer ended up custodian of all the unchallenged classics, breathing only the profound air of Beethoven and Goethe. In less exalted ways, this is the pattern of progress of the brilliant lecturer who is happy enough at the beginning of his career to review modern poetry, but as soon as he becomes a professor, turns his attention wholly to Tennyson or Keats.

Thus did the 'monasteries' preserve their virtue and remain insulated against the 'front line' news of those metropolitan journalists into whose company Porter himself was moving.

The 'metropolitan', non-university based critics mentioned by Porter included Randall Jarrell, Al Alvarez, Cyril Connolly, Clive James, Geoffrey Grigson and, further back, Eliot and Auden. Such names were contrasted with the relatively barren academicism of New Critics such as Yvor Winters, F. R. Leavis, I. A. Richards and neo-Leavisites such as Ian Robinson. The cleavage between town and gown seemed to Porter to be widening and culprits could be found on both sides of the divide, either popularizing banally for a wider audience or obfuscating obtusely for small bands of initiates. More threatening, however, was a developing movement among certain critics in academe, represented by George Steiner at this time, who strained towards 'International Significance'. Porter's publicly expressed suspicion of 'prophets of Regeneration', in poetry as well as in prose, may be seen in part as a siding with England against further cultural takeovers by America. This should not be taken as a reversion to the 'little Englandism' which he had criticized in the Movement poets; Porter was in favour of international influences in his universe of poetry. However, the residual 'Australian democrat' in him vigorously contested both the stand-offishness and self-righteous assertion of moral probity of England's 'New Model Army of Leavisites' and the evangelistic internationalism of America's 'Hermetic Avant-Garde'. The age, Porter concluded, was competitive and materialistic, and its discourses were in disarray. But he could see no reason why the modern writer, 'buffeted by commercialism, bad journalism and clichéd speech, should not make from these ingredients a work of the purest imagination'. His avatar here was the American businessman-poet, Wallace Stevens, whose art had 'a shining relevance for the age he lived in'.

In this article, then, Porter displayed a vigorous polemical style and few illusions about the jungle of literary journalism he had entered; he applauded cosmopolitanism against narrower provincial exclusivism, but entered a special plea for the registering of a missing element, pleasure in writing. While encouraging scepticism in the face of competing critical orthodoxies, he mapped out a role which was congenial to his own beliefs and skills: 'to watch over the talents which we have and hail them when they reach the point at which they might attain real standing'. While the article was too prodigal in its ideas and examples to be decisively influential, it nevertheless set a path and a purpose for its author in literary journalism.

Porter's ideas in 'Grub Street versus Academe' represented an important personal advance since the 1950s, when an immoderate admiration for certain 'great' artists of the past (Shakespeare, Donne and Pope, for example) and of the twentieth century (Auden and Stevens particularly) had threatened to stifle his own creativity. A recognition of a neurosis, excessive literary ambition, was expressed by Porter in an unpublished BBC script about an invented figure, a minor English poet, 'John Woodby', whom Porter alleged he had met in the 1950s.[2] 'Woodby' ('Would Be') had asserted in some hard talking to the poet Porter that his extreme admiration for various artists was really a form of aggression towards them. His alter ego was crucifying his own talent as a way of punishing himself for not being Auden, Stevens, whoever. If he were serious in his wish to write, he should come to terms with his own 'torque towards mediocrity'; he might then write something bearable and learn to live with himself. 'Woodby' approved strongly of keeping part of oneself at the early Freudian stage of self-love. This recognition enabled Porter to write with a less extreme anxiety about inferiority or influence; and even, on occasion, to expose a sense of vulnerability and 'ordinariness' as part of the rhetoric of his verse, thereby contributing to a more human and accessible poetic voice.

Another analysis of Porter's early 'condition' might emphasize the autodidact as overreacher; that he felt required to prove his knowledge and understanding. This theory probably has some validity, but it fails to explain the source of Porter's enormous and continuing curiosity, and his thirst for knowledge, which had as its accompaniment a repertoire of rhetorics and a vocabulary as prodigious as that of Dickens or Furphy (both themselves self-educated). Porter's attack on academic learning within the universities should therefore not be seen as a case of sour grapes but as part of his own developing consciousness as poet and critic, which favoured metropolitan inclusiveness and an engagement with contemporary literary and cultural practice rather than retreating to 'safe' authors or remote theories.

◆

The critical reception of Porter's two books of poems, *After Martial* and *Preaching to the Converted*, published simultaneously by Oxford University Press in 1972, is interesting in the light of his distinctions between Grub Street and Academe. In general, critics with a social mission or a specialist interest gave particular attention to the free-ranging translations of Martial's epigrams, while those with more existential interests (particularly university academics) preferred *Preaching to the Converted*.

Porter's forty-seven 'remodellings' of Martial (fifteen of which had been previously published in *Poems Ancient and Modern* and *A Porter Folio*) received praise from the teachers' journal *Classical Outlook*,[3] which found the volume 'representative [of Martial's work as a whole] ... intriguing, entrancing and exciting' to read, but unsuitable for classroom use because the poems were 'too bold for youngsters'. Porter's relish in finding equivalents for Martial's argot of lustful sexual activity was noticed also by other reviewers, usually with less apprehension than *Classical Outlook* had shown. At a more technical level, poet and classicist Gavin Ewart compared Porter's translations with James Michie's and found Porter's more 'open-ended and fluent, skilful rather than neat', and using more ambitious rhyming schemes. 'The boy from Brisbane', Ewart said, had obvious affinities with 'The Kid from Spain'.[4] Alan Brownjohn, like Ewart an Oxford-educated reader of Latin and a poet, observed the various kinds of 'liberty' taken with the originals, and commented on Porter's range of sympathy and poetic invention, ranging from tenderness, lyricism and delicacy to an ebullient, obscene humour.

A more socio-literary approach to Porter's work at this stage was taken by Oxford academic and Marxist critic, Terry Eagleton, whose interest in outsider figures in modern literature had recently been demonstrated in his book, *Exiles and Emigrés*.[5] Eagleton proposed that 'the unchallenged sway of non-English poets and novelists in contemporary English literature points to certain central flaws and impoverishments in conventional English culture itself'.[6] Eagleton acutely observed that Porter's art, which he praised, 'thrives on disillusionment. The more depressed he gets, the more poetically inventive he becomes'.[7] Eagleton contrasted the decline of Donald Davie's poetry following the latter's expatriation to the United States with Porter's more lively critical engagement with British cultural philistinism and political dogma; from Eagleton's point of view, the adversarial role for the poet in contemporary culture was the only appropriate one.

By far the most perceptive Australian reading of Porter's work to this point was Andrew Taylor's review of *Preaching to the Converted* and *After Martial*.[8] Taylor, an academic and poet from Adelaide, was interested primarily in Porter's art in relation to Europe, and his existential concerns. Taylor observed something which English critics seemed not to notice: that Porter had 'preceded the UK into Europe by a few years'. However Taylor's contention that Porter had, like Henry James and Ezra Pound before him, made himself 'an imaginary European' doesn't quite ring true, for in Porter's work there is a continuing and sometimes abrasive scepticism about all people in relation to the places they inhabit; and the strategic and only viable personal view of himself is as a visitor.

Porter's fellow expatriate, Clive James, neatly observed the ways in which Porter's knowledge of a European cultural past had also cut him off from contemporary Europeans. Porter, he declared, had shown himself in *Preaching to the Converted* to be culturally deprived through *having* culture: 'the savage paradox of being disqualified from modern society through appreciating its past is at the base of Porter's satiric position'.[9] In assuming this uncomfortable, and discomforting, status as outsider wherever he went, Porter was developing a capacity to reveal his own, and western society's, vulnerability to contemporary change, which he could both revel in and castigate. He was adopting a role as critic of genteel niceties who could entertain, but who refused to charm, since this would reduce some harsh truths he felt bound to impart as, for example, in 'The Isle of Ink' (*PPTC, CP* 185−6):

> . . . a million toddlers ride the bike
> Of Hardly Art. The rest of us will stay
> On the Isle of Ink, postmarked *Totenreich*.

'Totenreich' here takes the poet not to geographical Germany but to the sentiment of Strauss's one-act opera *Ariadne auf Naxos*, wherein the kingdom of death ('Totenreich') is perceived as the only place where everything is 'rein' − pure in spirit or radically innocent. This was the central drama which Porter discerned behind all cultural game-playing: the individual's hand-to-hand combat with death; artists who denied this were doomed to superficiality.

This feeling provided the primary impetus behind Porter's piling on of 'fuel for the dark' in the early 1970s. He responded as if in a race to death, and the race thrilled him. Sexuality may be enhanced in such situations, and a number of poems in *Preaching to the Converted* and *After Martial* express sexual desire. As Clive James wrote, 'Porter's is the brave version of shamelessness: he puts up no front, declaring his lusts as they occur to him, which means constantly'.[10] 'Affair of the Heart' (*PTTC, CP* 169−71) is a self-mocking confessional which exults, in a hyperbolic Martial manner, in sexual energy and desire. The unrhyming couplets link and unlink, gymnastically. The poem opens:

> I have been having an affair
> with a beautiful strawberry blonde

and concludes:

> only memory is like your tunnelling tongue,
> only your fingers tinkering tell me I'm alive.

Porter is never content with sensuality alone, however, and the poem contains some clever parodies of sexual representations in fiction as well as an autobiographical hint of the middle-aged male fear: a heart attack. But the poem essentially celebrates energy, desire and inventiveness while death lurks around the corner. A more mordant humour flickers around the spoof 'Sex and the Over Forties' (*PTTC, CP* 160–1), in which the poet observes the sexual problems attributed by a younger generation to himself and his age group:

> It's too good for them,
> they look so unattractive undressed —
> let them read paperbacks!

Adultery, in such a light, is boring:

> More luncheons than lust,
> more meetings on Northern Line stations,
> more discussions of children's careers.

Sexual inventiveness, the lore of the 1970s, becomes strained and faintly ridiculous:

> Trying it with noises and in strange positions,
> trying it with the young themselves,
> trying to keep it up with the Joneses!

But the concluding stanzas characteristically shift from this dismissive view of a younger generation to a more bemused view of the sense of fragility which comes with ageing. This does not eliminate the earlier view but lifts it to a level where ridicule is subsumed in pathos:

> All words and no play,
> all animals fleeing a forest fire,
> all Apollo's grafters running.

Back to the dream in the garden,
back to the pictures in the drawer,
back to back, tonight and every night.

The last line is the opposite of a clincher, but it has the capacity to chill the reader or listener with its semi-humorous, lightly serious portent of separation, an end to loving.

Preaching to the Converted is a distinctly serious book, but for Porter this does not connote earnestness. The mannered epigram, the hyperbolic phrase, the pun at a serious moment, are reminders of the Toowoomba schoolboy's expressed wish to have Oscar Wilde at his dinner table. David Selzer went further than to call Porter's fifth collection serious. For him it showed Porter to be a 'major' poet, 'that is, whatever you think of his politics, his religion, his aesthetic, his poems, he has made an indelible mark on our culture. All major artists make such marks, eventually. We measure by them — though they are not always self-evidently important to their immediate contemporaries'.[11] For Selzer, Porter's importance lay in an unusual combination of tradition and originality:

> ... his conception ... of the world is expressed in most eclectic terms. He draws metaphors and subjects from the spread of European culture — spread in time and distance. His conception is traditional in essence — the universals of love, insubstantiality, death — and original in appearance — violent, absurd, grotesque, anachronistic particulars.[12]

Selzer recognized the hunger for knowledge and experience which intensified this book, sometimes to bursting point, but did not notice a new development in the *placement* of ideas and images associated with the author's travels in Britain and Europe in these years. Indeed, it is in *Preaching to the Converted* that a sense of place emerges as an important anchor in poems as various as 'Fossil Gathering' (*PTTC, CP* 158–9), 'Evolution' (*PTTC, CP* 159–60), 'Seaside Resort' (*PTTC, CP* 163–5), 'On the Train between Wellington and Shrewsbury' (*PTTC, CP* 171–3), 'Delphi' (*PTTC, CP* 182–3) and 'Thomas Hardy at Westbourne Park Villas' (*PTTC, CP* 187). The other significant development, which had begun to emerge in *The Last of England*, was what Terry Eagleton called an 'ability to shift levels from low-keyed social observation to ambitious metaphysical pronouncement without a sense of unease'.[13]

The title poem, 'Preaching to the Converted' (*PTTC, CP* 157), might be compared with a comment which Porter made in an interview about poetry in the 1960s: 'I think it's bad to go about continually giving poetry readings which are preachings to the converted'.[14] In this piece, Porter's persona stands in the nave of the church of San Lorenzo in Florence, chosen because Savonarola, the greatest preacher and reformer of his time, preached there. Here Porter's own baroque imagination and sardonic irony rise to the occasion. He sympathizes with the dead, who are no longer revivified by their 'territorial imperative', and have no prospect of resurrection: all they can do is to suffer sermon after sermon — 'to huddle close/Under the sermon shower'. There is no good news here: 'God will be verbose'. The poem has some wonderfully chilling effects, such as the animation of Latin letters on the memorial tablets, lending an air of macabre comedy to the scene. It ends with the speaker, while claiming to be critical of evangelicism, entering his own resonant warning. But his message is secular and sceptical, the parameters are drawn up — death, not eternity, is our future:

> . . . the base is fixed:
> A *Grundrisse* for the dead — have marble fun,
> Put on your shoes and walk into the sun.

The allusion to another prophet, Karl Marx, is instructive. The mention of Marx's *Grundrisse*, the ground-plans of his work, suggests that his philosophy too, however powerful, is undermined by mortality. With this knowledge, what forgiveness? Porter offers no comfort; even pleasure must be 'marble fun'. An accompanying piece in the *Collected Poems*, 'Timor Mortis' (*PTTC*, *CP* 157–8), presents a similarly disillusioned view, ending with a mutter from Schopenhauer, the philosopher of pessimism:

> Man is ridiculous; if
> it weren't for his death,
> he'd have no value whatever —

Yet, in the very next poem, 'Fossil Gathering' (*PTTC*, *CP* 158–9), as in the earlier 'Seahorses', death is perceived to confer significance and value, a recognition that, in the great chain of evolution, 'every feeling thing ascends from slime/To selfhood and in dying finds a face'.

Questions of death and resurrection in the context of the evolutionary cycle were raised again in 'Evolution' (*PTTC*, *CP* 159–60), a poem written after Porter had visited Israel with a group of writers in 1971. Others in the party were Ted Hughes and his wife Carol, D. J. Enright, Jeremy Robson, Dannie Abse and Charles Osborne. Readings were given in many places to different groups. The Australian-born director of the Literature Department of the Arts Council of Great Britain, Charles Osborne, who organized the tour and was its 'sheepdog', has recalled the setting which Porter used in his poem:

I shall not easily forget my first glimpse of the walled city of
Jerusalem from the top of the Mount of Olives, surely one of
the world's most impressive sights. Those primitive notions of
the Last Trump are given 'a local habitation and a name' by
the green hillside liberally sprinkled with graves whose dead
shall rise and enter the holy city, Paradise, through that particular gate. And there, just outside the walls to the south, is
the valley of Gehenna, or Hell.[15]

In the poem 'Evolution', which was dedicated to D. J. Enright, Porter
turns this setting to his own designs, questioning in semi-jocular
fashion the speaker's Christian inheritance in relation to that of the
Jews:

This is the Jewish Cemetery at the foot of
    the Mount of Olives. From my point of view
(and I have been calling myself a residual Christian),
    a cemetery without crosses and headstones
looks like a stonemason's junkyard. But
    just beyond the cemetery is the Garden
of Gethsemane, the Vale of Kidron, the Dome
    of the Rock, and the Eastern Wall
with its Gate for the Messiah to come through
    (decently bricked-up). Staring in sunlight,
I am conscious only of the Jewish Resurrection —
    Christians can resurrect anywhere,
but every Jew has to go by underground:
    out of Cracow or Lodz, with his diary's
death, from the musical comedy stage of
    New York, from the pages of fiery
Medieval books — he dives into oblivion
    and comes up here to join the queues
as the millions swapping stories enter Heaven.

A collision of values and expectations is expressed humorously and
ironically, with Porter's comic book images of talkative, joke-swapping
Jews in the queue for Paradise (a Woody Allen touch) setting the scene
for a more mordant image of the individual salvation required of the
inheritor of protestant Christianity. The poem jangles with ironies of a
kind which would be appreciated by Enright, who has himself written
a book about irony.[16] When an autobiographical note enters the poem
towards the end, it adverts to one of Porter's recurrent concerns since
his departure from Australia, the question of metamorphosis. If rapid
mask-changing or dramatic transfigurations, such as those suggested by
Ovid and Apuleius (entertained elsewhere in Porter's poetry), were put
aside, evolution might provide an answer:

> Now we are changed: I haven't an atom
>   in my body which I brought to Europe
> in 1951. How beautiful is evolution,
>   that I am moving deeper into my own brain.

The sense of initial wonder here changes to deep situational irony: in the context of the poem the poet's praise of evolution is both a statement of change and an implied plea for company, if not community in the full Jewish sense. The condition of aloneness proposed in the process of 'moving deeper into my own brain' sounds ominous, not only contrasting radically with Jewish notions of a chosen people in their chosen place, but also revealing a vulnerability in the loss of identity. If national or ethnic identity is insufficient, what may replace it? If the individual must lose a sense of self in order to gain one, what would that rediscovery involve? Such problems would be re-examined after Porter's return to Australia in 1974 and subsequently, most notably in the poem 'On First Looking into Chapman's Hesiod' (*LIACC, CP* 210– 12). In 1971, however, twenty years after his first departure from Australia, in confrontation with Jerusalem as place and symbol, Porter's tone reveals both fear and hope:

> stay with me, my friends; truth and love,
> like miracles, need nowhere at all to happen in.

The vulnerability is patent and the final assertion has none of the force of a manifesto. Its status is that of the individual voice in retreat from all communities save that of friends, a reminder of E. M. Forster's rejection of piety and patriotism in favour of individual friendship.

As he began to touch base at certain places in Europe, and read and think more deeply about what Australians used to call 'the Continent', Porter's poetic and personal strategy was to seek out exile figures and relate them to a developing sense of his own spiritual and emotional exile. Thus Giotto's portrait of Dante in the Bargello in Florence told him that 'we never get home but are buried in eternal exile' (*PTTC, CP* 194). 'James Joyce Sings "Il mio tesoro"' (*PTTC, CP* 195–6) recalls Joyce being beaten by McCormack in a singing competition and reproduces the famous Irish exile's buoyant egotism, concluding with a boast which extends well beyond any claims made by Joyce's brother Stanislaus in *My Brother's Keeper*:

> If I had been christened Stanislaus
> I'd have claimed the throne of Poland.

Two composer-exiles were also invoked: Stravinsky, whom Porter regarded as the greatest composer of the twentieth century, and Scarlatti.[17] Displacement here became a major topos as Porter restlessly interrogated the relationship between creativity and place. The notion of Ireland as a place which evoked sentimental piety and patriotism was mocked in 'The King of Limerick's Army' (*PTTC, CP* 180—1). The marvellously disconcerting satire on racism and the language of prejudice, 'Mort aux chats' (*PTTC, CP* 184), attacked the smugness of those who, through becoming too settled, had lost a capacity to appreciate difference. Thomas Hardy, so often represented as the archetypal poet of place, was uprooted from his native Dorset and placed in a London which Porter viewed as a natural habitat for this 'watchful conspirator against the gods' ('Thomas Hardy at Westbourne Park Villas', *PTTC, CP* 187). Nor did extensions of territoriality by Americans and Russians to the moon necessarily fire the imagination: indeed it 'lowers/ A lamp inside Mankind's memorial dark.' ('A Hint from Ariosto', *PTTC, CP* 174—5). Imagination, it seemed, was out on its own, without the comfort of *terra firma*.

A tension between Porter's increasing prepositional reliance on places and his apparent rejection of their clinging demands, in the early 1970s, is evident also in 'Delphi' (*PTTC, CP* 182—3). In this piece, the place is much less than the dreams it has evoked. Modern Delphi is a convention centre, an ambiguity which Porter emphasizes by capitalizing the term:

> . . . The many Conventions we have!
> Addressed by bishops and conjurers —
> The ladies with streaky hair who study gin
> Through 2 p.m. light, the serene hum
> Of vacuum cleaners always in another room,
> The documentary film makers downing vodka
> And lime, shouting how marvellous it is
> In such a god-forsaken place!

These modern prophets bespeak a lack of knowledge, memory and imagination. Locked in present time, they are deaf to the voices of oracles. Once considered by the ancient Greeks to be the centre of the world, its very omphalos, Delphi is now 'drowsy D.... i,/The happy haven, a good place to come to die'. Delphi's modern 'tone of prophecy' is diminished to the discourse of vacuous tourist brochures.

But Porter's interest was in the vital ideas which such places may keep alive in the bustling commercial present. Appropriation into symbol was insufficient by itself. A place must build stories in the mind and stimulate the pleasure glands of ideational thought. Such is the case in 'Seaside Resort' (*PTTC, CP* 163—5), which Damian Grant considered to be the best poem in *Preaching to the Converted*, for its incorporation of two of Porter's 'most recognizable' ideas: 'an intense and complicated nostalgia for a vanished way of life on the one hand, and on the other a simultaneous mockery and weary acceptance of what modern life has to offer in exchange'.[18] 'Seaside Resort' offers these ideas and more, but its singular distinction is in its anchoring of experience in a place and time while it flashes out with characteristic brilliance towards a wider moral and political universe. The poem is based on a visit by the poet to Hastings St Leonard's, on the south coast of England. Its centre-piece is the statue of Queen Victoria, 'the small black rained-on Queen' who 'stares with disciplined eyes at the sea'. Clever switches of viewpoint move the speaker's perceptions in and out of an imperial outlook and an ambivalence is stated early: 'To be in two minds about her is right:/gone and for good riddance the Raj — / . . . but you regret Mr Mendelssohn/and the plant collectors of Udaipur'. The pronoun 'you' here has the status of a generalizing 'I'. Other poems (including 'Delphi') also reject the superannuated colonialists of Empire days and their way of life, but Porter was inserting here a less popular notion that the British Empire was also a bearer of people and styles which do not deserve outright rejection. The plant collectors of Udaipur, who lived in the walled 'city of Sunrise' in Rajasthan, for instance, represent decent people of the British Raj; and Felix Mendelssohn, Queen Victoria's favourite composer, emerges favourably from a comparison with the Eastbourne Brass Ensemble which 'showers its *Showboat*/*Medley* on the American-owned air'.

Empires come and go, and in Porter's view they are a recurrent feature of history, whatever one's moral objections; and their consequences may include a firing of the imagination beyond what is possible in the provinces. Thus, as in 'Vienna', the Austro-Hungarian empire, along with its perversions of power, also functioned to gather to its centre a powerhouse of intellect and creative thought. The looming American empire of the 1970s was however more threatening to intellect and creativity, because it 'coca-colonized' culture. Commercial popularization of thought had thinned culture out, making critical inquiry and genuine leaps of imagination increasingly rare. Against this background, what might be revealed in a small seaside resort town in the south of England?

> A new Ice-Cream Flavour or a boom
> in Rabbit's Ears, the visit of a hundred
> skinheads or a Television Team,
> a Conference of Escapologists,
> Christ walking on the water to claim
> the record Channel Crossing?

Where everything is sensational, nothing has value or priority. The more calmly authoritative voice of disillusionment takes over and encapsulates the prevailing mood:

> Nothing but the calm
> of history dying, the beautiful
> vulgarization of decay —

Yet the poem turns back at the end to its principal focus, the statue of Queen Victoria, and adds the poet's remarkably personal identification with this figure:

> I am almost in love
> with the small black Queen in the wind
> and I will not notice that the beach is full
> of mussel shells and crab claws
> and the smell is unimaginable
> yet like your mother's corpse,
> that the torn feather is a terrible
> catastrophe, and I am cold
> and lonely on an unimportant strand.

The mood here recalls Matthew Arnold's 'Dover Beach', but disrupts the euphonies of the Victorian poet's emphasis on the mind and ideas with anti-climactic elements such as the images of smell and the disturbingly personal reference to 'your mother's corpse'. The mixed feelings engendered suggest nothing as euphoniously distant as armies of ignorance clashing by night. Nor do they place the individual respondent in such a prominent, even Romantic, light as Seamus Heaney's later poem 'North', ('I returned to a long strand')[19] in which Porter's contemporary articulated his relationship to Ireland.

For all its discordant elements, the conclusion to 'Seaside Resort' revealed the poet, as Damian Grant indicated, 'in a mood of sober self-pity which is neither avoided nor indulged but fully expressed as a dramatic element in the poem's whole context'.[20] Edward Neill, who also considered this poem to be one of Porter's most important, viewed the conclusion as 'a statement of feeling which relates to his own sense of self and role (or rolelessness)'.[21] Both comments contribute to an understanding of a poem which, like 'The Last of England', shows Porter's nostalgia in the early 1970s for an idea of a liberal Victorian England which could be incorporated into his identity. There is no reason to disregard the comment early in 'Seaside Resort': 'gone and for good riddance the Raj'. But the whole era which saw Anglo-Saxon settlement of Australia, India and other countries was a past of dreams which the expatriate now perceived as a temporary anchor. The stillness and persistence of the iconic queen in 'Seaside Resort' are as important to the poem as its changes of spatial perspective from an outward-looking gaze to sea, towards a now disbanded Empire, and the closing long-shot of the poet-observer, lonely and insignificant among the holiday crowd on the beach. An autobiographical link with Porter's image of his grandmother in Sydney is difficult to avoid:

The old lady died, a dejected figure, dressed perennially in hard black, ankle-length dresses, with stoles of slightly less dark lace circling the neck of her blouse, and a penchant for fine lace visors like maiden hair hanging from her hats. She remained chatelaine of my grandfather's suburban castle to the end. I can't recall whether her severity relaxed into concern and kindness, but I do remember that my mother was her favourite daughter and I became, for a while, the grandchild she cared for most.[22]

There is nothing so simple here or in the poem as a transferred dependence to a 'mother culture' of Victorianism: Porter's adult restlessness of intellect and feeling had required a wrenching from all such dependencies, and alternative guides and authorities had been sought across a range of European literatures and lives. But 'Seaside Resort' exemplifies, in its contemplation of the still figure of the most recent Queen of Empire, the pathos of provincialism, with its story of separation from some central source of wisdom, knowledge and emotional fulfilment; a story which parallels the loss of parents, grandparents and others who also represent a source of potential stability.

Yet the personal and even sentimental dimensions of Empire are held stoutly at bay in other poems in *Preaching to the Converted*. In 'The Great Cow Journeys On' (*PTTC, CP* 181), for instance:

... Along the Gulf
of Time huddle the former empires, cold states
now peeling frontages for the tanned police —

Terry Eagleton praised here the deftness with which Porter used 'typifying abstractions' which 'organize but don't appropriate the sense of something actually observed'.[23] In these lines 'Empire' has been internationalized and historicized. Political tyranny is put in its place in 'The Dust' (*PTTC, CP* 176–7) not by images of countervailing violence but by the long view of history, and reminders of mortality:

The Saviour of the State was accoutred
in his people's love, but a detachment of dust
rose from Spanish wheels.

Here is the chief use of the image of death in Porter's fifth collection: as a closure upon heroics, the limit against which all prophets and their Empires test themselves and are ultimately found wanting.

◆

The most important artistic breakthrough for Porter in the early 1970s was his partnership with the painter Arthur Boyd in two books, *Jonah* (1973) and *The Lady and the Unicorn* (1975). Music had led to previous collaborations with David Lumsdaine, Christopher Whelen, Nicholas Maw and Don Banks, and would do in the future with George Newson, Ronald Senator and Geoffrey Burgon.[24] Porter's interest in visual art developed steadily through the 1970s, with London's National Gallery as the principal source of pleasure and instruction. The gallery's outstanding collection of early Italian painting and its special attention to Poussin and Claude were drawcards for the poet, who schooled himself in the iconography of European painting through the centuries and 'its modes of mechanical invention (perspective, geometric shapes, numerology and proportion) which seemed to serve human concerns'.[25] Porter was interested in a humanist aesthetic with origins more steeped in art history than modernism seemed able to offer. While not wishing to place Arthur Boyd, or himself for that matter, as complete opponents of Modernism (both have clearly contributed in different ways to Modernist practice), Porter sought deeper roots in the past. For Porter, Boyd's singular achievement was to extend the European tradition of the figure in the landscape to the quite different light and conditions of Australia. Although they had met at the Royal Court Theatre and other places in the 1960s, it was the English publisher and critic T. G. Rosenthal who brought the pair together in the early 1970s. Boyd's previous work had included several series of pictures on Biblical or religious themes, including Nebuchadnezzar and Francis of Assisi. Another noticeable development in Boyd's work since he had moved to London in 1959 was fantasy, with a basis in metamorphosis, in paintings such as *Nude Turning into Dragonfly* and *Bride Turning into a Windmill*. Like Porter, Boyd was an eclectic borrower from earlier artists, turning their images to his own purposes with playful ingenuity.[26]

Boyd initially responded to the 'powerfully pictorial' qualities he observed in Porter's work. He was also impressed with its emotional range and inventiveness:

> His language is universal ranging from humour to great tragedy so that my response to his work was automatic. I have always felt that as far as I was concerned there was no stage at which I ever felt stuck for a stimulus and I've never known this same rapport. His work seems to me to combine delicacy with strength. The whole of each group of poems always gave me all I needed and more to create the visual contributions to those joint ventures.[27]

After rejecting as possible subjects for a first volume the Book of Job, which was daunting to Porter because he would be following in Blake's footsteps, and aspects of the New Testament, Rosenthal insisted that the partners keep the project 'Jewish'. Porter suggested the brief Book of Jonah, whose neat structure of two parts with two chapters each — the Whale and Nineveh — appealed to him. He divided each part into ten sections, each one bearing on the story in some way, and planned his work appropriately. These plans were exceeded as the project went forward and thirty-five poems were produced, which were finally cut back to the fifteen in each part which appear in the published book.[28] The prodigality of Porter's response to his subject was out-done by Boyd, to whom Porter had sent poems in batches of threes and fours as they were finished. Rosenthal recalls that Boyd 'simply flooded my office with hundreds of drawings in line and wash. We almost drowned in his cornucopia and it took months to fine down to the selection that eventually appeared in the book.'[29] This method made possible what Porter has described as his most fruitful kind of collaboration, which requires 'each artist to go his own way, the resultant works being counterpointed rather than harmonized'.[30]

For Porter, the collaboration had a major impact on his methods of treating the past. The earlier experiments in historicism were given a new impetus by analogy with painting, which seemed to invite a kind of 'complex anachronism', whereby styles and subjects of all periods could be presented as if they were contemporaneous.[31] Boyd's work was rich in such transpositions and Porter was emboldened to go beyond the more strictly apposite parallels between past and present apparent in Auden's *Sea and the Mirror* and *For the Time Being*.[32] Critics, however, were slow to recognize the unique qualities of *Jonah*, or indeed of *The Lady and the Unicorn* and later collaborative efforts by Porter and Boyd, although these represent Porter's major contribution to the form of the long poem, and they led Boyd to new heights of erotic fantasy and experiments in visual metamorphosis. A problem for critics was how to describe the collaborative interaction between words and image. The *Times Literary Supplement* reviewer responded to the rapport which both partners conveyed by using a musical analogy: 'each is a descant on the other's theme; now one leads, now the other'.[33] Roger Garfitt took a more thematic tack, suggesting that there was in Boyd's 'scribbled homunculi, incessantly squirting at the elements or burrowing into compromising positions, a certain foetal vulnerability which is basic to Peter Porter's concept of Jonah the man, who is small, very ordinary and very afraid, taking every chance to dive back to the womb'.[34] But the neo-Romantic Australian poet, Robert Adamson, considered the book 'adolescent', contrasting it with Porter's previous volumes, which in his view had followed 'rational lines' and kept 'well within either an historical or social dimension'.[35] Adamson seems self-contradictory in criticizing the excesses of the Porter/Boyd book. More understandable is his objection, as a neo-Romantic, to its exuberant, profligate mixing of comic and serious modes.

Jonah does perhaps assume too much knowledge of the story, in its playful indulgences in parody and variations on the Biblical narrative. It is Porter's first construction of a significantly 'unstable text' and both he and Boyd seem to delight in the opportunities it provided to upset applecarts. Some of Porter's most interesting experiments were in prose, for example 'Jonah's Contract with the Whale' (*J* 54):

> ... that the party of the first part, hereinafter known as *The Whale*, shall undertake to deliver the party of the second part, hereinafter known as *Jonah*, by means of vomiting, expelling or otherwise voiding the said *Jonah* onto a beach, promontory or any such part of the adjacent coast at the earliest opportunity before the New Moon, always allowing for local impediments and hindrances offered by tide, weather or any other natural hazard.

The second section, set in Nineveh, the sinful City which Jonah is charged by God to reform, provides Porter with further opportunities to test the language of sermon and prophecy. This he does in 'Jonah's First Nineveh Sermon' (p. 78), as well as relishing the scope for anachronism:

> Whom the Lord chastises floats in fire
> To serve again as his scorched phoenix!
> City and people, seller and buyer
> Are thrown away like used Kleenex.

An Epilogue concludes the volume, recalling both the Bible and the Brothers Grimm, and drawing attention to Porter's new-found confidence in the capacity of poetry to bring together myths and history, word and image, past and present in the one song:

> For us the Garden is set in its clockwork glory,
> Nineveh sits by the river forever in the story,
> No-one is old and God listens on his throne,
> At the end the bone sings: He hears the singing bone.

The second of Porter's collaborative books with Arthur Boyd, *The Lady and the Unicorn* (1975), is a more strikingly handsome volume than *Jonah*. A larger format and fewer pictures give scope to Boyd's white figures with delicately etched details set against a deep black background. Ursula Hoff has described the essential elements of the myth which Boyd and Porter reinterpret and elaborate upon:

> According to myth the unicorn is an emblem of purity; it is trapped and betrayed by the woman it loves and made over to the hunters ... Porter and Boyd transport it into the fallen

world, where it becomes the touchstone of vulgarity, pride, possessiveness, lust and treachery.[36]

This project brought forth some of Boyd's best work. In Hoff's view, 'The contrast between the idyllic scenes of the unicorn in the woodland and the stark and grotesque images of the unicorn betrayed, imprisoned and crucified are the core of Boyd's finest graphic series to date'.[37] Porter was more than happy that his poems should be the 'onlie begetters' of such outstanding images:

> Each picture is white on intense black and the mastery of sheer line and complexity of drawing is virtuosic ... The celebration of both the fertile and terrifying aspects of sexuality, which is Boyd's constant preoccupation, makes the book as disturbing as it is beautiful. Nets, mirrors, dogs of various sizes and shapes, helmeted soldiers (looking forward to the *Mars* to come), spears and humans with aboriginal features are all dominated by the image of the Unicorn itself, which Boyd draws with Protean splendour.[38]

With no firm Biblical text to destabilize, Porter needed to create the events in *The Lady and the Unicorn*. It was this precondition which made for a less iconoclastic and frenetic approach to material drawn from a variety of sources, including the minor Victorian poet, Sabine Baring-Gould.[39] Porter made up much of the story himself. In this book, more than in *Jonah*, the poet, like the painter, was engaging with the pleasures and pitfalls of love, also a topic of pressing importance in *Preaching to the Converted* (1972) and more strongly still in *Living in a Calm Country* (1975) and *The Cost of Seriousness* (1978). The virtuoso poet reined in his energies in *The Lady and the Unicorn* and a controlled lyricism informs passages of both prose and verse. 'The Unicorn's Love Dance' (*TLATU*, X), for instance, is spritely and musical:

> I lay my horn in my lady's lap
>   To bring mankind to innocence:
>   We come from darkness and go back thence
> And only love may hold a map.

'The Death of the Unicorn' (*TLATU*, XVII) is a prose lyric which embodies a dream of the poet since his mother's early death:

> It can never have been so quiet on earth and yet all space is a Cantoria for angels, masking the finite shape of time. As fruit to the sun, as sky to the fountains, love is stepping from its robes. When the threads meet, blessed by the textbooks of despair, the

single palpable shape you know, ensexed by its own imagination, will press its starved fig against your lips. 'I have come home, mortal creature, the Muse of Waiting; this is my love which I give you freely. Now you may love everything'. And in that salt instant when creation is all sex, she enters your body in a kiss. 'Mother, Mother', you say, at the second of death. The Mother has come home to her son.

These examples indicate the strength of the lyrical impulse in *The Lady and the Unicorn*, allied with a deep nostalgia, which is held in check but not superseded by irony. A sense of ceremony prevails, from the opening Invocation and procession into the Ark (from which the Unicorn, another exile figure, is excluded) to the epilogue, in which the creatures of the tale are returned to the tapestry from which they came.

◆

Arthur Boyd left England to return to Australia in October 1974. By May 1975, he and his wife Yvonne had moved into their new home, Riversdale, beside the Shoalhaven River in New South Wales. This would be their main base despite periodic returns to Ramsholt, in Suffolk, where Porter would visit them from time to time. Peter Porter's returns to his native land in 1974 and 1975 were more complicated and emotionally fraught than the Boyds', and it is instructive to compare his expressed feelings about Australia immediately before and after these visits. In an interview with Elizabeth Riddell at an international poetry festival in London in July 1973, Porter had asserted that roots in a place were not necesarily an advantage: 'They may be twisted underneath the surface, and they may hold you there when you should go'.[40] Although he claimed not to feel *déraciné* in London, he supposed he was 'an air-plant, without roots'. London, he had said, provided his 'map of ordinariness' and was 'an ideally artificial place to live', somewhere he could 'feel at home in all the better for lacking the roots I had in Australia'.[41]

Porter's recollections of Australia were of 'fear, penury and heat'. So much so, that he had some difficulty believing other people's accounts of happy childhoods spent there. In a talk on BBC Radio after returning from a visit to his native country, however, he spoke of an 'Arcadian Australia', and confessed surprise at the distortions of memory — a memory which was to become increasingly the subject of his poetry:

In the past, I completely misunderstood Australia: I saw it through eyes blurred by personal experience and soured by the traumas of childhood. Now, for seven weeks, I glimpsed the real land, and it was as if a mask had fallen from a handsome face.

I don't want to overdo the eulogy, and Brisbane is still itself to present the other and barbarous profile of Australia. But I was profoundly disturbed to discover that I had roots after all. That I was a son of the country and might even have managed to grow up into some sort of an artist, if I had never left it.[42]

Porter was careful to distinguish in his developing personal mythology between Sydney, which was 'the great good place, the unconscious Arcadia from which I was expelled', and 'the sub-tropical city of Brisbane, with its everlasting suburbs, serpentine river and unpainted houses on wooden stilts', which had been his Inferno. As he implied in the poems 'Sydney Cove, 1788' (*PAAM, CP* 50−1) and the later 'Spiderwise' (*TAO* 56−9), Arcadia was preferable to Utopia because not too much was expected from it. Arcadians therefore do not suffer as Utopians must, from the inevitable collapse of high ideals. Something of the sardonic Australian myth figure of the 1890s tradition may be discerned in these sentiments expressed by a person who in many other respects seems set apart from it. Yet Porter also remained a defender of the English, who were being derided by some Australian nationalists in *Nation Review* and elsewhere as 'beat-up Brits'. Nationalism had its good points in reawakening awareness and even love of the natural scene, in its encouragement of local talent and its liberalizing influence in the Pacific, but its big drawback was an isolationist ockerism, which asserted that 'all the rest of the world is out of step with Australia'. Nevertheless, a buoyancy of tone informed Porter's reappraisal of Australia in 1974 (he even discerned a 'love affair' of Australians with their past), and a new appreciation was evident of a country which, in the early Whitlam years, he found interesting and emotionally compelling enough to argue with at close quarters.

An argument about national origins and their significance had been proceeding in Porter's literary journalism before his departure for Australia in 1974. The notion of 'Englishness', for instance, was explored in an essay on Byron who, after Auden and Browning, induced the feeling in Porter that poetry had more to do with pleasure than edification or mystery.[43] Porter's liking for Byron was due at least in part to the latter's position as a self-exiled figure in Europe, who was at his most 'English' when he was criticizing and satirizing the country he had left. In Porter's canon, you criticize only if you care. He discerned Byron's 'English' bias in Canto One of *Don Juan*:

Happy the nations of the moral north!
    Where all is virtue, and the winter season
Sends sin, without a rag on, shivering forth;

With his forthcoming visit in mind, Porter must have perceived his role in relation to Australia in a similarly adversarial light, but one which might connect him vitally to his country of birth. The Byronic model which appealed to Porter was not the earlier heroic Childe Harold phase but that of the author of *Beppo*, *The Vision of Judgment* and *Don Juan*. Byron, like Pope, Rochester and Auden, was part of a tradition of wits in English poetry of which Porter was a fully subscribed member at this stage. This tradition was, in Porter's view, set against the 'feeling' tradition of Keats, Wordsworth and the new American aestheticism. Byron also appealed to Porter because he was adept at *talking* in verse in a way which was provocative, amusingly serious and highly readable. Rather than straining after pure form, Byron's verse was eclectic and various, engaging with a wide range of ideas from the past and present; in Porter's words 'the quintessence of impurity'.[44] Discursive impurity of this kind increasingly attracted him.

'Americanness', on the other hand, was encountered by Porter mainly through its contemporary manifestations in the poetry of Robert Duncan, Charles Olson, Michael McClure, Charles Bukowski and others. His criticism of the Black Mountain Group, projective verse and other elements of contemporary American poetry was not that of a knee-jerk neo-phobe, although this may seem the case from occasional anti-American asides in certain poems. The rage for experimentation and the new in the United States had led to a situation in which, Porter acknowledged, American poets wrote 'more audaciously and more commandingly' than those in Britain. Yet the poetry of some seemed 'too aesthetic by far, lacking in the taste of life and prone to attitudinizing'.[45] While adopting anti-academic postures, they seemed in thrall to the universities. Porter found his preferred American tradition in Wallace Stevens and T. S. Eliot, rather than in the currently fashionable imagism of Ezra Pound and William Carlos Williams. The thinking person's discursive verse which Porter felt had helped him find a voice and manner for his talent appeared to be less favoured, in Britain as well as America in the early 1970s, than a self-conscious romanticizing of 'the poet'. In spite of such general reservations, Porter could be remarkably open to the work of individual American poets, and indeed be influenced by them. In a review he praised Robert Duncan's 'Imitations of Gertrude Stein'[46] and later wrote 'Gertrude Stein at Snail's Bay', *Three Transportations*, (*TCOS*, *CP* 256−60), in a similar vein. He found Olson's *Maximus* poems 'an astonishing stew which sometimes catches the light in very beautiful colours'[47] and, as Jonathan Raban noticed,[48] he used some of its techniques in his ode 'Europe' (*TLOE*, *CP* 127−9). Those whom he most severely attacked, on general theoretical grounds, Porter could also listen to, and even adopt in part, for their individual and particular qualities.

Before Porter's crucial return visit to Australia in 1974, he had reviewed volumes by Australian poets David Campbell and Les A. Murray, but had given more space and attention to the unusual collaboration of D. H. Lawrence and Western Australian prose writer Mollie Skinner in *The Boy in the Bush*.[49] In one of the most insightful reviews of this underrated novel, Porter described it as a 'fantasia' on another person's theme, which was 'the more enjoyable for being devoted to realizing somebody else's imagination'. Skinner was praised for her 'almost obsessive love of the facts and surfaces of beleaguered living which is characteristic of early colonial writing' (an acute observation), although the novel's ending was criticized as 'fairly unpleasant, bullying, self-worshipping Lawrence'. In her memoir, *The Fifth Sparrow*, Skinner had quoted Lawrence approvingly for his remarks that a gum tree 'seems to sweat blood' and that 'these hills frighten me, as if dark gods possess the place'. Porter was more critical: Lawrence was working these comments up from his reputation rather than from observation, and even he should have known why gum trees are called gum trees. Porter was not expressing any specialist knowledge of Australia here, rather, he objected to a kind of literary attention to a place which rejects intelligent observation and knowledge in favour of a self-regarding mysticism. Lawrence's air of self-importance, in fiction as in life, and his gospel of primitivism, offended Porter's Europeanized Australian sense of proportion. While he could mock himself as an alleged sufferer of 'Porter's Complaint' — 'an inability to feel anything when feeling/is the whole point of the operation' — he was not about to relinquish his intelligence to Lawrentian vitalism or primitivism.

◆

Straddling as it does the catchment areas of before and after Porter's first visit to Australia, his sixth volume, *Living in a Calm Country* (1975), shows to some extent the mental and emotional vacillations of this period, and provides a deeper source of insight into attitudes, values and feelings than the literary journalism. Jannice Porter's death occurred after the book was prepared for publication and is not referred to in any of the poems. A late dedication was however entered in the book: 'In Memoriam S. J. H. (1933–1974)'. The prevailing mood in *Living in a Calm Country* is not the untroubled calmness proffered by the title, but a profound disquiet, a lull before a storm. The energies involved in 'piling on fuel for the dark' still surge up from time to time but, as in 'The Storm' (*LIACC, CP* 217–18), their countervailing forces can also be seen. Out of an increasing sense of imprisonment, the poet sings of storms and lost rebellions:

> . . . The heart of the storm is now,
> God wrecking man on death,
> the absurdity of pitching his tone so high —

Even the act of writing poetry, which is always an act of disturbance, a rebellion against stasis, must be questioned:

> Why write poems.
> Why, for that matter, march on Moscow
> or ask your daughter if she loves you?

Such startling compressions of historic and personal moments indicate their interpenetration in Porter's imagination, which continually juxtaposes domestic ordinariness with the long vistas of history and space. While a Beckettian nihilism is only just held at bay in this piece, other literary connections operate: Thomas Mann's image of the sanatorium in *The Magic Mountain* and Dante's storm which sweeps up Paolo Francesco are faintly discernible background presences. Rather than a visually developed metaphor, 'The Storm' offers a more conceptually derived, 'force field' metaphor, which is strengthened by allusions to music and a verbal music of its own in alternating melodies:

> The storm will return
> but before it claps down on the foreshore
> and harbour, put out the lights, the nightlights
> and phosphorous and turn the sea upwards
> inverting the stars — see the long winking banks
> are like Mozart or Nature
> . . .
> awaiting
> the tempest, the null epicentre.

As the title poem in *Living in A Calm Country* (*LIAAC*, *CP* 199) notes quite explicitly, the deceptively 'calm country' in which the poet subsists is not England but himself:

> Living in a calm country, which is me
> Is not like architecture or the sea

First published in *The Critical Quarterly* in 1972, this poem mimics a mood of disappointment in which the poet's persona is 'calm as a cup', while underneath the temperature heats up:

> Playing with selfishness, I propose
> Rules for the game, the most outrageous clothes
> On truth, a cunning heart
> Pumping in praise of time, as the world goes.

Such acknowledgements of the complexities, deceptions and

betrayals of the heart make Porter a very different poet from D. H. Lawrence or Ted Hughes, who share a principal purpose of stressing the pre-conscious elements of human nature. The true wildness of nature for Porter is evident not in an abeyance of the mind but when mental and physical domains clash. This volatile mix is reminiscent of Marston and more particularly Webster, of a post-lapsarian world of deception and intrigue from which there is no withdrawal except death.

Anger is a part of this world, though the poem 'Anger' (*LIACC, CP* 198−9), in an uncharacteristic minimalist technique, suggests a deeper sense of hopelessness which the energies of rage cannot eradicate:

> Only to sit
> with one's back to music
> and no forgiveness
>
> Colouring
> the air and scattering
> flakes of fury·

The short lines, lack of punctuation and soft alliteration effectively silence the anger, recalling the mood of 'The Sadness of the Creatures' (*TLOE, CP* 136−7).

While recognizing that 'real pain is never art' in 'Ode to Afternoon' (*LIACC, CP* 200−1), Porter nevertheless attempted to expose some of its sources in 'That Depression is an Abstract' (*LIACC, CP* 215−16). First published in *Overland* in the summer of 1973−4, this poem reflects some insights following Porter's visits to a psychoanalyst in London, both with Jannice and alone. The poem is presented as a colloquy between doctor and patient and is packed to overflowing with mythic and personal references, ranging from Ovid's *Metamorphoses* to a nostalgic image from childhood which had recurred in his work — 'seeing my father in the rain trying to clip the bougainvillea'. (This image, deeply nostalgic, hints at the pathos of an ordinary man, the suburban gardener, setting limits to the scarlet bush which 'goes mad' in sub-tropical Australia.) But a know-all perverseness in the poet/patient is also explored in this poem, including his tendency to write polemics against avant-garde artists while adopting their techniques. Porter can be merciless in such self-examination. Indeed, the poem verges on self-defeat; both doctor and patient are mocked and the irony and clever evasions on the patient's part give neither any respite:

> None of this is necessary
> he says in his subtle lower case,
> you have landlines to ten capitals.
> Good thinking,

bring the calabashes of iced wine
and the little sausages, I reply.

A passing allusion to Leopardi mid-way through the poem signals
its overall mood of dark pessimism, which is summed up in an image
towards the end: 'terror pressed into depression'. A related poem,
'Down Cemetery Road' (*LIACC*, *CP* 226−7), reinforces this mood but
offers two possible escapes from the psychological morass — the
avenues of religion and music:

I think I was six when first I thought of death.
I've been religious ever since. Good taste
lay in wait and showed me avenues of music.

These were avenues which Porter would explore further in the late
1970s and 1980s, but for the present all ways seemed to cross and
cancel each other out.

An interim tone, *entre deux guerres*, of darkly jocular apprehension,
enters 'The Story of my Conversion' (*LIACC*, *CP* 204−6), a poem which
'celebrates' the poet's turning forty-five:

I have begun to live in a new land,
not the old land of fear
but the new one of disappointment.

This disarming expression of personal vulnerability masks a cunning
metaphoric play on the word 'land', which in the next stanza transforms
to 'my new country'. Here, and elsewhere, Porter reveals how countries
are in large part inventions of the mind. In anticipation of his return to
Australia, he reinvents from memory the Queensland of his childhood,
'the sun-slatted boredom, castles of the kingdom of bananas'. For the
moment, he seeks diversion in 'my students' albums,/girls' addresses,
hours of drinking, anger'. In self-mockery, he pictures a version of
himself with 'The Book of Useless Knowledge' in his hand, saluting
'the thousand green horrors, the self'. Considering the accidents of
birth, his persona ponders how things might have been worse: 'I might
have been born in Galicia,/in the poet-killing provinces'. (Porter's
Galicia is the district in Poland which has been the homeland of
generations of Jewish poets, psychiatrists and intellectuals.) Porter
here seems to echo Seamus Heaney's 'The Tollund Man': 'Out there in
Jutland/In the old man-killing provinces,/I will feel lost,/Unhappy and
at home'.[50] Porter anticipates instead the plain verandahs of vernacular
Australian architecture, which lead him to consider his own perversely complicating habits of mind (which, however, he shares with
some powerful allies):

1 Porter, aged almost nine, with his mother in Martin Place, Sydney, 1938.

2 'The Nook', 51 Junction Terrace, Annerley, Brisbane, 1990 (now 'Hillcrest').

3 New arrivals in
London: Roger Covell,
Porter, Brian Carne,
Stanley Gardens,
Notting Hill, 1951/2.

4 Porter in London,
1953.

5 Porter and
Jill Neville, 1953.

6 London party, 1958.
Porter (left) with
Diana Watson-Taylor,
Jill Neville and friends
at a basement flat,
93 Wetherby Gardens.

7 Jannice and Peter Porter after their wedding in 1961, with Jannice's
parents, Richard and Jean Henry.
8 Porter with his wife Jannice and daughter, Katherine, Charmouth,
Dorset, 1966.
9 Among the graves, Blythburgh Church, 1966.

10 Writers' tour of Israel, 1971. From left: Charles Osborne, Dannie Abse, Porter, D. J. Enright.

11 With Sally McInerney, Mt Victoria, Blue Mountains, 1980 (and Harpy with breast plates).

12 Reading *Island Magazine*, London, 1981.
13 Porter and cat Flora in the kitchen, 42 Cleveland Square, London W2, 1982.

14

15

14 Punting on the Tas, Low Tharston, South Norfolk, with Christine Berg, 1984.
15 Judging the BBC Africa Poetry Prize with Musaemura Zimunya (Zimbabwe) and Kofi Anyidoho (Ghana), 1988.

I am fond of the overdone. Of Luca Signorelli
and Castagno. I'll never learn simplicity,
I don't feel things strongly. I was never young.

In this *mea culpa*, Porter lays charges against his poetic personae
which echo those of some critics, and which he himself considers
justifiable. When pursued by Eumenidean Furies, or critics, Porter's
tactic is to invite them in. However the tactic is more cunning and
perhaps less self-destructive than it may seem, for the harassed poet
then shares his burden with other Fury-driven artists. In this context,
the early Italian Renaissance painters Luca Signorelli and Castagno
are no mean allies. But the major spiritual ally invoked in 'The Story
of my Conversion' is none less than Sigmund Freud, whose ghost
arises sporadically and with some significance throughout Porter's
work. Freud enters this poem through an anecdote to the effect that,
in the 1880s, the young Viennese doctor considered emigrating to
Australia.[51] In the personal allegory of Porter's work, this semi-
humorous alliance with Freud signifies an acceptance of neurosis
and nightmare: it is a pre-emptive exploration, however comically
expressed, of the poet's capacity to return and face the sources of his
sense of exile.

By contrast, a less self-defeating, more engaged tone enters certain
poems written after Porter's first Australian homecoming, in particular,
'On First Looking into Chapman's Hesiod' (*LIACC*, *CP* 210−12) and
'An Australian Garden' (*LIACC*, *CP* 208−10). Scottish poet and critic
Douglas Dunn commented that Porter's temporary return to Australia
appeared to have been 'a necessary catalyst to his imagination'. Going
home, Dunn averred, was part of a 'personal realignment', and 'no
discovery in poetry is ever as important as a re-discovery'. In Dunn's
view, Porter had 'recovered respect for his background and nation-
ality'.[52] Major Australian poet-critics, Chris Wallace-Crabbe and Les
A. Murray, in their turn, recognized the strengths of the new book,
Wallace-Crabbe advising readers who considered Porter 'too cool and
disengaged' to read poems such as 'Ode to Afternoon', 'The Storm' and
'Anger'.[53] More enthusiastically, Murray observed that this was Porter's
best volume, and the best book of verse he had seen for a long time.[54]
For Murray, these comments represented a turn-around. As he admitted,
he had vehemently disapproved of this Australian who had 'betrayed'
his birthright:

How dare he run off and court the supercilious literary circles
of London. How dare he look back at Australia with nothing
but scorn. I even used to make snide remarks about selling out
one's country to curry favour with the metropolitan enemy.

Murray now conceded that his prejudice was 'silly, heartfelt and deep', and had been dissipated by his meetings with Porter in Australia and by this new book.

The friendship which developed between the two poets, based on respect for each other's work in spite of opposing views on poetic inspiration and purpose, led to an important exchange of ideas on poetry and culture in Australia. The central document in this public debate remains 'On First Looking into Chapman's Hesiod'. In his review of *Living in a Calm Country* Murray referred to the poem as:

> . . . a superb piece of cultural analysis which gets the relationship of quintessential Australia to the metropolitan civilizations Over There dead right, and deserves to be in every Australian anthology from now on, as an angel we will have either to acknowledge or wrestle with.[55]

As subsequent events showed, Murray chose to wrestle with the evil angel of Porter's metropolitanism, and especially with the role of poet and his chosen community, which Porter offered as a credo in the concluding section of his poem:

> Sparrows acclimatize but I still seek
> The permanently upright city where
> Speech is nature and plants conceive in pots,
> Where one escapes from what one is and who
> One was, where home is just a postmark
> And country wisdom clings to calendars,
> The opposite of a sunburned truth-teller's
> World, haunted by precepts and the Pleiades.

Typically, Porter entertains his bardic opponent's case so well that he almost gives his own away: the 'sunburned truth-teller's/World' seems doubly blessed with its access to both the stars of the Pleiades and the group of poets who went by that name and brought about a Renaissance in French poetry. The speaker's own position seems understated and self-deprecatory by comparison. But this is not surprising: many of Porter's most heartfelt statements of belief or understanding emerge from positions of apparent weakness, as if the acknowledgement of weakness gives access to strength. Although the source of this habitual tendency in Porter's work appears to lie deeper than any mere rhetorical trick, it can nevertheless work rhetorically too. Here, for instance, the escapee from the fixed identity which may be imposed on a person by small communities seeks instead 'The permanently upright city where/Speech is nature and plants conceive in pots'. Humour, self-deprecation and serious statement combine fruitfully in Porter's expression of purpose and direction. Its basis is nothing as traditionally wholesome as pastoral nostalgia, yet the image of the 'permanently upright city', emerging to replace the earlier metaphors of horizontality in 'On First Looking into Chapman's Hesiod', achieves an air of quiet dignity and hints at its own tradition and aspirations. As Porter has said, his city is:

> ... not London; it's the ideal city state which doesn't exist but which has lived in men's minds since social order first existed. But perhaps it's just the republic of the imagination ... where you are made welcome as a confrère. To put this up against a real place — Australia, America, anywhere — is unfair, but tends to occur to people like me who have lived their lives in self-imposed exile.[56]

As a place of belonging, Porter's metaphoric city appears precarious, even to himself, when compared with Murray's earlier images of the rural community which he poetically represents:

> Like a Taree smallholder splitting logs
> And philosophizing on his dangling billies,
> The poet mixes hard agrarian instances
> With sour sucks to his brother.

Porter's non-rhyming talking verse here incorporates a Theophrastian comic 'character' who combines elements of the biographical Les A. Murray (schooled in Taree in northern New South Wales and living on the family farm in the district) and an agrarian literary type which crosses national and historical boundaries. While the portrait suggests something of the comic condescension of the metropolitan dandy's view of the countryman, this is balanced elsewhere in the poem, as has been suggested, by exposing the vulnerability of the city-dweller's position. The different modes of wisdom are thus held in a productive tension, as in a disputation where the two opponents seem to offer each other genuine alternatives. Porter's desperate malcontent of the 1950s and early 1960s, whose satire could be tinged with bitterness, has weathered here, through his London winters, to a more balanced philosopher of humankind.

That 'quarrelsome vitality' which Sylvia Lawson noted in Porter's first book could still flare up, but was contained in a more generous spirit of comic acceptance. Critical of simple-minded patriotism, Porter lampoons it in the second section of 'On First Looking into Chapman's Hesiod': 'A long-winded, emphatic kelpie yapping/About our land, our time, our fate, our strange/And singular way of moons and showers'. In Porter's world view, nationalism and patriotism have much to answer for. But is Porter's 'upright city' a preferable model of civilization? The debate broadened when Les A. Murray honoured

Porter's poem with a full-scale commentary in a book of essays on major Australian poems,[57] and again when an ABC interview between Sydney academic Don Anderson and Porter was published as 'Country Poetry and Town Poetry: A Debate with Les A. Murray'.[58] Academic articles which refer significantly to the topic have followed,[59] reinforcing its relevance as a debating point and a source of ideas and images relating to changing social and cultural conditions in Australia in the 1970s and 1980s.

What emerges from these discourses and analyses is the fertility of Porter's poem as a basis of cultural ideas and hypotheses. In re-readings by various hands, the different qualities attributed to Porter's urban Attica and Murray's rural Boeotia have been extended across national boundaries and historical periods and assumed the dimension of symbolic ideal communities. It has also emerged that Porter's ideal city and Murray's agraria, which were posed as opposites for the sake of debate, are not as exclusive as they first appeared. The concept of cultural relativity in Porter's poem is significant: 'The apogee, it seems,/Is where your audience and its aspirations are'. For London readers and critics, a provincial aspect was often perceived as part of Porter's poetic persona although, to most Australians, he seemed thoroughly metropolitan. Similarly, Murray could not be categorized as a laconic bushman; as Porter himself pointed out, Murray's style was at times 'high baroque'[60] and even his 'Boeotian rituals' were characterized by a 'Byzantine complexity'.[61]

◆

Porter's emotional return to Australia was signalled most definitively in the love poem 'An Australian Garden' (*LIACC, CP* 208–10). The garden, based on one at 'leafy Lindfield' on Sydney's north shore, where Porter stayed with Sally Lehmann and her husband in 1974, is a symbolically appropriate setting for this writer's reconciliation with Australia, recalling the images of a prelapsarian Eden in the childhood garden in Brisbane. But how does an apparently anti-Romantic city poet incorporate nature in his work? Porter was by now a self-confessed child of the Enlightenment, whose poem on a bird at this time was designated 'a study' which presented its images not through immediate sense impressions but through the eyes and images of other artists ('A Study of a Bird', *LIACC, CP* 218–19). In a similar way, Porter's garden is established within a framework of art: it begins not in personal communion or pantheistic contemplation but ceremonially, as in drama or opera: 'Here we enact the opening of the world'. This 'opening' is both sexual and personal; but its recreation of an original beginning in the radical innocence of the Eden myth is ambitious. In spite of this, Porter's assuredness never falters in the poem. Unlike Laura and Voss in their garden in Patrick White's *Voss*, Porter's lovers are sexual rather than spirit-beings, but in both the poem and the novel a journey within is implied: 'like entering/One's self, to find the map of death/ Laid out untidily, a satyr's grin/Signalling "You are here" . . .'.

The Miltonic Garden of Eden has been Australianized by Porter, but unlike the seekers after bareness and simplicity in Australia, such as White or Stow, Porter's garden is crowded with noise, movement and variety:

> . . . Such talking as the trees attempt
> Is a lesson in perfectability. It stuns
> The currawongs along the breaks of blue —
> Their lookout cries have guarded Paradise
> Since the expulsion of the heart, when man,
> Bereft of joy, turned his red hand to gardens.

In keeping with his hunger for variety and difference, Porter's trees are positively multicultural: 'there shall be Migrants,/Old Believers, Sure Retainers' and 'Spoiled Refugees nestle near Great Natives'. As Russell Davies remarked, 'Eden has seldom been spoken of so un-naively'.[62] If trees in this symbolic garden are personified and capitalized, so too are the 'Unknown Lovers', whose inward journey echoes James McAuley's poem 'Terra Australis'.[63] Clearly, the lovers in Porter's poem are partaking of more than casual lust:

> The act is canopied with stars. A green sea
> Rages through the landscape all the night.

Few of Porter's other poems so unequivocally invoke 'love' or 'joy': '. . . What should it be but love?' The 'thousand green horrors, the self' have been temporarily put aside. But the warnings of the Christian myth remain potent: 'The past's a warning/That the force of joy is quite unswervable —/"Out of this wood do not desire to go"'. The journey inward discovers both love and death, described by Michael Hulse as 'the twin powers of Porter's universe'.[64]

The concluding section of 'An Australian Garden' is a decisive stage in the evolution of Porter's autobiographical persona. The 'monster' of early poems such as 'Beast and the Beauty' and 'Metamorphosis' remains an element in the self-perception here, but the barefooted girl finds him lovely in spite of this. In place of gentrified England where 'the inheritors are inheriting still', Porter's current lovers find themselves the inheritors and proprietors of a dream garden where it is high noon and the sun shines continually. The European mind is still present, but here:

It would be easy to unimagine everything,
Only the pressure made by love and death
Holds up the bodies which this Eden grows.

'An Australian Garden' is the ultima Thule of Porter's romantic
quest for reconciliation, and it could not last.

8

THE EASIEST ROOM IN HELL

◆

In the construction of any retrospective account of a literary life and its products, even when it is ventured while that writer is in his early sixties and still in full flight, a temptation exists to identify a central crisis, a single event and make it the fulcrum upon which the story turns. Many reviews and commentaries have presumed that Jannice Porter's apparent suicide in December 1974, when Peter Porter was forty-five, constituted this central crisis. The literary and biographical evidence supports this view, to a large extent, but there are complicating factors. First, as was pointed out in Chapter 1, Porter's mother's death when he was nine presented the boy with a template of loss, which he would superimpose on subsequent situations and events: death had been a sudden visitor before and the aftermath of each death would have recurrent features. A second complicating factor was Porter's reconciliation, only nine months before Jannice Porter's death, with the country of his birth after a separation spanning two decades. This reconciliation was closely related to his love affair in Australia with Sally Lehmann, which had thrown his sense of identity, loyalty, purpose and place into new turmoil. This second factor is reflected in a duality in the composition of Porter's most praised volume, *The Cost of Seriousness* (1978). The book is dedicated to Sally McInerney (Sally Lehmann's single name which she resumed after separation from her husband), and concludes with two poems which celebrate the possibility of a new life with her, 'Roman Incident' (*TCOS, CP* 269–71) and 'Under New Management' (*TCOS, CP* 271–2). But the book's central poetic focus is a renewed acquaintance, in memory and imagination, with the wife who, like the mother, had preceded him across the river of death.

The historical Jannice Porter is an elusive figure. In Porter's verse, she appears in some of her aspects through an authorial screen of deep nostalgia mixed with guilt, fear and loneliness. Images of her are dredged from the routines of domestic life and holidays in Europe, especially Italy. Romance, realism, tragedy and nightmare provide filters through which she is realized. She appears in one of her guises as the speaker in 'The Delegate' (*TCOS, CP* 249–51); and again, mythologized as the self-sacrificing figure of Alcestis, in 'Alcestis and the Poet' (*ES, CP* 292–3). 'The Delegate' is a monologue in the form of some of Peter Redgrove's longer blank verse pieces, such as 'The Case' or 'A Friend',[1] and shares with his work a Jungian dream-like mix of

worldly and unworldly elements. As a whole, the poem sensitively
dramatizes the psyche of a woman who makes no great demands on
life for herself, and who, even before her death, was 'always receding',
her ambition being to 'accomplish/non-existence, to go out and close
the door/on ever having been'. The image here of the recessive self
(which recalls Jannice Porter in the later rather than the early years of
her marriage) differs radically from the defining qualities of other
women who had figured significantly in Porter's life and writings —
the more self-assured romantic bohemianism of Jill Neville, for instance,
or the socially defined energies of Diana Watson-Taylor.

The 'real' Jannice Porter is no easier to pin down than her literary
counterpart. Porter's 'character' is significantly complemented by the
perceptions of others who knew her. One such is Susanna Roper, who
lived in Cleveland Square with her young son and became friendly
with Jannice when they took their children to the local school. According to Susanna Roper, the major turning point in Jannice Porter's
married life was a severe attack of meningitis in September 1967 (she
was already afflicted with liver disease and had been born with a
congenitally malformed kidney):

> That's what changed Jannice. It was like hippy, trendy times
> and Jannice was not touched ... She was fantastic. I was
> always changing boyfriends or depressed with this or that,
> and she was very supportive. It was what you wanted. She
> had this amazing card in the kitchen — Goya's picture of the
> firing squad — and she said this is what it (the meningitis)
> feels like. Only afterwards did she seem to hit the gin.[2]

In Susanna Roper's eyes, her neighbour was a living reproach to
flightier women of her time, such as herself. Jannice was not judgemental, however, and was warm and supportive to anyone in trouble.
Instead of rebelling and breaking away, like some other daughters of
Marlow doctors, Jannice married and, after the birth of Katherine in
1962 and Jane in 1965, seemed to become a model mother. Deeply
attached to her father, she seemed dismissive of her mother but imprisoned in her proprieties: 'Don't you tell my mother that Peter and I
lived together before we were married!' she exclaimed to Susanna
Roper when she was in her mid-thirties. This was 'ancient' as far as
her younger neighbour was concerned. Jannice was not one to 'let it all
hang out'. From the early 1970s, while her husband's career flourished,
Jannice Porter's life seemed to diminish ('I must reduce, reduce ...' is
the way her self-perception is presented in 'The Delegate') into a hell
of hidden bottles and private drinking. Her situation may have been
exacerbated by the treatment she received at St Thomas's Hospital in
London, where she had been trained as a nurse; there, the doctors

prescribed a supply of drugs which Porter likened to the bombs being dealt out by American B−52s.[3]

The question of suicide was quite widely discussed in the 1960s and 1970s and Susanna Roper has recalled that Jannice Porter talked about it as well. She had also been close to some who had failed and others who had succeeded in their attempts at suicide in those febrile times. Roper herself, for example, had twice overdosed and survived. Sylvia Plath's successful suicide in the freezing winter of 1963 was also etched in her memory. During this period Jannice and Peter Porter used to hear about Plath and Ted Hughes through Al Alvarez, who was Jill Neville's lover while she was living in the flat above the Porters. It was Alvarez who told the Porters about Plath's death. His subsequent book, *The Savage God*,[4] placed her suicide at the centre of his diagnosis of contemporary society. The book contributed to a public fascination with the sources and motives of suicide; even while denying he was doing so, Alvarez glamorized it in ways which Peter Porter rejected as irresponsible. With her experience in nursing, Jannice Porter was unlikely to have been swayed by the 'glamour', but the prevalence of talk about suicide, and the actual examples around her, may have contributed to her seeing it as a possible conclusion to her later troubles. In this, one particular example seems to have been more powerful than others. It was that of Assia Wevill, the woman who replaced Sylvia Plath in Ted Hughes's life. Jannice Porter and Assia Wevill became close friends, and Jannice helped her after an abortion in 1965. When six years after Plath's death, Assia killed her young daughter by Ted Hughes and then herself, Jannice Porter was severely shaken. She may have had this incident in mind later, when her own difficulties increased and her depression became chronic.

Disturbed by the reliance on drugs and lack of humane psychotherapy at St Thomas's Hospital, Porter arranged a visit in 1974 for Jannice and himself to a psychoanalyst, Dr Wolf, to whom Porter has attributed a Viennese (and presumably Freudian) background. The visit was a disaster: Jannice refused to go back, referring to the psychoanalyst thereafter as 'the big bad Wolf'. Porter returned for one more visit on his own, at which he was confronted with some harsh and discomforting perceptions of his marriage. Porter has recalled the diagnosis in the doctor's own words:

From what I have read of your wife's history, it looks to me as though when you got married in 1961 you were both on the rebound from other sexual disasters and misalliances. You were both highly neurotic, highly disturbed, highly nervous, probably to the point of near psychosis. You couldn't cope, you thought there would be a kind of collective support in this. In practice, what has happened is that you have healed

in the marriage and your wife has got worse. She has enabled
you to grow strong and become healthy. You have caused her
to be destroyed.[5]

Within a few weeks Jannice Porter was dead. Much of the guilt
which inheres in poems in *The Cost of Seriousness* may derive from
Porter's sense that there was at least some accuracy in the analyst's
comments. In a later poem, 'Disc Horse' (*TAO* 49–50) he recreated,
and partly reinvented, a visit to Dr Wolf in the context of a broader,
ironic questioning of truth, illusion and the discourses through which
they operate:

> Enter a hot room, past the potted plants
> and whispers of Vienna; listen to the Big Bad Wolf
> of Parsifal Road — the jockeys of our joylessness
> are liberators really...

Somewhat ambivalent about the value of psychoanalysis, Porter
became more stoically Protestant in his belief in individual responsibility after his wife's death. But he also believed in an attempt to trace
the sources of one's personal misery, to recognize these 'jockeys of our
joylessness'. The notion of 'liberation', through psychoanalysis or
other means, has always represented a mixed blessing for Porter,
reducing inevitable complication to enforced simplicity; 'letting go' in
ecstasy, trance, despair or even psychoanalysis seemed a futile evasion
of the torments which comprise the normal human condition.

To see oneself as the 'destroyer' of a partner, as had been implied,
needed to be confronted. Harsh self-assessments of this kind recur in
Porter's writing, making a study of his literary ego an entirely different
proposition from those of Robert Lowell or John Berryman, for instance, for whom a romantic attraction exists in 'letting go', in indulging
their sense of 'manic power'.[6] The Furies are ever-present in Porter's
work, but their purpose is never to romanticize him as a post-Byronic,
ravaged hero. Rather, they are allowed in to reveal the tension and
conflict in a sane man's struggle for self-control and a sense of purpose
in living.

Before the last few desperate years, Jannice Porter seemed to
visitors and friends to find satisfaction and pleasure in the life of wife
and mother in London. Roger Covell, who visited the Porters quite
often, en route from Australia to musical assignments in Europe, remembered Jannice as 'very English' with a 'slightly flat' voice and a
figure like a ballet dancer.[7] She enjoyed cricket and would sometimes
take the girls to a Test match (unaccompanied by their father, who
preferred listening to music). The protective and caring side of Jannice's
nature was noticed by Ian Hamilton to whom she seemed a 'nurse by
instinct' and 'glad to have an occasion to rise to'. When Hamilton's

first wife was very ill psychologically, Jannice had gone out of her way to help.[8] Trevor Cox, another friend and regular visitor, enjoyed Jannice's company and rejected any notion of her as victim:

> Jannice was no sucker. She was the other side. Through Jannice you got to know Peter better. She was very very nice. Very good company. Sometimes she had an acid tongue. Now and again, in a couple of sentences, she'd say something that was abrasive and terribly right, to do with Peter — either contradict his point or give a point of view of her own. Together, they were very right for each other in so many ways. In other ways they were quite wrong for each other. I don't think he loved her enough ...[9]

By contrast, Jill Neville considered Jannice to be tight, withheld and very 'proper' in an 'English' way; but this may have represented a withdrawal from Neville's more extrovert manner and her tendency on occasions to appropriate Peter Porter in conversations on their shared past. Neville also felt that Jannice Porter was not interested enough in literature to get really close to her husband.[10]

Porter's verse after Jannice's death contains vignettes of their marriage, relating mostly to its latter years. This represents a writer's attempt to outface the misery of his loss, and its implications. The marital quarrels which are alluded to appear to have been chiefly verbal, perhaps once or twice physical. Each partner had had affairs outside the marriage. Each realized that, to some extent, they had married on the rebound from previous deeply emotional relationships. Yet it was in the nature of both to take marriage, and especially the responsibilities of children, seriously. There were happy times, especially on holidays together in Italy, or with the children on the south-west coast of England. There is little evidence to support Porter's harsh judgement of himself as an ogre. Roger Covell recalls Porter's unsuccessful attempts in the early 1970s to reduce Jannice's dependence on gin and other spirits by buying Campari and wine and drinking with her. But her strategies became more subtle, her dependency stronger. Porter would accompany her to the clinic where she was an outpatient and wait with her. To his friends, he seemed a devoted father, did most of the cooking and kept the house going during the latter years of Jannice's life. After her death, Porter remembered his own childhood misery, exacerbated by being sent off to boarding school when his mother died. He rejected the apparently easier option which his father had chosen, and decided to bring up his daughters alone. Friends noticed both the stoicism of his grief and the love and respect between father and daughters. In poetry, where the more buried aspects of his nature sometimes surfaced, he paradoxically held more closely, in his imagination, the woman and wife he had lost, so that fifteen years after her death, it could still seem like the day before.

◆

While much of Porter's finest poetry in *The Cost of Seriousness* seems to offer personal, and even confessional moments, it is important to stress that the poems which refer to the writer's relationship with Jannice Porter and her death should not be considered as strictly autobiographical statements. Throughout, a contemplative retrospection calls up different frames, phrases or fragments of memory which are given ceremonial, colloquial or allusive force according to the requirements of poetic form or imaginative appropriateness.

'The Easiest Room in Hell', for example, represents both a personal and a more generalized experience. On the one hand, the 'I' is a projected image of Peter Porter; the 'we', himself and Jannice Porter. On the other, the story told is one of human partnership and loss, simplified and clarified in the manner of parable by eliminating the dross of detail which had sometimes clogged earlier work. Lying behind the piece are the circumstances of Jannice Porter's death. With or without such information, the poem works, although the biographical context contributes to the appreciation of the ways in which personal experience may be transmuted into something broader and deeper in the process of composition.

Jannice Porter died in the upstairs attic of her parents' large house by the Thames in Marlow, Surrey. Feeling tired and jaded after a number of visitors to the flat in Cleveland Square, and having looked after someone else's child as well as her own two, she told her husband and her friend Susanna Roper that she would go down to her parents' place for a few days in order to be alone. (Her parents had gone away for the weekend.) The last Porter saw of her was when he waved her off at Paddington Station. On the first night away, she rang, as arranged. The second night, when she didn't ring, Porter tried to ring her, but there was no answer. The third day, when her parents arrived back, he asked them by telephone about Jannice. Assuming that she had gone to an aunt's place in Rye, however, they did not go upstairs to check until the next day, when they found their daughter's body, and a note which Porter has never been shown. A reconstruction of events would suggest that, on her second night away, Jannice Porter retired to the attic room, where she had slept as a child, and where she, her husband and children had often stayed. There she composed a note to her parents (there may also have been another, but only one was mentioned at the coroner's inquest). She then took a number of sleeping tablets (not a lethal dose in themselves) and followed these with a bottle of gin, to which she had become addicted in her by now

severe state of depression. The coroner's 'open verdict' indicated that
suicide should not be presumed, but Porter himself has never doubted
that she intended her death.[11] Porter chose not to see the body of his
dead wife, which had been trepanned for the coroner's inquest. Some
of the rituals of closure which might have been confronted in the
mortuary or before the cremation were acted out instead in his poetry,
in a continuing affair of the heart with the woman to whom, in some
respects, he had never felt closer than in death.

'The Easiest Room in Hell' is a poem which gets very close to this
pain. Its short, unrhymed lines arranged in two, three or four line
stanzas seem uncontrived, the equivalents of breath pauses in a muted
monologue, as the speaker recalls the site of his wife's upbringing, and
her death. The room in which she died not only marks out her life
span, but also assumes an emotional and temporal expansiveness as
the place where wife and husband have been lovers ('on a clandestine
visit') and parents ('We sorted out books and let the children/sleep
here away from creatures'). But neither the place nor the marriage is
idealized. Porter has said how he dislikes the ethos of the town of
Marlow,[12] which represents in some respects the gentrification and
settled upper middle-class values which he had satirized in the symbolic Haslemere of 'John Marston Advises Anger'. Porter's father-in-
law was a doctor, and perhaps this is why Marlow is associated with
this profession. In 'Death in the Pergola Tea-Rooms' (*OBTB, CP* 21−2),
the 'local doctor' fails to comfort the dying man; and in 'The Easiest
Room in Hell', doctors and ballet dancers symbolize the social ethos of
this 'heart of England/where the ballerinas were on points/and locums
laughed through every evening'. The mention of 'locums' here relates
also to Neil Micklem, who was Dr Henry's locum in Marlow and with
whom Jannice Henry, as an eighteen year old girl, had fallen in love.
The conclusion of her protracted affair with Micklem, whom she had
hoped to marry, occurred just before she met Porter, when she was 'on
the rebound'. She never ceased writing to Micklem and Porter came to
believe that he was the only man she ever fully loved. An unpublished
piece, 'The Doctor's Story', clearly refers to Micklem, presenting him
as cold and manipulative; this could hardly be considered an objective
view. The 'games' which husband and wife play in this house were
thus coloured by the past, and are an extension of those in the wider
life of this 'home counties' society: '*Inconsequences, Misalliance,
Frustration*/even *Mendacity, Adultery* and *Manic Depression*'. One
effect of placing such adult 'games' in the context of both the 'heart' of
English society and the childhood home is to demonstrate innocence
subverted. The shifts in physical perspective in 'The Easiest Room in
Hell' reinforce a distinction between the murderous games of adults in
closed rooms and the more expansive view of a childhood by the
Thames:

> From its windows, ruled by willows,
> the flatlands of childhood stretched
> to the watermeadows.

Here, as elsewhere in his work, Porter's tone is gentle and considerate of a childhood innocence which in his own life he felt deprived of too early.

In this poem, too, the recurrent concern with 'home' resurfaces. The arguments and betrayals of a determinedly ordinary marriage are not rationalized away but are taken up in a movement which has both autobiographical and parabolic force. As in Donne's 'The Sunne Rising', the world is contracted to a room inhabited by a particular emotion, though in Porter's poem this is of a more problematic kind. Nevertheless, the room chosen by the wife for her death, the poem's epicentre, is in a sense sanctified:

> all along it was home,
> a home away from home.

These lines, signalling a hushed awe in the speaker's tone as he contemplates the site of a chosen death, also suggest a psychological realism which recognizes a retreat adults make to whatever remains of childhood security. The room, held firm in the speaker's contemplation, is marked out for memory, as a reminder until his own death, of a space shared by a man and a woman. In his despair, he strikes out for hope:

> Having such a sanctuary
> we who parted here
> will be reunited here.

That this hope is illusory does not deny its psychological force. Towards its close, the poem moves into remembered dialogue, the wife quoting Portia in Shakespeare's *Julius Caesar*, who also felt excluded from her husband's confidence. The poem ends with Brutus's enforced confrontation with his aloneness:

> You asked in an uncharacteristic note,
> 'Dwell I but in the suburbs
> of your good pleasure?'

> I replied, 'To us has been allowed
> the easiest room in hell.'

> Once it belonged to you,
> now it is only mine.

The poem, stripped to the bone, does not reach out for validation in any wider social or cultural reference. Rather, it evokes an ordinary tragedy, and hints at the imaginative and emotional consequences of a belief that 'the easiest room in hell' is the poet's eternal home, and in some senses the reader's as well. There is a concentration of vision here, a stillness, which had not occurred in Porter's poetry before.

When *The Cost of Seriousness* was published in 1978, four years after Jannice Porter's death, a number of reviewers praised the book by comparing its tone and outlook to Thomas Hardy's 'Poems of 1912 and 1913', which were written largely in response to the death of his first wife, Emma. Patricia Beer, for instance, commented: 'After years of perfecting his poetic skills Peter Porter has been overtaken, much as Thomas Hardy was, by a subject from which he cannot escape — haunting and being haunted; the result is his most brilliant and moving book yet'.[13] In a perceptive extended commentary, Herbert Lomas noted differences between Hardy's 'uninhibited' and Porter's more constrained self-dramatization, and compared the 'atheism' of the two writers: 'Porter's atheism — that *is* the word, though he once claimed to be a "residual Christian" . . . is occasionally, as with Hardy, like a war on "God". His deprival's genuine, though: the ultimate power in life is death, and so death is God: Porter's poems are the prayer of the bone on the beach'.[14] In subject, tone and, to an extent, philosophical outlook, Porter's elegies are close to Hardy's. But the differences, both biographical and textual, are important. When Emma Hardy died in 1912, Hardy was seventy-two; when Jannice Porter died in 1974, Porter was forty-five. Although feelings of deep loss, remorse for recent neglect of a sick spouse and a search for redeeming moments of love and intimacy beyond the indifference which marriage had bred, were common elements in the experience of the two writers, the difference in their ages accounts in part for their quite distinct representations of that experience. Hardy's 'The Haunter', for instance, is a more euphonious and sentimental evocation of the wife's ghostly return to haunt the poet than Porter's 'The Delegate' (*TCOS, CP* 249—51), in which dissonant elements bring concreteness and conflict to a drama which embeds itself in a mid-life consciousness.

The concentration of feeling in 'The Haunter' is reinforced by its regular rhyme scheme and elimination of detail, such as other characters and social or physical surroundings. Although the ghost of a dead wife is first-person narrator in both poems, there are important qualitative differences. Hardy's dutiful haunter, with her coy playfulness, is in a tradition of Victorian ghost stories. Porter's poem, three times as long as 'The Haunter', achieves psychological complexity in a modern manner, by drawing on motifs from the previously established mythology of Porter's poetic life, together with some new elements. The artistic danger which the poem courts is a dissipation of impact.

This is avoided by a skilful juxtaposition of vividly realized 'scenes' — recalled incidents or fragments of speech — rather than by an attempt to maintain the tone of a single ghostly voice. A cumulative sense of a full, value-laden and therefore conflict-prone relationship is conveyed, replete with deep feeling, humour, and shared pleasures.

The garden which comprises the opening frame of 'The Delegate' is an image superimposed on all the previous gardens of Porter's writing and thought — from the back garden of his parents' home in Brisbane to the Cleveland Square garden in London to the recent garden of love in Sydney. The opening lines establish the distinctiveness of the present garden, based on the Slough Crematorium where Jannice Porter was cremated. The voice is hers, recalling the nadir, the 'worst hour' of husband and wife, at the moment prior to cremation: 'the wreaths/and the mis-named name competing with/the other mourners' flowers upon/the crematorium slabs'. The speaker dissolves into 'an infinity of myself, pieces/for everywhere'. Memories of her are likewise dissolving: it will be the husband's duty (he is still her husband), conveyed by the voice of responsibility which comes to him in his wife's tones, to give form and precision to their shared life: 'I cannot forget unless you remember,/pin down each day and weighted eye/with exact remorse'. The fragments which are shored up against their joint ruin recall the intimacy of children, holidays, bedrooms, drinking, shared superstitions. Regret at the loss of these moments is patent in the 'never more' refrain:

Never to puff up those sloping headlands
watching the children ahead negotiating
the lanes of the wide bay: never
the afternoon sun straining
the bedroom light to a tint distinctly
like gin: never more the in-flight panic,
refusing to see omens in our food
or the number of letters in the month.

Such images, recalled from different angles and in different moods, would indeed be pinned down with 'exact remorse' in a significant number of Porter's poems of the late 1970s and 1980s.

By including the dead wife as narrator in 'The Delegate', Porter was able to explore in a different way his chief obsession, death. Porter's fascinated engagement with the signifiers of contemporary living has never entirely diverted him from his deep thraldom to death. Through the literary 'character' of his dead wife, he approaches most closely in all of his work the transition from life to death. The notion of metamorphosis is crucial to this. Thus his 'delegate' at the conference of death likens her changing state to adolescence: 'such changes/as the schoolgirl saw in her body/are metamorphoses to the gods'. She warns that she is no longer what he remembers: 'I am not what you remember,/the snapshots in time and sunshine,/nor even the angry and accusing face/at breakfast, the suddenly delivered tone/of hope along a Venetian calle on a Sunday'. Nor has she been transformed to a goddess. The truth she offers is atheistic, quashing the illusions of optimistic believers in a deity and a happy after-life:

> What we do on earth is its own parade
> and cannot be redeemed in death. The pity
> of it, that we are misled. By mother,
> saying her sadness is the law, by love,
> hiding itself in evenings of ethics,
> by despair, turning the use of limbs
> to lockjaw.

The abstracts here ('pity', 'sadness', 'law', 'love', 'ethics') seem not to be imposed; they are validated by an intensely felt personal situation and from this gain their plangency, penetrating the reader's consciousness.

What hope is there, what purpose then, in living? The delegate offers one principal justification for her husband's life, and by extension, that of others: the capacity to live as an artist, to be a mediator of truths, as she herself has been in this poem, is reason enough for living. In a voice weighted with dignity, ceremony and tenderness, she offers her husband a benediction, but with it an exacting duty:

> And in this garden, love,
> there will be forgiveness, when
> we can forgive ourselves. 'Remember me,
> but ah forget my fate'. Tell me like music
> to the listeners. 'I would not know her in that dress.'
> The days I lived through change to words
> which anyone may use. When you arrive
> I shall have done your work for you.
>   Forgetting will not be hard,
> but you must remember still. Evenings
> and mad birds cross your face,
>   everything must be re-made.

The injunction to remember (quoting Dido in her lament before immolating herself in fire in Purcell's opera, *Dido and Aeneas*), and to reconstruct the world in the light of this memory, became a major obsession in Porter's life and poetry from 1975 to 1980; and remained a continuing force throughout the 1980s. The command of his imagined interlocutor, 'remember me', was the soundest artistic advice this poet could have been given. It drove him to face again the fundamental reasons for living against the certainty of death, and gave the exercise of memory an urgent personal purpose.

The poem most widely regarded as Porter's masterpiece among those written after his wife's death is 'An Exequy' (*TCOS*, *CP* 246–9). Like 'The Delegate', this poem recalls the sentiment, if not the style, of Hardy's laments, as in, for example, 'Woman much missed, how you call to me, call to me,/Saying that now you are not as you were' ('The Voice'). Significant differences exist, though, in the modes of self-dramatization employed. In 'The Voice' Hardy presents his first-person speaker alone with his obsessional haunting, in an outdoor setting with traditionally symbolic autumn leaves falling around him:

> Thus I; faltering forward,
> Leaves around me falling,
> Wind oozing thin through the thorn from norward,
> And the woman calling.

Poignant as these lines are, they call up images of Victorian melodrama. Porter's self-dramatization in 'An Exequy' is of necessity backward-looking, but it leaps across the Victorians to find its verse form in Bishop Henry King's early seventeenth-century lament of the same title, written after the death of his young wife. Porter's use of King's model, a threnody in octosyllabic rhyming couplets, and in simple iambic tetrameter form, brings to the fore the artifice of this and all expressions of grief, emphasizing their conventional and ceremonial aspects. As Herbert Lomas noticed, a more contemporary verse model was Auden's *New Year Letter*[15] but this, like King's poem, was soon transcended by the individually realized grief of the protagonist.

As is often the case in Porter's work, time and place in 'An Exequy' are registered precisely — it is 'wet May', a season of spring-time volatility (not the falling cadences of Hardy's poem) only five months after the death of his wife. The settings in this poem are generally interiors, the speaker is inside looking out. He is pictured first in troubled sleep, disturbed by self-accusation and guilt, which later modulates into a consideration of his 'map of loss'. After many shifts and turns of image and emotion, mixing memory and desire, the resolution arrived at is not unlike that in 'The Delegate' which was written seven months later: the wife has preceded and will pave the way for her husband's entry into death. The last words in the poem, spoken by her across the borders of death (and therefore a prelude to the full-scale dramatic monologue, 'The Delegate') are in German: 'Fürchte dich nicht, ich bin bei dir' — 'Fear not, for I am with you'. Appropriately, these words come both as a reminder of Christian solace and of the comfort provided by certain kinds of art; the words are from a Bach motet. Representations of numinous experience in Porter's work are often conveyed, as noted, by a fusion of Biblical phrase and musical allusion. Hence the last line of what is one of Porter's most powerfully emotive poems does not fizzle into obfuscation or suggest that German is somehow the proper language of the dead; rather, it is a ceremonial benediction in which music is the only true ratification of, or support for, a quasi-religious feeling.

Between the nightmares of the opening of 'An Exequy' and the benediction at its close, Porter's persona is shown surveying his 'map of loss':

> No one can say why hearts will break
> And marriages are all opaque:
> A map of loss, some posted cards,
> The living house reduced to shards,
> The abstract hell of memory,
> The pointlessness of poetry —

Unlike Hardy, who discovered in Emma's diaries some harsh comments on her husband's conduct, character and treatment of her,[16] Porter was left, like many who are suddenly widowed, with little 'hard' evidence of the marriage (lasting thirteen years in Porter's case) from the point of view of the dead partner. There were occasional notes and letters between the Porters but, as in most marriages, little expression of intimacy occurred in writing. He was left with remembered incidents, phrases, photographs. There was, of course, for Porter (unlike Hardy) the major comfort, duty and reminder also of children shared.

The image of the stairway in 'An Exequy', as in 'The Easiest Room in Hell', echoes from Dante, Eliot and Yeats. It also carries Porter's personal signature — his literal attic room carries particular reminders, of ends and beginnings, of the creativity and destruction of the human spirit. In this upper room, many imaginative revisitings will occur, as preludes to the definitive reunion:

> ... one day
> The time will come for me to pay
> When your slim shape from photographs
> Stands at my door and gently asks
> If I have any work to do
> Or will I come to bed with you.
> *O scala enigmatica*,
> I'll climb up to that attic where
> The curtain of your life was drawn
> Some time between despair and dawn —

The italicized line in the midst of this scene of tenderness and love is typical of Porter. Like the voice exclaiming 'There is no God' as if from nowhere in 'Death in the Pergola Tea-Rooms', '*O scala enigmatica*' ('O mysterious stairway', or 'ladder to where?') signifies a sudden shift of register, as in opera. The effect is to radically dislocate the scene of peaceful intimacy, propelling it into a different key wherein grief, apprehension and the sense of lost love can mingle. Porter is almost never a poet of uncomplicated emotion and such sudden transitions are a hallmark of his work. The effect is not a self-conscious grab for attention, as Desmond Graham suggested when he criticized this 'rhetorical leap' as a 'jolt' aimed at drawing attention to the poet's 'dramatic coup'.[17] Rather, it produces an effect which is both emotionally appropriate and a reminder of the necessary artifice of all verse, especially that which deals with pain and grief. The Australian poet Bruce Beaver recognized the deft way in which Porter's mixed discourses operate when he remarked that the contrast between 'ironically conceived intellections' and low-keyed understatements make the latter 'more movingly apt'.[18] The same effect may occur in reverse. In a scene of gentle understatement, like this one, a sudden upward spiral in diction may provide an authentic and dramatic urgency.

Following the recollection of domestic intimacy, 'An Exequy' offers another sphere of togetherness in the experience of Italy, which Peter and Jannice Porter visited several times during the early 1970s. In some respects, Jannice was the teacher and Peter the learner there; as mentioned, the two first met after her return from living in Rome for six months. In the poem the speaker depends on his companion's 'nurse's eye' and capacity to 'bandy words' (his Italian was not good) while the couple sought out the artworks they had read about:

> I think of us in Italy:
> Gin-and-chianti-fuelled, we
> Move in a trance through Paradise,
> Feeding at last our starving eyes,
> Two people of the English blindness
> Doing each masterpiece the kindness
> Of discovering it — from Baldovinetti
> To Venice's most obscure jetty.

The tone here is mobile, shifting in a few lines from the high Romanticism of the phrase 'in a trance through Paradise' to the playful rhyme of 'Baldovinetti' and 'jetty'. Indeed, Italy appealed to Porter for its potential for rapid transfer from ordinary living to high art, from humour or vulgarity to the most exalted experience. Such mobility, allied with Porter's insatiable thirst for knowledge and ideas, can cause problems for the reader, as we have just seen. Often, however, the seemingly obscure references, once identified, are illuminating. Baldovinetti, for example, has more than a rhyming significance in the above lines. Spoken of condescendingly by one of E. M. Forster's clergymen in *A Room with a View*, Baldovinetti's frescoes were nevertheless enjoyed by Porter as the work of a minor artist who is important for a full appreciation of a culture.[19] A false restriction to 'the greats' eliminates whole levels of understanding, as Porter had argued in challenging the doctrines of F. R. Leavis.

One of the pleasures for the reader of Porter's mutable poetry is observing his tussles with form as well as his cultural ideas. From the semi-humorous picture in 'An Exequy' of husband and wife with their 'English blindness' enjoying the pleasure of cultural tourism, we are taken closer again to the poem's emotional core:

> And, oh my love, I wish you were
> Once more with me, at night somewhere
> In narrow streets applauding wines,
> The moon above the Apennines

The 'oh my love' phrase, virtually proscribed in modern poetry, seems at home in this piece where an easy commerce between old and new poetic conventions is established. (The generalization about 'opaque' marriages epitomizes the contemporary discourse.)

'An Exequy' is notable for the deft way it ranges across past and present and draws attention to its poetic form and conventions, without its central drama ever being lost. The dramatized persona returns again and again to the death of a particular woman and its painful correlative, the upstairs room:

> The rooms and days we wandered through
> Shrink in my mind to one — there you
> Lie quite absorbed by peace — the calm
> Which life could not provide is balm
> In death . . .

The process of spatial contraction in these lines continues in the poem's concluding movement, funnelling in on the core drama and its consequences for the person who remains:

> I have no friend, or intercessor,
> No psychopomp or true confessor
> But only you who know my heart
> In every cramped and devious part —
> Then take my hand and lead me out,
> The sky is overcast by doubt,
> The time has come, I listen for
> Your words of comfort at the door,
> O guide me through the shoals of fear —
> 'Fürchte dich nicht, ich bin bei dir.'

This conclusion, though it echoes Bach and the Christian gospels, offers no religious hope of an after-life. In this respect it parallels Hardy. It does suggest, however, that the poet is willing to 'share' the experience because of the depth of his attachment to its victim — that abandonment to despairing love, to the lost image, is what will enable him to lose his fear of death.

. The foreshadowed end to the speaker's own life echoes, with some irony, the exit of Adam and Eve from Eden ('Then take my hand and lead me out'), but is doubly moving because, as Lomas remarked, the speaker 'has no shadow of hope in her being alive in some way, only in the days, places, people and poems of remembrance'.[20] For Porter, religious feeling does not portend salvation. The task (a compulsion for this writer) of making memorials in verse is a contest with the forces of amnesia which is analogous to death. While seeking, from the depths of personal loneliness, a continuity through memory and art with his partner of fifteen years, the poet shares the late twentieth-century inability to make the leap of religious faith. The chief alternative role to that of believer is exile, and this is the role which occurs most naturally to Porter. In dramatizing the fate and sensibilities of an exile who longs for reunion with beauty, intimacy and truth, Porter invents a persona reflecting a wider allegorical range than merely autobiographical reference would indicate.

The constructed 'I' of Porter's elegies has clearly allegorical elements in some of the shorter poems, including 'Scream and Variations' (*TCOS*, *CP* 253–4) and 'Non Piangere, Liù' (*TCOS*, *CP* 267–8). As James Olney suggested, 'the great virtue of autobiography . . . is to offer us understanding that is finally not of someone else but of ourselves'.[21] The allegorical tendency in Porter's verse is allied with a wish to extend personal experience to a more public domain — the personal and particular is not enough. Stephen J. Greenblatt has pointed out that 'allegory arises in periods of loss, periods in which a once powerful theological, political or familial authority is threatened with effacement'.[22] If the project is driven, as Greenblatt suggests, by a sense of 'the painful absence of that which it claims to recover', then Porter's loss of his wife may be the catalyst for representations of the self which simultaneously exhibit a broader loss of faith, purpose, and desire.

The tragic perception which informs 'Scream and Variations' extends its range with allegorical elements. The poem opens with images of horror:

When I came into the world I saw

My face in my mother's blood
and my father crying where he stood.

The personal-sounding nightmare image, which the author has said is imaginary, expands to an allegorical image of perverted nature: 'the lark ascending a poisoned sky'. The images which conclude the poem are from a nightmare by Kafka out of Graham Greene, with music by Benjamin Britten:

> My coffin in an unknown place
> among enemies and the police;
> my daughters trying to please
> me in dreams with their replies;
> a tear belonging to my wife ablaze
> in the darkness of widowhood; peace
> in her voice but only death's applause.

> Myself in the world is all there was.

For a poet who sought as his abode one of the most heavily populated cities, this isolation is horrific, and recalls the round zero at the centre of the figure in Münch's painting, *The Scream*. An apparently personal and private poem, then, expands outwards, through its placement in a wider cultural context and by its powerful projection of an iconic figure, alone in the world.

'Non Piangere, Liù' demonstrates the way in which personal grief may be communicated while speaking of something more than just regret. The fourteen-line poem is composed in short lines, stark diction and relatively flat rhythms, minimalist in style. There is a sense here, as in the earlier 'Anger', of great emotional stress between the lines. In some poems emotion can be obscured by a garrulous rapidity of idea, image and statement; here it is confronted more directly. The poem's title, 'Non Piangere, Liù' ('Don't cry, Liù'), is from Porter's favourite Puccini opera, *Turandot*.[23] It alludes to Prince Calaf's words to the loyal slave girl, Liù, who has followed him to Peking and who tries to prevent him striking the gong which signals his candidacy for Turandot's fateful test — 'Non Piangere, Liù', he sings. Like many of Porter's borrowings from other artists, this poem identifies a keynote of the original: Porter heard in Puccini's music, especially in *Turandot*, a 'massive directness' in confronting difficult emotions.[24] (A later poem, 'At Lake Massaciuccoli', celebrates Puccini as the eccentric, amoral composer who conveyed the sounds of creatures, animal and human, 'abandoning life for love'.) Puccini's opera both moved and challenged Porter when he wrote 'Non Piangere, Liù'; at this time he was struck by the composer's inclination towards 'frail and doomed heroines', like Liù, and this resulted in 'an unhappy insight' into himself.[25] Porter has not explained further, but it seems likely that he was referring to a similar tendency in his own characterizations of Jannice Porter in poems such as 'An Exequy' and 'The Delegate,' in which an assumption is made about the wife's loving compliance in leading her husband towards death. In 'Non Piangere, Liù' the possibility that she will reject him is confronted. Indeed, the poem moves towards an awareness that all previous meaning is lost with her death and cremation.

The motif of tears is used with delicate insistence in 'Non Piangere, Liù', concluding with the stanzas:

> Do not cry, I tell myself,
> the whole thing is a comedy
> and comedies end happily.
>
> The fire will come out of the sun
> and I shall look in the heart of it.

Nowhere in Porter's verse is the switch from the rationalizing human intelligence to a direct confrontation with feeling better conveyed. The poem's final recognition, which emerges as an epiphanic moment, is a tragic vision; but 'tragic' is itself a human construction and is imaged here in the element of fire, which will consume all human life as it has consumed the loved one. The speaker's stoicism, and the directness of his gaze, will themselves be reduced to ashes in the consuming fire which awaits us all.

As if in response to this vision of annihilation and despair, some of Porter's poems of the late 1970s take a new grip on place and history. The evocative setting of 'An Angel in Blythburgh Church' (*TCOS, CP* 245–6), for instance, is the Holy Trinity Church in Suffolk, known in the region as the 'cathedral of the marshes'.[26] The central image of this piece appeared as the cover illustration on *The Cost of Seriousness*. It shows, as Pamela Law saw it, 'an angel ... smiling slightly squint-eyed and enigmatic into eternity and suitably pitted with (beetle?) holes'.[27] The holes are actually bullet holes; in the sixteenth century this church was raided by Cromwell, whose men tied their horses up in the church and shot at the angels.[28] The shot-down angel, resting against a wall of the church when Porter visited it in early 1976, provides a focus for the meditative flow of this poem, which is lent an air of quiet formality by its arrangement into eight quatrains of predominantly iambic metre and alternate rhymes on the second and fourth lines in each stanza. The speaker in the poem is less trigger-happy than Cromwell's soldiers, but he implicates himself in their activity through his own quiet iconoclasm: 'What is it/Turns an atheist's mind to prayer in almost/Any church on a country visit?' The questions intensify:

> Greed for love or certainty of forgiveness?
> High security rising with the sea birds?

A theology of self looking for precedents?
A chance to speak old words?

The questions are part of a personal drama located in place and time. The icon of the angel, shot down from the high wall of the church, has eyes which, though marred by time and circumstance, signify a faith both 'wooden' and 'certain'. Such certainty seems impossible, the poet broods, in the present age of doubt. Returning to his most potent source of loss and pain, he thinks of 'a woman' who stared for hours at a ceiling, with death as her 'only angel/To shield her from despair'. Illusions are stripped away again, recalling the personal statement in '"Talking Shop" Tanka' (*TCOS*, *CP* 265–6): 'I share death not faith with Donne'.

Although visits to the English countryside became part of a recuperative process, the poet never seemed quite at home there. Trevor Cox, whose properties in Devon Porter often visited, retains an image of his friend standing with briefcase in hand in an open field, dressed in smartly casual city clothes, seeming nervously anxious to return to the town. Cox has nurtured a small grove of trees at his property near Crediton commemorating each of Porter's visits — a Porter Grove — but the person honoured in this way soon becomes restless for the city and its diversions.[29] Fellow poet Charles Tomlinson, in a poem called 'Hedgerows',[30] dedicated to Porter, lectured him and fellow 'townsmen' about the 'urbanity' (and therefore the value) of England's disappearing hedgerows which 'contain, compel as civilly as stanzas'. Porter himself tends to present the English countryside only in terms of art. In 'At Ramsholt' (*TCOS*, *CP* 238) the setting is Arthur and Yvonne Boyd's Suffolk property. Porter's restlessness in the setting, which he renders in terms of a quiescent Dutch realism, is evident in his disruptive allusion to the greatest source of international unrest in the mid-1970s, Vietnam:

This is the Deben, not the Mekong, but a sail
curves round a copse; masts for Woodbridge
crowd three degrees of the horizon, edging
a painterly Dutch sky. Clouds are curdling.

From Dutch realism to Vietnam in a sentence — Porter's imagination continually pushes against the boundaries of what he feels to be rural and regional constraint. In 'Waiting for Rain in Devon' (*TCOS*, *CP* 239), he reflects on the drought in England in 1976, but does not take up the concerns of farmers, as would Les A. Murray, for instance. His perceptions are those of a metropolitan aesthete: 'Come back, perennial rain,/ stand your soft sculptures in our gardens/for the barefoot frogs to leap'.

Italy as subject and setting began to gain prominence in Porter's poetry in the mid to late 1970s. In some ways, Italy was a half-way house to Australia, travel to the southern hemisphere being prohibitively expensive for an author who lived by his hand as a freelance writer in London Porter understood the British need to escape to the Mediterranean, for the freedom, landscape, wine and art which novelists from E. M. Forster to John Mortimer have praised, with comic pleasure. For Porter, however, it also carried correspondences to the Australian landscape and light (especially in the north), linking this country with the prelapsarian Paradise of his early childhood. But Italy, through its art, also gives access to a 'high' cultural past. A habit was established, first with Jannice and later on with friends, of holidaying in Italy and making pilgrimages to galleries and churches there.

Australians in the 1950s and later used to deride those with an excessive interest in the 'high' arts such as opera, music and painting as 'culture vultures'. Porter's fascination with these arts took him well beyond social pretentiousness: those who have travelled with him to Italy have commented on his encyclopaedic knowledge of Italian painting and opera, and his tendency to find interpenetrations of these arts in everyday discourse. (He has opposed that brand of socialist thinking which sets folk or mass art as cultural norms — ordinary people are complex, he believes, and deserve access to 'high art' as well as popular culture.) Nevertheless, Porter himself has recognized a predatory element in his visits to Italy: 'I come to Tuscany for what I can get out of it, an exploitation fortunately which does the environment no harm'.[31] Such enthusiasm derives from a powerful emotional attachment: for Porter, Italy, and especially the Italian influence on the Germans, seems the great and central European tradition. In this, as in other areas of cultural thought, Porter is in agreement with W. H. Auden, whose notion of 'Italia in Germania' proposed similar artistic preferences. France, Spain and even England were marginal, in this view.[32]

Porter has never presented himself as other than a tourist in northern Italy. Unlike other Australian expatriates who have lived there, such as David Malouf, Germaine Greer and Jeffrey Smart, Porter has never been able to afford a house in Italy, or anywhere. Poetry does not provide a sufficient income, even for a best-selling poet. But Porter's 'rootlessness' on his visits to northern Italy give piquancy to his attachments to a region which he inhabits with knowledgeable love. The Italy with which Porter has been concerned in his verse has little to do with politics or economics, and everything to do with the creative arts. As an Australian, Porter feels that northern Italy holds special lessons:

> Here man himself is the measure of beauty, and here, despite
> despotism and, of recent years, reaching for easy commercial

success, people and landscape remain in agreement. In
Australia, we have hope, Protestant mercantile optimism, and
the despair which comes from fighting Nature and losing. In
Tuscany a peculiar truce still remains.[33]

The diagnosis of Australia here seems lacking in contemporary awareness. Judith Wright, among others, has shown the fragility of nature in
Australia, and there is now a wider recognition of this, though the
battle for environmental awareness has barely begun. At the same
time, the notion of a viable 'truce' between nature and human culture
in Tuscany suggests a worthwhile basis for ecological criticism. For
Porter, such a truce between people and place is a necessary precondition for a civilized intercourse between art and everyday living. In
'Frogs at Lago di Bolsena' (*LIACC, CP* 212), for example, the human
value of this lake in Umbria is attested by the ways in which its raw
wildness has been 'squared into art' by painters and writers. But in
this poem, like others set in Italy, Porter's assertion of the value of the
limits set by art on nature is partially undermined by a music of
dissonance which attests to wilder forces and feelings abroad in the
world. Thus the frogs at Lake Bolsena boom out a pain which is
inconsolable in spite of the best efforts of artists and tourists to 'square
it off'. Similarly, the images of Venice in 'To Venetia's Shores Have
Come' (*TCOS, CP* 262–3) are overtaken by the melancholy prospect of
this art-city becoming 'lost on a dark sea'. In 'Evensong' (*TCOS, CP*
261), which recalls a visit with Jannice Porter in 1973 to a church in
Verona, the tombs of the despotic della Scala family, placed outside
rather than inside the church and named after dogs, are reminders that
death and its horrors cannot be ignored. Porter's perception of a truce
between humans and nature in Italy is therefore problematical. The
savagery of battle, and its attendant spirit death, lurk just beyond the
corners of paintings, the walls of churches, the limits of verse; the
romantic spirit paces restlessly within classical confines.

◆

The Cost of Seriousness ends with two poems which attempt, as Porter
himself pointed out, 'to look forward to a renewal of life'.[34] Both
'Roman Incident' (*TCOS, CP* 269–71) and 'Under New Management'
(*TCOS, CP* 271–2) speak of a growing affection for another woman
following the poems on Jannice Porter's death. They recall, with particular poignancy, Yeats's question: 'Does the imagination dwell the
most/Upon a woman won or woman lost?' The subject of these pieces
is Sally McInerney, whose presence may be felt, along with the more
ghostly presences of Jannice and Marion Porter, wife and mother,
through a number of poems in this and the next volume, *English
Subtitles*.

The relationship between Porter and McInerney began in circumstances of abnormal stress and difficulty. Soon after his return to London came the unexpected and, for Porter, cataclysmic death of his wife, for which he felt not only sorrow but guilt and remorse. To further complicate matters, Sally McInerney's separation from her husband Geoffrey Lehmann was a bitter affair, and she had three children to look after, while Porter had two. A relationship which had begun at the Adelaide Arts Festival was darkened and complicated by these events and also by the tyranny of distance. Although Porter returned to Australia in 1975, 1976, 1977, 1978 and 1980 as writer-in-residence at universities and colleges, and Sally visited London and Italy in 1976, much of their relationship was inevitably, and frustratingly, carried out by letter. Ian Hamilton has recalled tortuous conversations in which Porter considered a possible future in Sydney with her.[35] A major barrier to this, in many ways highly desirable union, was the problem of children. After his 1975 visit to Australia, the response of Porter's thirteen year old daughter Katherine, who had stayed with Susanna Roper in London, was obvious: she feared desperately that her father would go to Australia and take his daughters with him. She stopped eating and, in Roper's words, 'really freaked'.[36] The daughter's terror at the prospect of leaving home and friends in London was matched by Porter's horror at the state she was in. These and other problems remained unresolved. The end of this love affair, when Sally McInerney and then Porter himself moved into new relationships, was felt by some of Porter's friends to be a tragic conclusion. Certainly, her influence on Porter's emotional reconciliation with Australia was enormous.

Ironies abound. In Les A. Murray's prothalamium upon Geoff and Sally Lehmann's marriage, 'Toward the Imminent Days',[37] the lines spring with celebration and hope, concluding with the poet's blessing: 'For your wedding, I wish you the frequent image of farms'. Porter, a principal agent in Sally Lehmann's departure from the life wished for her by Murray, was characteristically more guarded, but she appears as his own muse of hope in the love poem, 'Under New Management':

> I dedicate my years in the field to one
> Body only, she of the catalytic eyes.
> Whether Dionysus or some other hard man
> Of the hills will lead us, we shall be
> Caught up in the evening's enterprises.
> As god of this place, my own savannas,
> I praise her and I lead her forth
> Into a published garden. She'll be silver
> By mine and by the stars' true light.

While the pun on 'catalytic' here is playful, the overall effect is ceremonial, a formal 'progress'. From the raging love-making of 'An Australian Garden' to the solemn act of faith in 'Under New Management', a journey towards commitment has been enacted. That it is not completed is less significant, poetically, than that the journey has begun. Ironically, however, like Murray's prothalamium, Porter's poem expresses a hope which was dashed by time and circumstance, subsequently confirming its author's scepticism about high hopes of any kind.

More interesting poetically, and psychologically, than 'Under New Management' is 'Roman Incident', which refers to a visit to Rome by Porter and Sally McInerney in 1976. The poem reflects a phase of confusion and uncertainty in the relationship. If its metaphysics reside in questions about the nature of paradise, its principal psychological focus is the speaker's subconscious grappling with the influence of different women in his life; and, more generally, with the feminine principle. The poem's apparently casual narrative structure presents a man recalling an incident in Rome when, after 'a night of quarrelling and making love', he and his companion have stopped in the Borghese Gardens. While she sleeps among the other lovers in the garden, he leaves her for the nearby gallery, where his attention to the paintings reflects a compulsive search for meaning among the women he has loved. He is transfixed before Dosso Dossi's painting of Melissa, 'apotheosis/of the watching female and her autumn shades'. He wills the enchantress to do her work and transform him, to 'uncork' his heart. The need for a deep emotional union is enacted:

> Standing frozen before her on a plinth
> of grief and awkwardness, I tried to cry,
> to force water from my eyes, so that
> Melissa might turn me back to manhood —
> that, I said, is what I want. Magic, fortune, love:
> the luck to be kissed and smiled on
> no matter what ridiculous wizard
> corks up my heart. To mother,
> wife and all the sultry dead I prayed,
> lead me to the enchantress whose one kiss
> undoes the tactless misery of self.

When he returns to the garden, his actual lover has gone. Desperately, he looks for her, then hears her call his name. What he sees first is her body, eased by sleep. They eat, and walk together, and ahead of them is the anticipated sexual excitement of 'the night and our hotel' — paradise regained.

The passive masculine figure seeking an intercession by women from his past, via the intermediaries of paintings in a gallery, is transformed to a more actively anticipatory masculinity. The poem suggests that sexual love is a practical answer to unrealistic expectations of intercession. However, this particular woman in the park, who awakens the speaker from his sense of loss and disorientation, might herself be discerned as the goddess in a new guise. 'Roman Incident' displays a narrative interest, and something of the enigmatic playfulness, of a Malamud or a Barthelme short story; its verse form, discontinuities, hints of mystery and its placement in Porter's canon, give it a force and urgency which goes to the heart of his search for meaning and purpose in these years of change.

Inevitably, Australia, as place and idea, was also quarried at this time for signs of purpose and hope. In 'The Orchid on the Rock' (*TCOS, CP* 240), the orchid, which has spent some fifty years sending its 'roots like rivets into the rock', represents a symbolic antithesis to the self-exiled poet who has found nowhere to belong. Porter has described the poem's genesis in his visit to Australia in 1975:

> The decade of the Seventies was a troubled time for me. My wife died at the end of 1974, just after my first return to Australia for twenty years. My daughters and I spent five months in 1975 in Sydney and we visited Arthur [Boyd] at Riversdale. While there we ventured inland, over many ancient fallen cedar trunks, to view a superb orchid growing out of solid rock in the rainforest. Back in London, I wrote a poem entitled 'The Orchid on the Rock', which derived from my Riversdale experience.[38]

More direct in its symbolism than most of Porter's discursive verse, 'The Orchid on the Rock' is also spatially precise and sensuous, though sound is the predominant sense. During the walk in the bush the speaker hears waterfalls and the sound of the walkers' feet 'unplugging from the grassy bog'. The orchid itself seems to offer a natural synaesthesia: 'not scent/But a colourless colour, so intense/It eats the light, brings us up close'. Botanic details are of less interest than the numinous possibilities which this orchid represents. By what accident, or hidden purpose, did it make this rock its resting place? In the closing stanza, an old problem is posed afresh, but not resolved:

> The orchid blooms in perfect nowhere;
> We go home to our electrified fort
> To create relationships.
> From the verandah, viewing the river
> On its civilizing course, a fever
> Of loneliness encroaches, grips
> The mind. A clarinet has caught
> The empty evening unawares.

The verandah — where much Australian writing has occurred[39] — is the speaker's vantage point, from which, while hearing the clarinet's note of loneliness (perhaps from a record player in the 'electrified fort'), he observes the 'civilizing course' of the river. The Australian environment is no longer the hostile, masculinist force it seemed in the early poems, 'Forefathers' View of Failure' (*OBTB, CP* 3–5) or 'Eat Early Earthapples' (*1962/3, CP* 44). The subtext to this new accommodation to the Australian bush is the capacity to see it, in sympathy with Arthur Boyd, as a subject of art.

The role of artists in Australia, and the country's resistance to art, are interrogated wittily in 'Three Transportations' (*TCOS, CP* 256–60). The three artists whom Porter imaginatively transports to Australia are the American prose writer Gertrude Stein ('Gertrude Stein at Snail's Bay'), the Italian painter Piero di Cosimo ('Piero di Cosimo on the Shoalhaven') and the Italian composer Luigi Boccherini ('The Boccherini Music Camp'). Each of these artists is in some respect a misfit, perceiving and rendering aspects of the environment as a visitor or exile would: the exile's capacity to disrupt or disturb 'normal' perception here is their defining characteristic. Boccherini, for instance, is humorously characterized as an Italian migrant:

> As I said to your Arts Reporter
> asking about myself
> and Scarlatti and other
> erstwhile exiles in Iberia:
> 'Out of Italy we come,
> great talent, little room;
> the hatred is for family.'

The great cellist's 'Australian' music moves into Porter's private domain, to give expression to moods of sadness or discomfort: 'The call of strings through upland wattles' awakens personal disturbances, such as 'Do my children love me?/I doubt that any of my quavers/ would bail me out — I have been far away/all my life'. In contrast with the cello's 'sounds of sadness', the painter Piero (whose work Porter deeply admires) inhabits a more wittily apocalyptic Australia:

> Neither Adam nor Jesus ever laughed
> But the serious earth is quite hilarious.
> This is Eden as the cattle go past
> The electric fence; the faces are so various
> Of flower and shadow, which will last?

The American writer Gertrude Stein, on the other hand, offers Australia some of its own characteristics magnified. In a brilliant parody of

Stein's style (which Porter had praised in Robert Duncan's imitations of Stein as 'the tone of commonsense raised to the power of "n"'),[40] he reveals Australian laconicism and an associated optimistic spirit in a new cross-cultural perspective. As in most of Porter's parody, the ridiculed object is not demolished, rather, it is incorporated into the author's world view. The arrogant assertiveness of Stein's first-person narrator as she appears in *The Autobiography of Alice B. Toklas* (1933) is registered in Porter's poem as a dogged optimism in the face of the world's misery; but her apparent stupidity and snail-like logic emerge as attractively translucent in their simplicity:

> I am an observer,
> I observe the blue and you
>
> I see an immense rain
> washing pebbles up the beach
> and evacuating misery
>
> The plane for America is a sort of star

No such 'immense rain' or evacuations of misery are possible in Porter's world view; and the sublime belief in 'America' as a model for western societies such as Australia's seems patently false. However, in spite of its humorous undermining of Stein's perspective, Porter's poem also accommodates something of its spell.

The Australia which Porter envisages is thus inhabited by artists of all kinds, those who are ridiculed as well as those who are praised. They will come from everywhere. This vision of Australia as a place where artists might thrive in the mid-1970s had its sceptics back in Britain, however. In a verse letter addressed to Porter in Sydney, Clive James humorously requested his friend to return quickly to suffering Britain ('the pit of desperation') and not be lured by Utopian Australia's health, wealth, and philistine values.[41] The empire's tables had been turned, James averred, and (with punning reference to Porter's first book), 'the biter's bitten'; but Britain, with its apparently incurable inflation and raging unemployment, 'where soon must fall a dark night of the soul', is the place where Porter should really be. Like other friends, James missed above all Porter's conversation:

> I miss your talk not just because of savouring
> Its bracing lack of artificial flavouring,
> But also for the way that Grub Street scandal
> Is spiced by you with thoughts on Bach and Handel,
> And whether the true high-point of humanity
> Was Mozart's innocence or Haydn's sanity.

James's affectionately slanderous diagnosis of his native country's unsuitability for the artist was being rectified by Porter's closer look at an actual revival of the arts in Australia, which he applauded. However, he foresaw problems in the prosperity and confidence of the mid-1970s. Australians, in his view, were principally concerned with 'getting on' in financial and worldly terms, robbing the ground of its ores and emptying the seas of their fish. This sentiment is expressed in ' "In the New World Happiness is Allowed" ' (*TCOS, CP* 254–5):

>   . . . Why did nobody say
> that each successful man needs the evidence
> of a hundred failures?

If the materialistic success ethic was perceived by Porter as a barrier to artistic development, (a somewhat puritanical socialism seemed to re-emerge at this time in his thinking), an even greater barrier was the artificially prescribed climate of happiness. In response to Les A. Murray's contention that, in contradistinction to Europe's inheritance of misery, happiness was allowed in the New World, Porter boldly asserted, 'No, in the New World, happiness is enforced'. Porter did not here claim superiority over European *Angst*; on the contrary, he contended that the suffering and unhappiness of ordinary Australians, and other New Worlders should be allowed as a valid element in their art. Porter's contentions in this poem and elsewhere should be seen not as impartial sociological diagnoses, but as observations relating to the reasons for his own self-exile. In retrospect, Porter mused, it was an earlier version of this ethic of imposed happiness, which 'leans your neck over the void', that had sent him and others of his generation off to:

>   . . .Europe and the crowded hearts,
> a helplessness of pasta and early closing days,
> lemons glowing through the blood of Acre.

In the context of other poems in *The Cost of Seriousness*,' "In the New World Happiness is Allowed" ' is deeply ambivalent about the poet's relationship to both Europe and Australia. The hybrid European-Australian is ironically serious as he considers the polarities (heart-head, home-away) between which he is tossed:

>   . . . Trying to be classical
> can break your heart. Depression long persisted in
> becomes despair. Forgive me, friends and relatives,
> for this unhappiness, I was away from home.

Neat aphoristic statement is torn into dissonance by personal confession: 'I was away from home' seems autobiographically true but
as explanation is ironically insufficient. 'Home' remains a mirage,
continually sought but always unreachable.

9

THE LYING ART

◆

In *English Subtitles* (1981), as in *The Cost of Seriousness*, one of Porter's recurring themes is disillusionment with words. Anne Stevenson has rightly noted a process of declining faith in the power of language.[1] She might have extended her observation, however, to comment on the ways in which Porter had used this process to notate a wider breakdown of faith, in the face of personal pain and grief. These issues are intrinsic to poems such as 'The Cost of Seriousness' (*TCOS*, *CP* 255–6) and 'The Lying Art' (*TCOS*, *CP* 243–4). In the former, Herbert Lomas observed two main 'strings': that art cannot be serious, and that art should not be serious.[2] To be totally serious is to face the fact of death and be overtaken by it. The artist must therefore 'play'. The nature of this 'play' is taken up further in 'The Lying Art': the playful deployment of words will always fall far short of the truth of emotions, especially pain: 'It is all rhetoric rich as wedding cake'. Furthermore, poetry seems irrelevant, because it doesn't make anything happen. Yet from these Audenesque negatives, Porter paradoxically creates a sense of pain and need; the reader is invited to respond to this with his or her own recognition of the inadequacy of words in the face of reality, but to recognize also the necessity of the gestures they can make. In the concluding stanza, dishonesty is accepted as necessary: 'all epithalamiums/are sugar and selfishness' — an allusion perhaps to Les A. Murray's epithalamium 'Towards the Imminent Days'. Faced with the knowledge of these failures of language and love, the reader might be excused for asking what is the point of living. Porter never answers such questions with a Romantic, intuitive affirmation of life: life and death are unchosen; and the paths between may be lit up temporarily only by the exercise of human wit, imagination and feeling.

The inadequacy of words to subjects as a topic is further developed in *English Subtitles*, in poems such as the title piece, 'A Philosopher of Captions' (*ES*, *CP* 279) and 'At Lake Massaciuccoli' (*ES*, *CP* 306–7). Porter here ventured into dangerous territory, seeming to undermine the claims of Donald Davie and others for the pre-eminence of language. If words were felt to be at best 'slim sentences', like English subtitles under a foreign film, the captions of a philosopher, or a mere code which always falls short of the writer's intentions, surely the poet should give the game away? Moreover, by admitting these things, wasn't Porter opening the gates to the philistines? If these were his sentiments, shouldn't he keep them to himself in the interests of solidarity?

Porter has never remained silent in the face of arrogant self-justifications, whether the subject is poets, critics or politicians. But the subject here was broader than individuals. *English Subtitles* was realistic in recognizing the growing dominance of film as the medium of the 1980s and pointing to the necessary modesty of wordsmiths in this new era. In an enthusiastic and wide-ranging review of the volume in the *Times Literary Supplement*, John Lucas directly addressed this issue and the disquiet which Porter's stance had caused in some literary circles:

> Peter Porter's remark that poetry is 'a modest art' has excited a good deal of comment, most of it unfavourable. There is a general feeling that he is somehow selling poetry short, and that while one might not actually want to mention the divine afflatus nowadays (although some don't mind claiming it for themselves) it's there right enough and to pretend otherwise is boorish or downright philistine. Yet in an obvious sense Porter's remark seems perfectly just... And as to the claim for 'inspiration', of the poet as *vates*, there are two good reasons for wanting to challenge that. In the first place to do so clears away the rubble of poseurs and hysterics; in the second, it creates a space in which art can freely stand. More particularly, perhaps, the call for modesty is the satirist's way of reminding us that *his* poetry matters.[3]

This was a shrewd commentary on Porter's strategy. Like his attacks on the pretentious poseurs of fashionable London society in the 1950s and 1960s, Porter's broadsides in the 1970s were fired by a spy within the ranks, a complicit outsider who has never been accepted wholly as either 'British' or 'Australian'. Strong admirer though he was of Porter's poetry, John Lucas did not include him in his critical book *Modern English Poetry*[4], presumably because he considered Porter an Australian in England, whose work therefore did not reflect an English tradition. Porter's access to language, and a range of subjects, was enhanced, in Lucas's view, by the outsider's perspective:

> Porter's language... is truly an international language, quite simply because it isn't the language of any native English poet and because he can do things with it that no English poet can. It's dazzlingly eclectic...
>
> Yet what I am attempting to describe is perhaps less a matter of language than of what such language can reveal. It is a habit of mind, an attitude, sometimes and quite properly a

pose even, that is extraordinarily generous, accessible and responsive to ideas, to art, and to music, to the things of the world; and it can therefore move at ease among a whole variety of subjects which English poets typically approach, if at all, with awestruck solemnity or rasping contempt.[5]

English Subtitles posed a problem for reviewers, then. While proposing in its title piece the extreme modesty of poetry's claims upon the world, the book's offerings were rich and various, making considerable intellectual and emotional demands on readers. As Lucas put it:

> The poems of *English Subtitles* have extraordinary verve and a quite new eloquence, measured but genuine, in which wit is poised against melancholy in ways that are often expressed through taut, graceful rhythms...[6]

One of the important poetic characteristics of this new volume, which its title pointed to, was a renewed interest in aphorism. Porter's ability as a phrase-maker and epigrammatist had been evident earlier, but now he emphasized their modern function and value. No longer could the writer get by with the leisured and apparently complete 'pensée', or the Wildean manicured epigram. As Claude Rawson suggested, the modern writer was more inclined to the 'stab of sudden comprehension, incomplete and fleeting, but capable of piercing deeper than discursive talk'.[7] The aphorism used in this way encouraged a poetry of surface disconnection, replacing logic with the language of imagination or feeling. In bringing this aspect of his verse to the fore, Porter was emphasizing a growing sense of affinity with Wallace Stevens and Stevens's American successor, John Ashbery (though he was less consistently lyrical than Stevens and brisker and more referential than Ashbery). Like Stevens and Ashbery Porter became addicted to a poetry of sudden changes of register and direction, in which the fortuitousness of insight or revelation is exposed. As Rawson observed, Porter's aphorisms 'readily merge ... into a surrounding context in a way which denies undue status, even of a momentary sort, to self-enclosed crystallizations of experience. But there is no doubting their importance in the poems, and in his ideas of how poems work'.[8] Denying special status for poetry on the one hand, and claiming it on the other through surprising flashes of brilliance is, as we have seen, characteristic of Porter — a reminder of his description of his family history, for instance, as a 'diaspora of the undistinguished',[9] claiming in what appears a casual remark special attention for the apparently ordinary.

The insights into poetry as fictional construction in 'The Lying Art' were extended in 'A Philosopher of Captions' (*ES, CP* 279):

> Here space a fear and there placate a pun,
> Or adjudicate through childhood, one noun
> Up and another down, with everything a fiction.

It would be mistaken to deduce from these lines a cynical or merely mechanical approach to composition, as did Laurence Lerner.[10] Like other poets, Porter claims an 'authenticity' for his work. But he refuses to claim this as a divine, or even a special gift: it is more an obligation, a compulsion to get things right:

> This special authenticity must grow on one
> After baffled if dutiful years putting down
> Some orders of words towards definition —

The poet's art lies in knowing the limits with and against which he works: puns are not simply to be indulged in but 'placated'; childhood memories 'adjudicated'; philosophizing reduced to 'captions'. None of this denies 'the power ... still somewhere in us, hovering/In the forehead auditorium of sounds'. But for Porter, in the late 1970s and early 1980s, as he confided in 'A Philosopher of Captions', poetic composition was a continual wrestle with 'pain', which he perceived as 'the one immortal gift of our stewardship'. And as he put it elsewhere, 'No one produces the art he wants to,/Everything that he makes is code,/To be read for its immaculate intention'. ('At Lake Massaciuccoli', *ES*, *CP* 306−7).

For some critics, Porter's wrestle with language was leading him into a cul-de-sac, diverting him from the grief and pain apparently more directly confronted in *The Cost of Seriousness*. Ian Hamilton, for instance, having welcomed the 'emotional power' of the 'almost heroically meditative personality' which had informed the earlier volume, was disappointed with what appeared to be a reversion to poetry in which 'the ironical intelligence is pretty well always in control'.[11] James Lasdun speculated that Porter was now at an impasse, wherein 'private melancholy finds itself at odds with a style that was developed for public vituperation'.[12] Having discerned a still centre in Porter's whirling world of images and ideas in *The Cost of Seriousness*, such critics wanted to keep him there, attending the inner fires. To a large extent, Porter did continue to explore the feelings of pain and loss associated with the death of his wife, but he also became increasingly fascinated with the verbal vehicles for such emotions, from gem-like epigrams to complex baroque structures. He admitted the 'rococo' intertextuality of a number of poems in *English Subtitles*, but insisted that 'the poems speak of plain human feelings, however rococo the references'.[13] Deborah Mitchell was one of a number of critics who observed a successful containment of both strong feeling and complexity of associations.[14] British poet and critic John Fuller considered

English Subtitles 'a first-rate collection from a sad, ingenious, compelling poet writing at the height of his powers',[15] and John Lucas
thought the book revealed 'extraordinary verve and a quite new
eloquence'.[16] If those seeking emotive lyricism were disappointed,
others who preferred eloquence and wit were not. Yet none of these
assessments was extended enough to show the interplay between
powerful feeling and ironical intelligence in this eighth volume.

Australian reviewers of *English Subtitles* again exposed their difficulty in accepting a writer who had lived elsewhere for thirty years.
Recent visits had strengthened Porter's bonds with his home country,
so that there was now no question of trading-in his Australian passport;
however 'European' his orientation was, he was also Australian in
ways which seemed more compelling than they had before his first
return in 1974. At last his work was being noticed more widely in
Australia, following visits, readings, press notices and broadcasts between 1974 and 1980. In welcoming his 'authentic' voice and recom-
mending it to Australian readers, Elizabeth Riddell commented that
Porter's poetry was still far better known in Britain, and conjectured
that if he had remained in Australia his work might have been set for
study in secondary schools, as had that of contemporaries such as Les
A. Murray, Bruce Dawe, David Malouf and others.[17] (Some of Porter's
work was beginning to be studied in universities, but not, except
for the occasional poem, in schools.) There were other barriers to
acceptance. Sydney poet John Tranter, who was discovering his own
cultural references chiefly in American film and the popular arts, expressed irritation at the foreignness of Porter's European 'cultural
baggage'.[18] Tranter's criticism of Porter's alleged cultural referents —
'a German opera, an Italian composer, a Greek god, a Renaissance
painter' — revealed not only the wider turning away from European
'high culture' in Australia in the 1970s and 1980s, but also the cultural
orientation of some of the 'new Australian poets' whom Tranter perceived himself as representing.[19] Some Australian poets had more in
common with Porter's discursive urban poetics than Tranter's attack
allowed; affinities existed with Melbourne poets such as Evan Jones,
Chris Wallace-Crabbe, Peter Steele and Philip Martin as well as
Sydney-based poets such as Laurie Duggan, John Forbes, Gig Ryan,
and certainly Tranter himself. Young Western Australian poets such as
Philip Salom and Shane McCauley also benefited from Porter's visits.
In spite of their professed preference for the American Black Moun-
taineers and French symbolists, Australia's 'new' poets were often
concerned to conceptualize and problematize ideas and emotions
ironically in their verse, and Porter's work indicated ways of doing
this. In spite of occasional critiques on the limitations of a modernism
which rejected the past, Porter's thinking and practice were well
attuned to contemporary movements, as his reviews in the *Observer*
and a number of Australian magazines show. The licence he found in

New York poet John Ashbery's work to 'resist the referent' through deferral of meaning, for example, may not have been the best direction for his own verse (it encouraged, at times, too much 'poetic talkativeness'), but his capacity to be influenced indicated an openness to movements and modes, which appealed to younger writers.

However, the problem of a perceived 'foreignness' in Porter's work persisted for some reviewers. In an aggressively critical review, Fay Zwicky remarked that, 'read in an Australian context' Porter's poems seemed 'marmoreal'.[20] Failing Zwicky's American/Australian test of 'visual concreteness' (except in an early poem like 'Ghosts' *1962/3*, *CP* 42–3), Porter was alleged to have turned away from the life-saving prescriptions of William Carlos Williams — 'No ideas but in things!' (Williams and his followers were a lodestar for many Australian poets at this time.) In Zwicky's view, Porter's rejection of Carlos Williams's prescription of idiomatic speech patterns and concise images of commonplace reality as the basis of poetry, was the cause of his damnation as a poet. By putting his faith in 'art' and not in 'life', Porter had lost energy and direction. Being Australian, however remotely, may have saved him however: 'Expatriate or not, Porter's energy does not submit easily to the dampening pressures of his adopted country's mores'. It is obvious that Zwicky's argument suffered from simplistic certitude and national stereotyping — of the English and of Australians. Bolstered by the growing acceptance of 'American ascendancy' in poetry, and within this, of Williams's imagist prescriptions,[21] Zwicky's attack on Porter suggested that poetry should be visually concrete, and should avoid reference to other works of art. Porter answered what seemed to be a popular case against his work:

> Most people seem to think about writing literature that one uses experience and that experience should be as fresh as it can be, and it ought to be, in most cases, new; and it ought to be about taking a car down to the beach, or about looking at a flock of galahs, or even about Sydney cocktail parties, but it should not be about other works of art, nor even about ancient Italian cities. This seems to me to be a pre-empting of human imagination. It suggests to me that if reading books is not one of the major experiences of your life, why do people bother writing them? For many people, what they read, or look at, or listen to, is primary experience.[22]

These comments, and those of some Australian reviewers, suggest that if he had returned to Australia Porter would have found himself swimming against fashionable poetic currents. But resistance to prevailing fashions had become Porter's way; wherever he lived, he would be a renegade.

Poets were among Porter's strongest supporters in Australia, indicating a powerful degree of professional interest in his work. Robert Gray, for example, found that Porter was 'the most intelligent poet, the most lively mind in the whole of the art, working in Britain at present'.[23] Gray placed Porter close behind Philip Larkin and Ted Hughes (and ahead of Heaney, Tomlinson, Thomas, Davie, Sisson, Hill and Graham) for his 'over-all achievement and memorability'. Jamie Grant, while acknowledging the difficulty of some of Porter's work, praised his adept use of a variety of personae and his sense of humour, suggesting that he should be as widely read as 'homebodies' such as Les A. Murray and Judith Wright.[24] Clearly, the difficulties of a writer-in-exile are not restricted to his divided personal loyalties and feelings; they extend to the feelings and loyalties of critics and readers and the cultural contexts in which books are bought, read and studied.

For all its concern with the processes of art, the overwhelming preoccupation in *English Subtitles*, as in *The Cost of Seriousness*, was with difficult personal emotions. Surprisingly, none of the reviewers of this volume, British or Australian, noticed that, together with a small number of major poems recalling his dead wife, *English Subtitles* was marked also by the return of a disruptive sexuality, more humorous and relaxed than in earlier phases. In the playful poem 'How Important is Sex?' (*ES*, *CP* 277), Porter answered his title question in the first line: 'Not very'. Yet succeeding stanzas demonstrate the endless fascination of sex for the young, the middle-aged and the old. Sex may have its comical side, but the speaker is a respecter of its power:

... having seen a skinny girl
Screaming in the playground, oblivious
Of boys, wake to her hormonal clock

As Juliana or as Mélisande —

The title is also taken up in the first line of another piece, 'The Need for Foreplay' (*ES*, *CP* 275), which:

Is felt by these trees
accepting the wind along
their winter arms.

In such poems humour and lyricism are nicely attuned; Porter's mix of tones is neither demeaning nor evasive. He is at a stage of acceptance in which, instead of remaining tied up in anticipatory theory, it is possible to 'love the mechanics of love,/go backwards into mystery'.

'Sonata Form: The Australian Magpie' (*ES*, *CP* 283−4) is one of Porter's most entertaining poems about sex and love. Although the manuscript at the Australian National Library contains a note in

Porter's hand, 'Poem in honour of an Australian magpie',[25] the bird is a vehicle for a playful yet serious consideration of matters arising from his relationship with Sally McInerney. It is one of the central poems of Porter's reconciliation with the country of his birth, none the less serious or important for its brightness of tone. Porter's sprightly sonata in short, unrhymed stanzas of three lines mixes thoughts, ideas and observations which do, in fact, honour the magpie's agile boldness, though not in any traditionally romantic way. What principally characterizes Porter's magpie is its predatory nature, which it shares with humans:

> Its opening theme is predation.
> What it scavenges is old cake
> soaked in dew, but might be eyes.

> Such alighting and strutting
> across the mown grass at the Ladies' College!
> Siege machines are rolling near.

Both the Australian magpie, which swoops down on passers-by in the nesting season, sometimes injuring eyes, and the human on a mechanical mower are, in Porter's view, part of a universal drama of predation: the tone may be light, but the game is survival.

The magpies are a reminder of the primacy of appetite. Humans are as greedy for sex and love as these birds are for food. The contrasting themes of musical sonata form chiefly involve an attempted differentiation of human behaviour from that of the magpie. The speaker presents his rather unmusical attempt at intellectual superiority humorously:

> You can upbraid the magpie,
> saying, 'What do you know of Kant?'
> It might shift a claw an inch or two.

This philosophical distinction is soon blurred, however, when human 'cries of love' and the noise of magpies 'sighting food' become musically indistinguishable. Questions of sex, love and their endurance recur towards the poem's end, in the third last stanza:

> The coda, alas. It can be Brucknerian.
> We say the end is coming.

The pun on 'coming' and the jocular allusion to Bruckner's codas (which, some critics have observed, go on forever) link human nature and art by analogy.

Finally, the poet is faced with a problem. In Australia, nature still seems to predominate over art. Where can an appropriate balance between art and nature, love and sex, be found? In a vivid surrealist image which concludes the poem, the poet flies 'to the top of a camphor laurel'; that is, escapes to his art. At this point:

Girl and magpie leave him in the tree.
Tomorrow a trip down the coast for her
and spaghetti rings left out for the bird.

At the dramatic level, this might be seen as another poignantly comic vignette of the poet as loser in the love stakes — 'high' on his art but deserted by the one he loves and desires. At the biographical level, it may refer to the end of his relationship with his Australian lover. For the reader, however, one of the main pleasures in this conclusion lies in its demonstration of the comic inadequacy of a 'high' art which remains out of touch with ground-level realities. Porter's magpie-like forays into questions such as Australian nationalism in this poem indicate that, whatever the disappointment, he would be unlikely to remain isolated in the tree of art, however tempting that might occasionally seem. In *English Subtitles* this temptation is most fully tested.

The buoyant tone of Porter's sonata for the Australian magpie was matched in 'My Old Cat Dances' (*ES, CP* 275), written early enough to be included in *The Cost of Seriousness* but held over for *English Subtitles* by the author and Oxford University Press's poetry editor, Jacqueline Simms, because it fitted the tone of that book better. 'My Old Cat Dances' is a lyrical celebration of Porter's cat's self-sufficiency in the face of human desperation:

In his lovely sufficiency
he will string up endless garlands
for the moon's deaf guardians.
Moving one paw out and yawning,
he closes his eyes. Everywhere
people are in despair. And he is dancing.

Such celebrations are not unusual in Porter's work. Since his early days in London flats he has kept cats which have appeared as 'characters' in his poems. In contrast to T. S. Eliot's somewhat condescendingly playful verse for children, in *Old Possum's Book of Practical Cats* (1940), Porter's cats inhabit an adult urban world of violence, anger, humour and fantasy. Aware of the role of cats in legend and witchcraft through the ages,[26] Porter also reveals a companionable understanding of their 'human' habits and ways; he never condescends to these animals, which have a fictional life of their own, as witnesses and co-respondents in the dramas of domesticity.

Cats had appeared in some of Porter's most successful poems since the early 1970s, including 'The Sadness of the Creatures' (*TLOE, CP* 136−7), 'The King of the Cats is Dead' (*PTTC, CP* 155−6) and 'Mort aux Chats' (*PTTC, CP* 184). In 'The Sadness of the Creatures' they were pictured as 'gentle predators' who witnessed the running down of a marriage and were resented for it: 'when the cats/come brushing for food their soft/aggression is hateful'. But 'The King of the Cats is Dead' presented an altogether different, more heroic aspect. Here the tomcat is praised in wonderfully lavish terms for his masculine powers and heroic feats:

> Between his back legs was a catapult
> of fecundity and he was riggish
> as a red-haired man. The girls
> of our nation felt him brush their legs
> when they were bored with telling rosaries −

And his heroic deeds against mice were the stuff of legend:

> When he stalked
> his momentary mice the land shook
> as though Atlantic waves were bowling
> at the western walls...

One can imagine the roll of drums off-stage as Porter winds up for the final lament:

> *The King of the Cats is Dead*
> and it
> is only Monday in the world

Such brilliantly orchestrated hyperbole works in two ways: to enhance its subject (the 'King of the Cats' may be perceived as an inspirational leader in any field), and to mock the excessive claims made for or by such subjects (fashionable poets, or politicians, could be in Porter's sights). 'Mort aux Chats', (*PTTC, CP* 184) on the other hand, is a satire of prejudice and the formulae it uses:

> I blame my headache and my
> plants dying on to cats.
> Our district is full of them,
> property values are falling.
> When I dream of God I see
> a Massacre of Cats...

The cats are here presented as a target for the old clichés of racialism, which are thereby ridiculed in a fresh and arresting way.

In a less rhetorical piece, 'Cat's Fugue' (*LIACC*, *CP* 228–9), Porter recalled Scarlatti's legendary cat, which the composer is said to have put on his keyboard, where it trod so gingerly that the sequence of notes was harmonically connected, giving Scarlatti the theme for one of his densest and most serious keyboard fugues. In this poem — which reintroduced Porter's favourite mad poet Christopher Smart and his cat Jeoffrey — Porter humorously observed the uses and abuses of cats in poems:

> Before you make your poem seem too twee
>   I'll warn you, said the cat,
> it's knowing when to stretto, how to keep
>   your counter-subjects simple,
> What to do when grandeur blows your mind —
>   also, you'll notice that my fur
> lies one way, so please don't brush it backwards
>   and call the act experiment.

If this is not quite Matthew Arnold's 'object in itself as it really is' hissing at the poet, Porter's poem does offer a good-humoured dialogue between poetic subject and object. However tempted he was at this time, it is difficult to imagine Porter disappearing into an experimental formalism which ignores 'the way things are' in the physical and solid world around him. Nevertheless, the cat's advice was a warning against the kinds of vanity and solipsism Porter explored in his next collaborative volume with Arthur Boyd, *Narcissus* (1984).

The two 'cat' poems in *English Subtitles* — 'My Old Cat Dances' and 'The Killing Ground' (*ES*, *CP* 288) — provide radically contrasting perspectives on a domestic scene. Whereas in the first, the cat remains oblivious, in his 'lovely sufficiency', of the unhappiness and desperation which surrounds him in the human world, the cats in these killing fields share a 'disputed territory' with husband and wife in their flat:

>   ...The air is lined
> with demarcations of despair.
> Each burly mote can tell
> how she sat there at last breakfast,
> night fumes in her dressing-gown,
> and how she said the sun upon the square
> was a massacre if only we could see
> the bodies. Then the cats walked straight to her
> with authority from the tomb,
> debating exits with their excellent tails.

The cats, witnesses of the woman's vision of a massacre in the calm London square, remain to observe the widower's visions of her returning ghost, bringing 'thunder to the keyboard' of his continuing life in this shared territory. Here they may observe, if they can, the speaker's chilling recognition that he has 'never shifted from the killing ground', a territory in which predatory humans and cats are at one.

The continuance of Jannice Porter as the key figure in Porter's poetic memory five to six years after her death is evidenced in major poems in *English Subtitles*, including 'Good Ghost, Gaunt Ghost' (*ES, CP* 281−2), 'Alcestis and the Poet' (*ES, CP* 292−3) and 'Talking to You Afterwards' (*ES, CP* 295−6). Such poems have an authority which gives them a dominant role in the book; they do not merely counterpoint other works which testify to a reawakening of sex and love. In 'Good Ghost, Gaunt Ghost', the immediacy of the ghostly presence is announced in the opening line: 'She is coming towards me'. Her approach invades all aspects of the speaker's life, including his dreams and the books he reads. No longer can she be externalized:

> . . .I see her as my hero-coward
> who has dared to be myself, erasing
> caution and suspicion. Soon I will be her
> and we shall keep creation to ourselves.

Gender differences are blurred in an imagined total merging of personalities through death. By contrast, 'Alcestis and the Poet' (*ES, CP* 292−3), like 'The Delegate', presents the dead wife as a separate identity speaking to her man through the filter of myth. The figure of Alcestis gives legendary shape to the contemporary story. In Euripides' play, Admetus, the husband of Alcestis, is presented first as an ingenuous egoist, who allows his wife to die for him. However, he returns from his wife's burial a changed man, having 'learned his lesson'.[27] In Euripides, Alcestis is a straightforward, unromantic woman who accepts as natural the duty of dying for her husband. In 'Alcestis and the Poet' the protagonist's questions are poignant and disturbing: 'will you/Find the inexplicable, out-of-reach-of-art/Intensity you mourn for outside Hades?' Death still beckons him with gently powerful insistence.

'Alcestis and the Poet' is interesting artistically, not only for its significant variations on the Alcestis legend but also for certain technical checks and balances. In keeping with Auden's dictum that 'every race should have its hurdle', Porter used nine words per line in this poem (with one exception), and his echoing of syntax and structures from Shakespeare and Milton broadened the poem's intertextual dimension. Such elements give the poem a baroque elaborateness, contrasting with the minimalism of 'Non Piangere, Liù' (*TCOS, CP* 267−8). But the poem's emotional impact, though broadened, is not diminished, as the following lines show:

> . . . Sited in great art, but tearful still,
> The creatures that we are make little gestures, then
> Go to nothing. The wind urges the trees to sigh
> For us: it is not a small thing to die,
> But looking back I see only the disappointed man
> Casting words upon the page. Was it for this
> I stepped out upon the stairs of death obediently?

The poignancy of the woman's imagined feelings is distanced but not lost in the echoes of Macbeth's nihilism and Milton's exit of the first lovers from Eden in Alcestis's speech. Art here builds upon art, expanding the range of reference without detriment to feeling, and the personal focus is expanded to include perceptions beyond those of the two private actors.

A more casually intimate set of effects was produced in the dramatic monologue 'Talking to You Afterwards' (*ES*, *CP* 295–6), in which Porter's speaker is presented conversing with his dead wife from a beach house in tropical Australia:

> Does my voice sound strange? I am sitting
> On a flat-roofed beach house watching lorikeets
> Flip among the scribble-gums and banksias.

None of the usual paraphernalia of traditional myth or spiritualist seance intervene here; the relaxed tone and syntax almost suggest a post-coital conversation. The poem was written in April 1977 when Porter was holidaying with his uncle at Noosa Heads on the Queensland coast and looking out towards Mount Tinbeerwah, and refers back to a stay at this house two years earlier when he was writing 'An Exequy'. As if in fulfilment of a pervasive sense of exile in his poetry, some of Porter's most successful poems were written at a considerable physical distance from the place in which they were experienced.

The Australian setting explicitly affected Porter's style of writing in 'Talking to You Afterwards': 'I can make words baroque but not here'. The admission is interesting in view of Porter's presentation of himself elsewhere as a homeless exile uninfluenced by place: his sense of place was in fact developing rapidly, due partly at least to his fresh encounters with Australia after two decades' absence. In the tropical beach setting, his persona is presented as relaxed and open; the source of his exile lies somewhere deeper within himself. His conversation with the wife who has preceded him to the other shore represents both a flirtation with and a challenge to the thanatophobia which surfaces recurrently in Porter's writings. Further back in his consciousness, as we have seen, is the recognition of himself as the sole survivor among his mother's multiple pregnancies, a person for whom life is fraught with reminders of extinction.

In mythical terms, Jannice Porter had left for 'The Winter Capital' (*ES*, *CP* 274), the dark underworld where Pluto took Proserpine from sunny Sicily. In the Australian sunlight, the departed wife and lover was wooed again in 'Talking to You Afterwards' in the medium of casual conversation, spiced with shared memories. Here a new aspect of the woman appeared: her English fear of the 'dreadful health' which, in her last months, Australia and Australians seemed to represent.

> Perhaps I had Australia in me and you thought
> Its dreadful health was your appointed accuser —
> The Adversary assumes strange shapes and accents.

She is also attributed with an instinctive recoil from names such as 'Mount Tinbeerwah', thereby casting her, perhaps unfairly, as brittle and 'English' in her rejection of 'uncultured' Australianisms. The speaker himself remains resolutely European in his perceptions of the landscape which he has seen from the top of Tinbeerwah:

> A plain of lakes and clearances and blue-green rinses,

> Which spoke to me of Rubens in the National Gallery
> Or even Patinir. The eyes that see into Australia
> Are, after all, European eyes . . .

The Europeanist perspective here is clearly Porter's, not his wife's, and so are the anti-nationalist gibes at subsidized Australian painters. Tensions remain in the poet's reconciliation with Australia, as he insists on the right to doubt and criticize from an outsider's perspective.

One of the achievements of 'Talking to You Afterwards' is its revelation of the ways in which words, silence and proximity may combine in a marriage to keep both partners at an emotional distance which neither has chosen:

> . . . We exchanged so few letters

> Being together so much. We both knew Chekhov on marriage.
> The unforgivable words are somewhere in a frozen space
> Of limbo . . .

The disenchanted realism of Chekhov's scenes of marriage, to which the poet seems attuned, contrasts with the popular prose fiction and diaries read by his wife. 'Carrington', whom the dead wife had 'loved', is Dora Carrington, Lytton Strachey's mistress, a self-sacrificing woman

who shot herself after Strachey's death. Jannice Porter had read Carrington's diaries avidly and she and Porter had seen a play at the
Phoenix Theatre in 1973, in which her relationship with Strachey was
dramatized. The concluding lines of 'Talking to You Afterwards' are in
a way Chekhovian. They suggest that all humans are predatory — like
the 'meat-eating bird' and the magpies in 'Sonata Form: The Australian
Magpie'. Life provides no second chances, so it is necessary to 'get
on with the business of living,/With only children for friends and
memories of love'. But the remote Australian setting has allowed him
the space to confront his loneliness and make new communion with
the one who has died; from here the 'party of the world' (a favourite
Porter metaphor), to which he is called, seems distinctly unattractive.

Porter's work from 1974 to 1981, after the initial intensity of
Jannice's death, reveals a persona torn between contending loyalties
and attractions: between prospects of a new love and memories of a
dead wife, between the country of birth and his adoptive country,
between living and dying. The predominant trope during this period is
that of returning. Porter's orientation as person and poet is to seek
sustenance in the past rather than place his faith in a brave new world
of the future. This orientation was evident even in his twenties, when
the past gave pungency to present desperation. In his late forties, there
were other futures to consider, and these could not be severed from the
past. In 'The Future' (*ES*, *CP* 290), written after Porter had seen some
frescoes by Luca Signorelli in Italy, he perceived the future as the
present frozen into art: 'We do not die, they say, but harden/into
frescoes'. Porter's poetic hypotheses of the self, the individual subject,
at this stage, are different from those of his twenties, when he could
indulge in rapid metamorphoses. This is not to accede to the notion of
a marmoreal self. Rather, in a period of critical reassessment of his
place in the world, Porter's contemplation of the future continually
drew him back to questions of continuity with an emotional past
involving, principally, his mother, father and wife. In this dynamic
process, the stillness of the fresco had an appeal not unlike that of
Keats's Grecian urn, in which the movement of the world was held
still in the contemplative act of art.

Coming to terms with the past was no mere intellectual exercise
for Porter at this time — it was a means of survival in a period of
disruption. In 'Returning' (*ES*, *CP* 280—1) he contemplated with wry
irony the tendency, after forty, to become a spectator of others' actions:

> Nobody feels well after his fortieth birthday
> But the convalescence is touched by glory
> So that history's truculent deeds of hate
> Are lived through in dreams, the story
> Followed to the investigator's hut, pain seen

> Through a window on its knees, late
> Help lost over marram dunes or never
> Felt at the deliverance on a screen.

Looking at one's own life:

> ...What you see
> Is coiled in an uneventful past, rough
> Justice of the body's failures, a commentary.

Lacking the audacity of childhood and youth, the middle-aged speaker ruminates upon survival in day-to-day living, surprising himself with an image of incandescence:

> ...He has learned how
> To live another day and wakes, ringed
> By the golden wallpaper of the sun.

This, then, would be the new territory of exploration, not apocalypse, but an incandescence of the ordinary.

The candour of Porter's ruminations on ageing is disarming. In 'Occam's Razor' (ES, CP 280), the imagery of departures, journeys and arrivals is contained in verse of such prosaic and relaxed casualness that the observations have entered the reader's consciousness before their full significance is felt:

> ...this long haul
> to a sort of maturity is nothing more
> than a persistence of arriving in someone
> with nowhere to go, rather than, as I once thought,
> an inveterate but soon to be vindicated
> earnest restlessness at the port of life,

The second stanza reflects on what is missing: madness (as in Ibsen) or magic (as in Haydn). However, what could lift the imagination from the salt pans of prose is a tantalizing apprehension of perfection, for which analogies can be found chiefly in a distant past. This notion was explored in 'The Imperfection of the World' (ES, CP 289):

> Yet we are haunted by our memory
> of perfection, of setting out among ferns
> for our father, the birds of the air
> moderating painful noon
> with their clamant cries.

The dreams which recur also recall the mother, who emerges in

'Myopia' (*ES, CP* 285—6) through the blurred figures of more recent
women, paintings, and mythical legends as the climactic emblem of
the grown man's search. In a poem which in some respects parallels
'Landscape with Orpheus', the lost mother in 'Myopia' returns, 'out of
sight of her son,/looking for the love which she was promised'. Such
dreams and fragments are hints of an unachieved perfection, and may
be awakened in unusual ways, such as by the sound of scratches on
a Bellini record failing to obscure the melodies behind them ('The
Imperfection of the World'). The tearful recognition that someone, or
some major element in one's life, is missing, hints at an underlying
dream of perfection:

> Therefore in the night when I cry
> that I have been deserted, I am
> once more made perfect.

It is difficult to conceive a reading of these lines which does not find
them laden with scarcely controlled emotion: rationalization is its
code.

Neo-romantics as different as Dylan Thomas, Robert Lowell and
Ted Hughes might have used the traumas and crises of this period of
Porter's life in different, more spectacularly egotistical ways. What got
in the way of 'letting go' for Porter, in the face of his grief, pain and
confusion, was an almost stoical sense of responsibility; and it was one
of the triumphs of his poetry at this time to allow strength of feeling to
emerge in the struggle with this barrier. Nevertheless, some dangers
are evident. If the danger of 'letting it all hang out' in poetry dissipates
its effect, concealing the emotions can sometimes smother them. Porter
confronted these kinds of problems in his poetic development in 'The
Werther Level' (*ES, CP* 296—7) and 'What I Have Written I Have
Written' (*ES, CP* 286—7). 'The Werther Level' alludes to scenes in
Goethe's popular *Sturm und Drang* novel, *The Sorrows of Young
Werther* (1774), in which the young hero kills himself out of hopeless
love. In one of his few autobiographical references to the 'brief Bohemian days' of his early years in London and his relationship with Jill
Neville, Porter recalls his 'tied-up, fitful twenties'. At fifty, elements of
this younger self persist, rehearsing a desperation not unlike that of the
young Peter Mainchance in 'The Losing Chance'. Women friends tell
him that his body is now a 'better prospect' than it was in his twenties,
and that he has every reason to compensate a neglected Romanticism.
But the young Werther's words are themselves a surprising check on
melodramatic actions: 'Do not escape responsibility/for living'. Though
drawn to such extremes, the speaker knows he must 'wear his suffering'
and make what style he can of that. He does not see himself as a star
turn of the *Sturm und Drang*.

The question of personal responsibility was broached most powerfully at this time in 'What I Have Written I Have Written'. Porter here seems to endorse the capacity to stand by a perceived truth under pressure, but the allusion to Pontius Pilate indicates a more problematical concern — Pilate condemned Jesus to death in spite of considering him innocent. While Porter's poem never refers directly to Pilate, the title, repeated as the last line, confirms this figure as a hidden presence in the poem, a truthful traitor with whom the poet identifies. Curiously, the other figure lying behind Porter's allegorical persona in this piece is Samuel Beckett's tramp, from the novel *Molloy* (1951).[28] In a brilliant scene which parodies mathematical reasoning, Beckett's Molloy picks up sixteen 'sucking' stones from the beach and considers different ways of managing them, 'not haphazard, but with method'. Finally, he throws all but one away, 'which I kept now in one pocket, now in another, and which of course I soon lost, or threw away, or gave away, or swallowed'.[29] Porter's speaker keeps his stone by him: 'It is the little stone of unhappiness/which I keep with me'. Whatever the allusions to other literary figures, Porter's own signature is firmly inscribed on this stone.

In a drained, mesmeric monologue, Porter's speaker in 'What I Have Written I Have Written' observes that the stone of unhappiness has been his companion since childhood. He put it in a drawer where, across the years of growing to young manhood, it was joined by many letters and poems. (Since adolescence, Porter has been a prolific writer, and recipient, of letters.) Then came the central crisis:

> I gave the stone to a woman
> and it glowed. I set my mind
> to hydraulic work, lifting words
> from their swamp. In the light from the stone
> her face was bloated. When she died
> the stone returned to me, a present
> from reality . . .

In the context of Porter's life, the reference here to his wife's dissolution and death are clear enough. But the device of the stone and the minimalist technique of pared-down vocabulary and simple syntax allow the poet to focus on essential elements in the speaker's life-history.

As in other poems, an aphoristic utterance in 'What I Have Written I Have Written' suddenly emerges as epiphany: 'Duty/is better than love, it suffers no betrayal'. The recognition stands out like a crag from the images of growing children and gardens which surround it in the poem. As Dorothy Green observed, the question of responsibility which was raised in the early poem about schooldays, 'The View from Misfortune's Back' (*PAAM, CP* 57–9), is answered here with 'a Roman

dignity'.[30] It would be unlike Porter to allow his alter ego an heroic
Roman posture without self-mockery however; hence the alliance with
Beckett's tramp and the self-disparaging image of the paranoid poet
who discovers that his name has been expunged from 'the Honourable
Company of Scribes', and who feels that the books in his room turn
their backs on him. The concluding movement recalls the poem's
opening in its mood of grave, though not earnest, dignity:

> Old age will be the stone and me together.
> I have become used to its weight
> in my pocket and my brain.
> To move it from lining to lining
> like Beckett's tramp,
> to modulate it to the major
> or throw it at the public —
> all is of no avail. But I'll add
> to the songs of the stone. These words
> I take from my religious instruction,
> complete responsibility —
> let them be entered in the record,
> What I have written I have written.

This piece is a moving testimony to Porter's growing commitment to
poetry which records the 'songs of the stone' of suffering, whatever the
cost.

Some readers and critics would deny that a poetry of suffering can
be as valid a form of art as that which pronounces optimism and joy.
Porter addressed this question in 'At Lake Massaciuccoli' (*ES, CP*
306–7), a poem which explores the writer's love of Puccini's operas
with a view to understanding better his own role as a poet. Elsewhere,
Porter has commented on the 'provincialism' of Puccini, with which
he identifies: 'I think of Puccini's melancholy and the formal perfection of his structures as characteristically Tuscan. He was a provincial and so am I'.[31] The designation of himself as a provincial
would puzzle many Australian readers who see Porter as a suave
metropolitan. But a growing recognition of himself as an exile of the
spirit led Porter to identify with other provincial eccentrics, including
writers from the provinces in Britain and elsewhere. Such artists find
their vision in their own surroundings. Puccini's vision is perceived
by Porter as influenced by the lake in north-western Italy beside which
he lived (Torre del Lago, which runs into Lake Massaciuccoli):

> He was not so soft: what he saw
> Was this lake made into the world — not to be
> Changed or pitied but crying through the night
> Abandoning life for love.

Out of such melancholy and intense need, Porter suggests, the most beautiful music may be made.

The concluding poem in *English Subtitles*, 'Landscape with Orpheus' (*ES*, *CP* 309–10) links music and painting with the dominant trope in this volume, that of returning. It is a baroque poem, richly layered and, as Evan Jones observed, 'loosely autobiographical' in its rendering of images from Porter's childhood experience.[32] As in 'Alcestis and the Poet', the framework of mythology allows personal experience to be generalized: Orpheus is a composite figure, of mythical and modern elements, and of the personal experience of the author. Less significant as a 'character' than as a structural device, Orpheus is seen as the figure through which dreams and visions can be filtered.

The modern story suggests the legend of Orpheus, whose descent to the underworld to bring back his lost wife Eurydice would be successful only if he did not turn to see her until they reached daylight. With many hesitations and reversions, and frozen 'stills', the spools of childhood are re-played as the reader of Porter's oeuvre is taken back to magic images of ferries on the Lane Cove River:

> The camera is rewound and there is the old latch,
> The gate, the pepperina tree, the ferry rounding
> Onions Point...

The ferry is ambiguously both an image of childhood pleasure and the legendary boat which takes passengers across the Styx to Hades. Will the boy-man take the ferry? 'The future must be crowded into now,/ Paradise and hell on deck'. But looking through the telescope, 'The Town Hall clock shows Orpheus looking back'. It is a reminder of Orpheus's fate, and Porter's fatal compulsion. As Lemprière noted, Orpheus 'forgot his promises, and turned back to look at his long-lost Eurydice. He saw her, but she instantly vanished from his eyes. He attempted to follow her but was refused admission; and the only comfort he could find was to soothe his grief at the sound of his musical instrument'.[33] This is the peril of looking back; the final and only pleasure is the soulful music. In this concluding poem in *English Subtitles* Porter again dramatized the limbo in which he found himself. Screening his imaginary film, he reconsidered his own past and looked towards a beckoning but unknowable future, with 'Paradise and hell on deck'.

10

BEYOND APOCALYPSE

◆

Peter Porter's star was ascendant during the 1980s. His *Collected Poems* (1983) won the Duff Cooper Memorial Prize. The publication of *Fast Forward* (1984), *Narcissus* (1984, with Arthur Boyd), *The Automatic Oracle* (1987) and *Mars* (1988, with Arthur Boyd), followed. An eleventh volume, *Possible Worlds*, was published in 1989, the year of Porter's sixtieth birthday, which was also commemorated by his publisher, Oxford University Press, with *A Porter Selected: Poems 1959–1989*. Porter was awarded the Whitbread Prize for poetry for *The Automatic Oracle* in 1988, and *Possible Worlds*, like a number of the previous publications, was a Poetry Book Society recommendation. Although overshadowed by the sales figures of some Faber poets, Porter was Oxford University Press's best selling poet, and the paperback *Collected Poems* was reprinted in 1988. Porter was awarded honorary Doctor of Letters degrees at the University of Melbourne (1985), and at Loughborough University (1987), in recognition of his work. Masters and doctoral theses at universities in Britain, Europe and Australia were being written on his work. Throughout the 1980s, Porter remained poetry reviewer for the *Observer* and was a regular commentator and critic on literary and cultural matters in Britain and, to a lesser extent, in Australia. He was in frequent demand at conferences, literary festivals and readings in Britain, Europe, North America and Australia. Porter strengthened his connections with Australia by accepting invitations as writer-in-residence at the University of Melbourne in 1983 and the University of Western Australia in 1987, and managed visits to his native country at almost annual intervals. In 1990, in his home city, Porter was awarded Australia's oldest and most prestigious literary award, the Gold Medal of the Australian Literature Society. Significant developments also occurred in Porter's personal life, and these will be considered later in relation to the ways in which his writing developed during the 1980s.

The publication of the *Collected Poems* in 1983 provided the first major opportunity for readers and reviewers to take stock of his remarkable and prolific achievement. The volume contained 316 poems — nearly everything Porter had published in book form from *Once Bitten, Twice Bitten* (1961) to *English Subtitles* (1981). Only twelve poems were omitted from the collection on the grounds that they had 'proved to contain mistakes or untruths' or had seemed to exceed 'even the most relaxed bounds of proprietary tolerance'.[1] Unlike Auden, Yeats and other predecessors who were concerned to rewrite their canon, Porter chose at this time to let his work stand in its virtual entirety, feeling loath to 'act the prig' to the 'gauche or cocky past self' he found in his earlier work.[2] It was therefore a large and inevitably uneven corpus that the critics considered.

In the event, Porter's *Collected Poems* received a very positive press in the newspapers and periodicals of Britain, Australia and, to a lesser extent, North America. Praise was especially strong from reviewers who were also poets, who together presented a picture of a writer of imaginative verve and poetic inventiveness, with an unusual capacity to open up the whole field of poetic activity for other writers. They perceived a poet with a strong grounding in traditional poetry and ideas, and an orientation towards the past, but one who was also fully alive to contemporary ideas and styles. Critics compared the magnitude and significance of his achievement to that of Ted Hughes, Philip Larkin, Seamus Heaney and Geoffrey Hill in Britain, and to A. D. Hope, Judith Wright and Les A. Murray in Australia.

The veteran poet and critic Stephen Spender, after noting the intoxicating power of Porter's language, acutely discerned in his poetry 'a wakeful deliberate dreaming, a controlled fantasizing'.[3] Spender's assertion that 'when Porter seems nearest to writing confessional poetry (which I am sure he despises) it would be wrong to assume that he is writing about himself and not inventing fictions to fit himself into', is a useful corrective to naive assumptions of poetry as straight autobiography. Having commented on Porter's 'fine lines' and brisk rhythms, Spender concluded with the quandary of Porter's attractive 'strangeness' for British readers:

> The feeling of strangeness underlying these poems is difficult to analyse. Perhaps it is Australian. I think of those Australian painters from Nolan on, lonely figures, painting landscapes in which wooden shacks float on light reflected from the desert as though they were standing on stilts. Perhaps it is the Australian light which keeps this lively, witty, allusive, sometimes mysterious poetry a few feet above the ground.[4]

The question of Porter's Australian origins and orientation which Spender's early review raised was taken up by other reviewers in conflicting ways, indicating a variety of critical predispositions as well as ambivalent feelings about the national 'ownership' of creative writers, especially expatriates. Unlike Spender, London poet and former Group member Alan Brownjohn located Porter firmly in a British-European setting, arguing that the image of an 'uprooted outsider writing somehow at a tangent to English literature and English life' was mistaken.[5] Porter had arrived in England, Brownjohn contended, with 'a much more informed, complex and ironical view of his native Australia' than the 'uprooted outsider' image allowed:

> No-one could have been less of an ambitious settler-in-reverse,
> coming with the intention of forcing his image upon the town.
> Porter came loaded with a whole cluster of ironies which he
> was to carry and refine throughout thirty years of an extreme
> commitment to the value and seriousness of poetry.[6]

As previous discussion has shown, this view was a fairly accurate
summation of Porter during the 1950s and 1960s, but it failed to
account for qualitative changes wrought by his return visits to Australia,
when he confronted old ghosts and revivified imaginative possibilities.
While Brownjohn observed in the *Collected Poems* a quick adaptation
to British and European society, Les A. Murray was struck by 'the
sheer number of Australian passages, inflections, references and
counterpointings, as well as the number of poems actually set in this
country'.[7] Murray correctly noted that such poems occurred in two
main groups — early in Porter's career and in the mid and late seventies,
when Australia had become 'much more congenial' to him. With all
the zest of an advocate, though Porter is a poet with whom he disagrees on
most issues, Murray urged Australians to take this expatriate more
fully into their national consciousness. But just as Brownjohn found
the British and European Porter easier to identify than the Australian,
so Murray fell into cliché when he described the 'English' element in
Porter's verse in terms of 'Sloane Ranger conversation' and 'a more
serious and intelligent Bertie Wooster in anguish'. Neither of these
flippant images does justice to the complexity of Porter's relationship
to England, which he has both loved and loved to hate.

Another Australian reviewer, the poet and academic Vincent
Buckley, who traced his own roots to Ireland, presented Porter as 'not
really at home outside England'.[8] Buckley failed to substantiate this
claim, suggesting an 'at home-ness' which was at best an ambivalent
metropolitan restlessness. The English landscape poems of the mid to
late 1970s, on which Buckley hinged his contention, were few, and
expressed a transient and by no means untroubled accommodation to
English landscape-as-history. Robert Gray's essay on Porter also tended
to play down the Australian element in his work, commenting that less
than thirty pieces in the *Collected Poems* were 'about Australia',[9] but
not taking into account, as Murray had, the many 'inflections, references
and counterpointings' which, like the Americanisms in Eliot's poetry,
characterize his work. Gray did however consider the reasons for
Porter's sense of incompatibility with pre-existing Australian poetic
traditions and conventions in the early post-war years. These traditions
in Gray's view were threefold: landscape or descriptive writing; pro-
letarian and socially critical verse; and 'nouveau-classicizing' (the
'vitalist' revival via Norman Lindsay and others). Gray was right in

suggesting that Porter was attracted to none of these — even the 'nouveau-classicizing', though seeming at first glance to be compatible, was, in its Australian variant, too muscular, optimistic and over-blown.[10]

These arguments do not dispel the possibility that, had Porter stayed in Australia, he may have developed in some different and important way as a poet. Porter may indeed have discovered a com-patibility with some of the urban poets who began writing in the 1950s, especially in Melbourne, including Buckley, R. A. Simpson, Bruce Dawe and Chris Wallace-Crabbe, whose later work he came to admire greatly. However, he would scarcely have developed in the directions he did. As Julian Croft has suggested:

> More than any other Australian poet, Porter combines with ease and fluency both the traditions of early modernism and the directions taken by English poetry in the 1930s and 1940s. His debt to W. H. Auden is evident, as is debt to the anti-romantic social commentary of the Movement, but in his com-bination of a sceptical distrust of any system and a sense of the individual's powerlessness he is an Australian modernist poet, as he is in his belief that there is a redeeming force in human affairs — and for Porter, that force is art.[11]

Porter's expatriatism, and exposure to a particular set of experi-ences and influences in Britain and Europe, contributed to his special role in Australian literary history as a 'late modernist'. Although some of his later work employs modernist and postmodernist methods (even while making cracks at them), his sense of what A. N. Whitehead called the 'witness of the past' links him with previous traditions in English language poetry, especially the Jacobeans, the Augustans and the Victorians. More significantly, perhaps, his debts were to individual poets of the past, including Shakespeare, Pope, Byron, Hardy and Browning. Porter's *Collected Poems* showed him to be one of the few contemporary poets who had learnt to play the whole poetic keyboard, but whose own preoccupations and obsessions gave his work its unique signature. If, for Australian literary history, he is one of the 'necessary exiles',[12] from another perspective he is out on his own, creating his own tradition as an émigré who mocks, mourns or exults in a fate which he perceives to be inherent in his nature.

The fate of such an exile at the hands of critics may be to fall outside all convenient schools, traditions or categories. The relative lack of attention given to Porter's work in North America for example, may indicate this. And Thomas D'Evelyn, who praised the 'gift of phrase' and ability to mix levels of discourse in the *Collected Poems*, found himself at sea with Porter's Australian origins and influences: 'Mr Porter is an Australian. What is the poetic tradition of Australia?

That's like asking, "What is the poetic tradition of Bakersfield?"[13] Sometimes Americans rejected all national affiliations as meaningless, as did expatriate Australian Patrick Hanrahan in his review for *The Daily American*: 'Porter sees the world in a Porter-like way because he is Porter. Also, when he writes on Australia, he sees Australia in a Porter-like way as well'.[14] American ignorance of Australian culture was only beginning to be rectified, chiefly at an academic level, in the 1980s.

Apart from the question of national or regional qualities in Porter's writing, the major critical issue addressed by reviewers of the *Collected Poems* was the presentation of self in his work, and its changes or developments. The issue was tackled in different ways. Peter Levi, for instance, seemed delighted to find that the characteristic quality in Porter's work was its humour, for which he conjectured two principal sources. The first was ancient Roman, 'full of grit and gravel and sudden baroque soarings and bespattered with vinegar'. The second source was Australian, 'the abrasive, ground-level humour of the Australian armies of 1914 and 1939 ... still alive and kicking like a mule'.[15] Levi contended that Porter had injected an 'alarming vitality' into the contemporary English poetry scene: Geoffrey Hill might be 'deeper', Ted Hughes 'more exclusively carnivorous', but 'there are more laughs in Peter Porter'. This, Levi asserted, was 'an aspect of his seriousness, his refusal of dishonesty'.[16]

A problem pointed out by some reviewers was the lack of a 'central self' around which they could hang their responses. Dick Davis, for example, complained of an 'absence of vision'. 'Where does the poet stand?' Davis asked, commenting that this poet was 'profoundly given to having his cake and eating it too'.[17] Christopher Reid was critical of the 'determinedly protean' scope of Porter's *Collected Poems*: 'The word "I" abounds in the work, but, with his fondness for dislocated scenarios and quirky narrative devices, he has consistently made it hard for the reader to identify a true confessional voice'.[18] Blake Morrison reiterated this line of cross-questioning in his long and overall sympathetic review in the *Times Literary Supplement*: 'Where is the true poet among his many masks? Canonisation demands a canon; recognition requires that there be something to recognise — a voice, a method or allegiance'.[19]

Australian reviewers seemed less worried about Porter's profligacy of talent and changeable personality. While conceding that Porter remained a protean figure in his work, Evan Jones noted an 'unabashedly personal voice' occurring first in 'The Sadness of the Creatures', written when Porter was about forty.[20] The key to development in Porter's poetry, in Jones's view, was a *loss* of self-consciousness, contrasting with the heightened self-consciousness in other major twentieth-century poets such as Yeats and Lowell.[21] Unlike those critics who praised Porter's 'public' verse, Jones drew a distinction between the poems whose structure was simply 'the rhetoric meant to wow an audience *out there*' and the poems 'crafted for himself with love'.[22] Jamie Grant also observed a development, beginning in *The Last of England*, in Porter's capacity to 'write about his own feelings'.[23] In qualitative terms however, Chris Wallace-Crabbe found little development in Porter's work: he was just as good in the 1980s as he was at the start of his career as a publishing poet.[24]

Change and development was observed by some critics in the ethical and 'religious' dimensions of Porter's work. Dorothy Green saw in the *Collected Poems* 'a religious debate that puts a convincing case for man', couched in fresh, energetic language modulating into 'a haunting, elegiac music'.[25] Douglas Dunn also discovered 'a near-religious feeling in Porter's seriousness which, if vexed and doubtful, is present for all that'.[26] Like Green and other critics, Dunn saw Porter's elegies for his wife as the apex of his achievement, but also noted a development through his work from 'satirical particularity' to 'a general tough-mindedness' and maturity of self and mind.[27] The issue of 'near-religious' belief in Porter's work and the question of changes of vision are particularly interesting in the light of his work in the 1980s, and are discussed later in this chapter.

While many critics and commentators preferred to concentrate on the self and private vision in Porter's work, there were some who emphasized a powerful interaction between the self and society, with an emphasis on 'the world' in and outside the poems. For John Lucas, the 'social poet' was at the centre of Porter's achievement, he was Britain's 'best satirist of *moeurs contemporaines*, with a 'disenchanted, tough relish' for the world he observed.[28] Martin Booth stressed Porter's continuing relevance as a socio-political poet in Britain. While disappointed at the number of 'poets' pieces' in the *Collected Poems*, Booth considered Porter 'a major figure and one to be read in order to see how society looks at itself in our present times'.[29] Alan Bold, editor of the *Penguin Book of Socialist Verse*, crystallized what he saw as Porter's social vision: he was a 'natural romantic' who had 'controlled his lyrical impulses with a satirical severity'. In Bold's view, Porter's vision was essentially reconstructive, representing a search for 'grace' in a 'grotesque world'.[30] The socio-political element in Porter's outlook was accurately discerned by such critics as democratic, but this perception was not followed through; while sharply critical of tyranny from above, Porter was also critical of the tyrannies of mass culture. He wanted democracy to be large enough to include doubters, sceptics and critics and to prevent the relegation of 'high art' to the status of an upper-class activity.

The present study aims to show that Porter's social poetry cannot be separated as neatly from his personal poetry as some of these reviews and commentaries might suggest — world, author and text are in a process of continual and dynamic interaction. Never one to baulk in his self-criticism, Porter has himself reflected on the tenuous and changing balance between these elements in his life and work:

> I may not know much about this pushy pronoun "I", the great asserter and maker of a way for itself in the world. And I can justify it still less readily. But its fellow pronoun "me" has received some exalting messages from the world beyond it in dreams, in other people's art and artefacts, in the assertion of its birthright of feeling and knowing ... I do not know who I am; I have poor recall of what happened to me; I have been in exile from myself all my life; but I am part of a world nexus which is extraordinary ... a given, not a chosen domain.[31]

Having taken stock of Porter's poetry, and critical responses to it up until 1981, let us turn to a consideration of the oscillating relations between author, world and text in Porter's career throughout the 1980s.

◆

The early years of the 1980s were punctuated for Porter by several important emotional breaks with the past, and so new beginnings. Since much of his best work deals with the perils and rewards of intimacy in human relationships, the significance of his father's death in 1982, and the end of his seven-year liaison with Sally McInerney in 1980, followed by a serious relationship with Christine Berg, should not be underestimated.

William Porter died at a Brisbane nursing home, three months short of his ninety-seventh birthday. In 'Where We Came In' (*FF* 29) Porter recalled visiting this home, shortly after his father's death, to collect his few possessions. It was a Brisbane summer, 'at the time of jacaranda falling':

> here, where I was born, the cycle
> not yet complete but estrangement
> Made absolute by time.

The cycle of personal history is shaken by no spectacular self-revelation; the son is afforded a typically low-keyed apocalypse:

> At last I was alone with incandescence
> and did not question the mystery,
> The son was now the father.

Porter's preoccupation here, as elsewhere, was not with material 'possessions' (these were meagre anyway), but with what he might be possessed by. His father's death represented 'another heat shield gone'; the cosmonaut had one less protective layer against the ultimate fire towards which he was proceeding. This realization itself provides a moment of 'incandescence', a lighting-up of the spirit on its flight towards death.

There is an ironic disparity between the emotional impact of losing one parent when nine years old and the other when fifty-three. Porter's mother's death clearly marked his emotional life more intensely than his father's much later one. Even in his fifties, Porter could make an unconscious slip when referring to his mother's death by saying, 'When I died ...'.[32] The young boy's strong identification with the mother seems to have created in his subconscious a sense that he was responsible for her death. To have expressed some anger towards the father, which he no doubt felt over subsequent events, might also, in a boy's subconscious, have compounded his felony in killing the mother. This anger may have been repressed and displaced in different forms of aggression. Irresistible as such psychological conjecturing is, it should be seen in the light of an ambivalently expressed love and respect for this father, estranged by distance and time, who had always, in the son's eyes, been 'kept waiting'.

If in his Brisbane garden the father was transmuted to an Adam figure, then his son was Cain. Unlike Byron, Porter does not flaunt this role, yet it is hinted at in the remorse at deaths and departures, remembered or imagined, and in the recognition of murderous impulses. William Porter had a restorative role in his son's cosmogony. He was the one who chose not to enlist in the first world war, and whose example might have accounted for his son's participation in anti-nuclear demonstrations in the 1950s. An imaginative fascination with and revulsion at war is apparent in *Mars*, one of Porter's major books of the 1980s. The father figure in 'Tobias and the Angel' (*OBTB*, *CP* 37) also exhibits the unusually 'ordinary' virtues of honesty, decency and undemonstrative love, which the son hears in the 'sad music' of his conscience during his first years in London. The incandescent recognition that 'the son was now the father' was therefore tinged with subtly conflicting emotions.

Another important ending for Porter in these years was the conclusion of his close involvement with Sally McInerney. The literary evidence, in poems such as 'Roman Incident' and 'An Australian Garden', indicates her emotional significance and role in Porter's reconciliation with Australia.

'The Decline of the North' (*FF* 11) hints at the difficulties in their relationship. The first of the two eight-line stanzas takes some of its details from 'Spring Forest' near Koorawatha, the family farm of Ross and Olive McInerney, Sally's parents, in central New South Wales. Porter had visited Koorawatha with Sally and his daughters in 1975, and without his daughters in 1977. The details of this place were employed by Porter to quite different effect from Geoffrey Lehmann's use of them in *Ross' Poems* (1978), in which Ross McInerney's memories and reflections merge with the poet's own. *Ross' Poems* contributes to a view of Australian country life which Les A. Murray has called the 'vernacular republic', in which Australians are bonded by their inheritance of rural attitudes and values. In contrast with this, Porter's bush in 'The Decline of the North' appears as a biblically derived exemplum; in a sense, it is a libretto for the desert, wherein silence, dust and sacrifice are evoked:

> Decibels of silence fill the day
> When belling dogs and creek are tired at heart:
> A kingdom comes with dust, a slaughtered sheep
> Hangs from a river oak, the text of life.

What this 'text of life' implies becomes more apparent in the second stanza, as the speaker addresses an unidentified other, suggesting a disputation within the self, as in some of Donne's poems or sermons:

> Your heresy is in new starts, that wheels
> And economics travel latitudes
> Across the bays of hope. Home might be anywhere,

The 'heresy' presented seems to be that the self is not bound by its past, and may be continually reconstructed.

In an interview in Australia in 1984, Porter commented more generally on the power of continuities in human nature, including his scepticism about 'new starts': 'My obsessions are with the impossibility of new starts, the conviction that the human spirit is likely to be what it is forever'.[33] This 'obsession' was tested, revised, challenged and reimagined in different ways in Porter's poetry throughout the 1980s.

In spite of such scepticism, Porter did experience a 'new start' of a kind in the early 1980s in his relationship with Londoner Christine Berg. Fifteen years younger than Porter, Berg was a divorced mother with two daughters, living in north London. She was a lecturer in the Extramural Department of Paddington College and had completed a Bachelor of Arts degree at Essex University. The two met at a party held by Porter's old friend Susanna Roper, who taught with Christine Berg at Paddington College. After several years of commuting between her place and Porter's, Berg moved into the flat at 42 Cleveland Square, which Porter still rented, in 1986. She accompanied Porter on travels to Europe and on his writer-in-residence assignment in Western Australia in 1987.

Christine Berg's influence on Porter appears to have been intellectual as well as emotional: her interests in feminism, psychoanalysis and literary theory (including deconstructionism and versions of

Marxist and Freudian theory) all seem to have been absorbed by Porter's 'oceanic' mind.[34] Berg's love for and loyalty to the two daughters she had brought up alone after her divorce deeply impressed Porter; her experience formed a continuity with his own. Berg was unlike other women in Porter's life. Vivacious and eloquent, she was from an Irish Catholic working-class family. She had left home at eighteen and married Michael Berg, an English businessman of Jewish background. She entered, for several years, the world of London Jewish intellectuals, including psychoanalysts in the Freudian and post-Freudian traditions. This interest persisted and in 1989 she commenced as a trainee child psychoanalyst at the Tavistock Clinic in London.

One of Porter's light verse pieces, 'Legs on Wheels' (*TAO* 27−8), affectionately recalls Christine Berg driving her Morris Mini, in the first year of their relationship, between her place in north London and Porter's in Cleveland Square. In this poem, the poet reflects with lightly satiric humour on the contradictions, the gaps between theory and practice, of those humans who like to consider themselves the 'Master Species'. This species:

> . . .lays out gardens,
> uses up hydrocarbons,
> has forgotten how to walk
> but stuffs itself with talk,
> jumps into the car
> if it has to go a yard
> to get the cigarettes
> and hoards like nuts regrets.
> A challenge to the seasons,
> it has its special reasons
> for poisoning itself
> with recipes for health,
> but though it screws the planet
> from Andaman to Thanet
> it dreams a Green Revival
> and sponsors the survival
> of all attractive creatures
> (O hide your ugly features,
> you vultures, snakes and rats
> and purr you pussy cats!)

The poem captures something of the good-humoured, witty argumentativeness in conversations between Porter and Berg on a wide variety of topics. Here, among other things, Porter questions the theories, motives and practice of proponents of the 'Green Revival': he is as vigilant an interrogator and critic of the intellectually lazy Left as of the rampant Right.

However, a larger contention lies behind this poem's tour de force of images and rhythmical variations. Throughout the 1980s Porter, like Berg, was a strong critic of the abuse of Darwin's 'survival of the fittest' theory of evolution by governments such as Margaret Thatcher's in Britain, to justify the devaluation of the weak, and helpless. Characteristically, Porter's poem does not remain at the level of polemical condemnation or indignation. The inventive adaptation of serious thoughts to the light verse form suggests a pleasure, and even a degree of complicity, in this game of survival in big-city living:

> May we avoid disaster,
> escaping ever faster
> from all the patent deaths
> (Aids, cancer, madness, meths),
> rejoice we swirl on wheels
> instead of our bare heels
> and changing down for climbing
> perfect our human timing,
> unnatural in our daring,
> quite naturally despairing
> of just our own devices
> but clever in a crisis
> show evolution that
> we are the where it's at.

The poem's 'message' — the hubris of individuals, governments or others who use simplistic myths or theories to justify despicable actions — is almost lost in the simulated 'game' of survival. From one point of view, this is a failure on the poet's part, an over-subtlety whereby he does not proclaim his beliefs directly. But Porter's subtlety is more insidiously effective than polemical directness, suggesting that only by allowing oneself an imaginative complicity in other people's games can we appreciate their content. Only in the awkward, skidding halt of the poem's final line are we fully reminded that hubristic games can have ridiculous ends. Thus, although the poem mocks the pretensions of humans who, lost in their own importance, place themselves above nature, Porter nevertheless relishes the opportunity to challenge the limits of possibility in his own way. There is something in this of the spiritedness of new challenges, a note which recurs throughout his first two volumes of the 1980s, *Fast Forward* (1984) and *The Automatic Oracle* (1987).

◆

Porter's first new book of poems after *English Subtitles* (1981), *Fast Forward* (1984), was dedicated to Clive James. This volume, coming

one year after the *Collected Poems*, appeared as a personal statement
about getting on with things after the years of introversion, as well as a
recognition of the urgencies of the video-technology world in which
Porter's friend and fellow Australian expatriate had made his mark.
More significantly in the *roman fleuve* of Porter's verse, the title of the
new book suggested a counter-tendency to the principal orientation of
'Landscape with Orpheus', the last poem of *English Subtitles*, in which
the Porter-Orpheus figure looks back towards a lost world. Yet looking
forward must always be an ambivalent act for Porter, as fraught with
threats as with possibilities. It is not surprising to find, therefore, that
although *Fast Forward* is packed with an awareness of contemporary
idioms, manners, mores and beliefs, Porter presents a humorously self-
deprecating image of himself as 'Stuffy, old signaller/of sickness' ('At
the Porta Humana', *FF* 9). In a present age of instant answers and
visions, the humanist is bound to seem 'stuffy', reminding people
of the imperfection of the species and the certain death of every
individual. Utopian futures will be brought to earth by such humanists,
in a world where the poetry of visions and the prose of reality are in
continual and dynamic contention.

The most significant 'new start' in *Fast Forward* was not, therefore, an engagement with technological progress, but with questions of
religious belief and their human consequences. In this volume the
'residual Christian' in Porter continued to employ the term 'God',
assuming that anyone brought up in a culture connected with two
thousand years of Christian inheritance could hardly avoid using the
word and feeling something of its force. However, he used the term
without believing either in a Creator who manifested Himself on earth
at one particular time, or that such a Being (if He/She did exist) would
answer particular needs and wishes.[35] But, as Chris Wallace-Crabbe
observed, Porter revealed another approach to religion in this volume,
'speaking repeatedly about "the gods", a tactic which whisks him
clean past Thomas Hardy junction'.[36] One stimulus to Porter's rethink-
ing of these matters was a conference he attended in Cyprus in 1981.
Taking into account his Mediterranean location and audience, Porter
gave a talk on literary representations of humans and the gods:

> I tried to run through various ways and stages of thinking
> about the gods, beginning with Lucretius's utterly indifferent
> gods to Auden's poem 'The Shield of Achilles' about the
> irrelevance of the gods because of the evil of man, and con-
> cluding with Hölderlin's 'Die Linien des Lebens', where the
> gods are transcendant. Between the Auden and Hölderlin, I
> slipped in one of my own called 'The Missionary Position'.[37]

'The Missionary Position' (*FF* 13) presented the gods, in Porter's
words, as 'decent, unimaginative, hardworking civil servants'.[38] In the

poem, they are in a queue, lined up and sighing for 'ordinariness,/To bring the brilliant seed of the future down/To an average hearth'. What these gods desire is not the exotic, transcendant moment, but all that was implied in the sexual discourse of the 1970s and early 1980s by 'the missionary position' — in the words of the poem, the kind of relationship which exists not with the stars but 'among the rugs and rules/Of generations, removed from graphs of passion'. Porter's poem indulges the humorous and ironic conjecture that while earth-bound individuals imagine that 'the stiff sky is always on the point of breaking/And intervention, above the roof of a shearer's hut/Or in a shower of gold' is awaited, the gods themselves provocatively prefer to 'Serve their chosen companionship face to face,/The decency and boredom of diurnal love'. The poem's spatial orientations of vertical and horizontal are nicely poised, denying neither the reality of apocalyptic hopes nor the necessity of ordinariness. The subtext is the capacity of humans to create gods according to their desires and needs.

A similar capacity is suggested in 'The Flock and the Star' (*FF* 12), in which the poet places himself and his readers as complicit artists in the recreation of paintings or engravings such as those by Blake and Palmer. In the active role he assigns to himself and his readers in relation to these emblematic paintings, Porter suggests an analogue of humanity's relationship to myths of the creator:

Take to this picture, God,
Consider the thing you are —
The unguideable flock,
The painstaking star.

Although the syntax and punctuation here suggest ambiguous possibilities, it seems that, as in other poems deriving from Porter's contemplation of the visual arts (for example, 'Paradis Artificiel', *TAO* 19), he was stressing the reader/observer's active role as creator of new meanings. Such an attitude may have been influenced in part by the rise of 'reader response' theories in literary studies, to which Porter would have been exposed through conversations with friends and colleagues in British and Australian universities. Their premise is that the poem is remade by the reader. But like 'The Missionary Position', 'The Flock and the Star' takes up a classic agnostic's position: fully aware that humans are continually drawn to the 'shower of gold' or the 'star', Porter is also painfully aware that most of us are denizens of an 'average hearth', that together we are an 'unguideable flock'. This recognition is not a cause of despair for the poet; on the contrary, it gives scope to the active powers of imagination in a middle realm between heaven and hell. Porter here assumes that God is an invention of the human imagination. As an artist who has no special access to the truth of human existence, Porter allows that there will be a plurality of 'gods' (or 'Gods') according to the needs and desires of the human subjects who create them.

What is rejected here is the kind of mysticism with which the young Porter had temporarily flirted in unpublished poems such as 'Some Questions for Thomas à Kempis'. (But even there, under Auden's influence, he was adopting a humanist posture: 'I am my own desire and find my Heaven so'.) Resisting mysticism, whose human exemplars he has generally found dull, Porter turned instead in the early 1980s to the anthropomorphized gods of Mediterranean culture, who first entered his consciousness in his ancient history classes at Toowoomba Grammar School. What was he turning his back on? Looking further east than the Mediterranean, this inheritor of Anglo-Australian Protestant pragmatism felt uneasy with religions such as Hinduism and Taoism, and was inclined to ridicule the fashionable western pose of the 1960s and 1970s of 'sitting around, in lotus positions, looking for inspiration or hoping that the divine afflatus will come fluttering down'.[39] This was not just a facile crack at a popular movement; it arose from a considered view of the value of rational intellection, however imperfect, and a fear of 'letting go'. Porter has recalled several blazings of temper at school, when he was teased by other boys, which drove him to the edge of 'homicidal rage'.[40] From these and other similar experiences in his twenties, Porter concluded that losing one's self-control was perilous, that 'ecstasies' (of joy, of religious illumination, of anger) were 'fearfully transfiguring and dangerous things . . . not very far from death'.[41] Death is of course the central and recurring fear which drives his imaginative impulses; and letting go, in mystical or other forms of ecstasy, is an analogue, and a prelude, to dying. At the same time, Porter recognized that non-rational activity was the stuff of life, that the imagination can soar beyond the facts, that the dream-life and religious experience can be major realities in one's life.

A possible resolution to this conflict was offered by the down-to-earth gods of Mediterranean culture, in which art and literature tell us that these gods and goddesses once walked. In 'Cyprus, Aeschylus, Inanition' (FF 39−40), for instance, a (Clive) Jamesian vision of Aphrodite at Paphos in Cyprus is evoked, at the spot where myth tells us she was born from the foam of the sea:

> Moon landscapes of kaolin hills, framed by
> Sexual froth where the light-stepping goddess
> Walked up the beach like a starlet at Cannes −

While modern 'miracles' of this kind can occur, so too can the terrible anti-climaxes of tourism, such as the tee-shirts at Paphos printed with 'postures of fucking' while 'the silent amphitheatre sings to the sea'. More lively than the somewhat predictable verse journalism of this piece was the 'found' poem, 'A Guide to the Gods' (*FF* 43), which begins:

> You must recall that here they are intense,
> Our gods. Each corner shop may need oblations
> So the genius rests content. Kick a dog turd
> Leftways to the gutter, but only if it be
> Dry and crumbling. Every third display of okra
> On the footpath must be stolen from ...

Porter's practice here suggests that gods which cannot be spoken of irreverently and humorously do not qualify for a place in his pantheon at this stage; and that their chief revelation is about the humans who construct them.

Yet Porter could not let matters rest there. Through the filter of other writers' styles and patterns of belief, he questioned further the consequences of the disbandment of traditional religious belief in his own times. Echoing Racine's long lines and rolling cadences in 'A Vein of Racine's' (*FF* 53−4), for example, Porter's speaker regrets the reductive ordinariness of life without sustaining belief and observes that, 'As in the spent Renaissance, life turns to the ideal'. Yet this gap, or need, is an outcome of the humanism which he himself has supported: 'Everything and everyone was clangorous for honesty,/The world became one Salon des Refusées'. As representative outsider seeking to belong somewhere in late twentieth-century society and culture, Porter observed the recurrent dilemma of his species, who were driven to reconstruct ideals as a reaction against the fallen world: 'Always there was elsewhere,/The Golden Age, the Innocentest Isthmus, a land on/Stalks beyond the eyes of youth'. 'Innocentest' here mimics the child's view without undermining it. The mature adult's perceptions are nevertheless corrosively honest. A democratic rationalist spirit, for which Porter's speaker and his kind are also responsible, has levelled the playing-field. One of the 'Proverbs of Democracy' reflects what has happened: 'You will put the gods inside you and make death their king,/You will know your nervous system a bland oracle'. But as the storm of destruction gathers, a small hope resurges in the impulse of individuals towards goals beyond themselves: 'The youth his girl friend, the warrior/His tall challenge, the poet his obsession in the head'. This is not a soft or gratuitous hope, for the poet also acknowledges the prevalence of power and circumstance, the destruction of the good and the great in a world, like Racine's, of tragic folly and blindness.

The dreamers in Porter's work are often exile figures, as in the title poem 'Fast Forward' (*FF* 59), which begins with an exclamation: 'The view from Patmos, the ghost inside/the module!' Patmos is the small

Greek island where John, presumptive author of the Fourth Gospel and the Book of Revelations, was said to have been exiled by the Romans in about A.D. 95. Using anachronism — as dreamers, visionaries and poets rather than historians do — Porter's opening lines link John's revelations with the kind of apocalypse which late twentieth-century observers have attached to moon travel. In doing so, Porter demonstrates what 'fast forward' can mean. It is one of several demonstrations in the poem of 'the tape so stretched it might at any second/snap to oblivion'. Here Porter indicates the pressure and urgency of his tight-packed imagination as it makes rapid 'future shock' connections across vast space and time, like some lightning-fast, inter-galactic communications network.

Another image from modern technology suggests the amazing capacity of the human intelligence to land where it wants to:

> Left to itself, the brain, circuiting the world,
> becomes a rapid deployment force
> and blasts ashore on any troubled sand.

The military image, and counters to it, are taken up elsewhere in this rapidly changing poem. For instance, the Greenham Common women who protested against nuclear stations in Britain are imaged in those who calm themselves by 'crooning/death songs to the mush of midday heat' as they sit with 'a little health-food lunch/among the flowers of a military estate'. (Porter's use of the first-person plural pronoun 'we' indicates an empathetic identification with their protestive vision of a world on the verge of destruction.)

The state of 'wakeful dreaming' which Stephen Spender observed in Porter's poetry is powerfully evident in 'Fast Forward', which qualifies as one of Porter's most 'difficult' pieces, with its anachronisms, rapid spatial dislocations, ellipses, syntactic contortions and sometimes abstract vocabulary. In addition, the poem offers no single personal voice for the reader to accept or reject. But these difficulties indicate a broader problem: a de-centred world out there of intolerably complex and contradictory forces. Is Porter right to include so much of it? Or should he have adopted the minimalist mode of some key poems in The Cost of Seriousness? It seems though that the difficulties of this poem contribute in the main to a rich and complex view of imaginative possibility, from which real politics and human weaknesses are not omitted. For instance, the poem suggests that early visionaries were right in assuming that power and madness are closely allied: 'that the countenance of state,/rock-featured and as bright as Caesar's eyes,/ floats on insane shoulders'. It is a view with which many Romanians in 1989 would have agreed.

How will this world end? Porter goes well beyond T. S. Eliot's formulation, 'not with a bang but a whimper', in his recreation of an

'Apocalpse Now' which emerges from the Vietnam conflict, and its representations in the media:

> Where will the end be staged, and whose hand,
> held against the light, shall glow, brighter
> than a thousand suns? Helicoptering
> in lightning numbers, the god of prose
> hears his own voice prophesying peace.

This build up to a modern, dystopic version of John's revelation of the Horsemen of the Apocalypse ends anti-climactically; it suggests nothing as magic as the world of Coleridge's 'Kubla Khan' (which is echoed in the poem's last line), but recalls instead the hypocrisy of declarations — whether by Lyndon Johnson, Ronald Reagan or others — of peace, when war is meant. The recognition of war as a site of revelation, which Porter explored later in *Mars* (1988), contributes to the poem's mood of fertile despair. The stripping away of illusions in Porter's work, here as elsewhere, was presented as a necessary pre-condition to truth.

Although poems about the human construction of gods, visions and beliefs represented a major advance in Porter's poetry in *Fast Forward*, a suite of more personal poems in the shadow of Jannice Porter's death also appeared. Completed some eight to ten years after her death, they represent a mix of moods and styles in her last major appearance in Porter's poetry, though her memory returns, on occasion, in later poems. 'Dejection: An Ode' (*FF* 30–1), working, like a musical score, off Coleridge's poem of the same name, dramatizes the process of seeking a purity of mood in dejection:

> Just turn the mind off for a moment
> to let the inner silence flow into itself —
> this is the beauty of dejection, as if our unimaginable death
> were free of the collapse of heart and liver

But Porter is not a poet to let the mind turn off for long, or to allow himself or his readers to sustain illusions about the unadulterated purity of visionary moments: this is his realism, which includes bowels, liver and the organic heart. Betrayals of truth, beauty and goodness are everywhere around us, his poetry suggests, not least one's own. Even truth must be questioned as an ultimate good:

> What has the truth done to our children's room?
> The toys are scattered, the pillow damp with crying,
> chiefly the light is poor and no-one comes
> all afternoon: *Meermädchen* of the swamp of mind.

Against the temptation to wallow in sentimental memories is set the much more sharply focussed 'Doll's House' (*FF* 23), one of the best poems arising from Porter's total revaluation of the purpose of living following his wife's death. Evan Jones has justly praised this poem for its 'masterful use of simple, strict form'.[42] The alternately rhymed quatrains in predominantly iambic meter, with a short last line, give a clipped precision to a story Porter's readers have heard before, but which gains resonance in this piece by its explicit parallel with Ibsen's play *A Doll's House* (1879). Porter takes up the feminist slant of the play thereby giving the aching memory of his wife's departure a new dimension: 'Its owner slammed the door and fled/Like Nora to the liberal hinterland'. The poem cinematically dissolves into images of children with their miniature house for dolls, which they (dissolve again) have outgrown. Eight years after Jannice Porter's death,[43] Porter had a clearer awareness that he could no longer consider himself the remorseful protector of wife and daughters: they would, and had, made their own decisions, chosen their own fates. Apart from its rhythmic control and precision of detail, Porter's 'Doll's House' is remarkable for its evocative spatial imagery, culminating in the recognition that love has 'Liliputianized the scale of pain'. However, the mood of bereavement still seems at times the most suitable accommodation for him. A mood which 'has no passage into afternoons/But through diminished doors and hush/Of darkened rooms' is a place of temporary belonging outside the 'liberal hinterland' whence Jannice/ Nora has fled, in her act of liberation from the restrictive house she had inhabited.

A reduction in the scale of pain through time and distance was recognized also in 'Elegy and Fanfare' (*FF* 27-8). The autobiographical speaker in this poem recalls a visit to a cemetery in Cambridgeshire after attending the races at Newmarket. He regrets that he can neither feel his wife's presence as a Rilkean angel, nor cry. Echoing Auden's famous phrase in 'The Shield of Achilles' — 'and no help came' — he recalls 'and no tear came/to swing the afternoon up to my face'. A conclusion of kinds is reached. In having relived the death of his wife for eight years, the poet feels the Rilkean angelic wings and a rhetoric:

> ... booming
> and blooming in himself, most precious
> of his properties, as, starred with vigilance,
> he looked down the unrecurring light
> to death. In his hell his heaven blazed.

These were the fanfares of his interiorized angels, the chief inspiration of his writing:

> For me they were the English ends of hope
> from which I now reluctantly retreat

This hesitant and reluctant retreat promised no 'new starts' or 'bays of hope', but signalled a necessary move forward, emotionally and imaginatively.

In this restless period when endings were occurring, the exile figure in Porter's verse explored the antithetical need for community, the scope for belonging. In 'Venetian Incident' (*FF* 25−6), the speaker's companion in Venice, based on Jannice Porter, makes herself 'at home' when she joins in the rituals of a high-church Anglican service in the foreign city. The speaker, based on Porter himself, recalls being pleased at her spontaneous capacity to let herself go and belong, however temporarily, to this English-language community: 'You were home,/a devoted unbeliever but English and real/I a frightened sceptic'. The play on 'home' here is a reminder of one of the most recurrent and potent words in Porter's poetic vocabulary. His persona decides, perversely and erratically, that God is pleased with his wife's actions, for 'even if he doesn't exist,/he likes us to try to belong'. He compares his wife's lightness of being at the church in Venice, as at her death, with his own obstinate self-control, and concludes that he and those like him are 'death's real retainers'. The poem concludes with a reflection upon the 'joy' (an unusual word in Porter's work) which this transient image of belonging has given him. The changes of mood are finely conveyed.

Most often, the representative cameo is one of exclusion. 'Santa Cecilia in Trastevere' (*FF* 44−5), for example, shows the speaker, 'a tired pilgrim', locked out of his favourite saint's chapel, where he hears only 'the music of exclusion,/gates locked upon flowers and fountain'. In 'The Cats of Campagnatico' (*FF* 41−2), it is the cats who mark out their territory in this small village in Tuscany, a hundred kilometres south of Florence, where fellow Australian expatriate author David Malouf lived, and where Porter visited him in 1981. More generally, Italy is presented in this poem as a country which may be claimed through the force of love: 'we are/As international as an opera festival,/We who love Italy. We have no home/And come from nowhere, a marvellous patrimony'. For one who felt his life was unchosen, the *choice* to love Italy, both rationally and emotionally, as a visitor, was significant: here, it was possible to have a 'vision of belonging', like the cats of Campagnatico. The concluding lines are marvellously playful and lyrically resonant, with that comic seriousness which is one of Porter's main legacies: 'between the fur of the cat/ And the cement of extinction, there are only/Cypress moments lingering and the long tray of the sky'.

The capacity to connect vitally to the world without final commitment to a fixed abode was persuasively dramatized in 'Marianne North's Submission' (*FF* 37−8), one of Porter's most successful monologues of the 1980s. The poem's speaker is nineteenth-century British botanist and diarist Marianne North, who travelled through Asia, Australia, North America and Europe making drawings and paintings of flowers, plants and bushes which were subsequently housed at Kew Gardens in London in 1881. Marianne North's journal was edited by her sister, Mrs John Addington Symonds, and published as *Recollections of a Happy Life*.[44] But Porter is typically less interested in botanic detail than in dramatizing the temperament and world view of this 'character', who emerges as one of the charitable, and in her way artistic, figures of Britain's Victorian empire. As Porter's persona declared in the earlier piece, 'Seaside Resort' (*PTTC, CP* 163−5), he is in two minds about the British Empire and does not condemn it outright, as has become fashionable.

Marianne North, like 'the plant collectors of Udaipur' in 'Seaside Resort', is treated sympathetically. Her tendency to view the world as a collection of flower gardens was epitomized in her description of Albany in Western Australia: 'The whole country was a natural flower-garden and one could wander for miles and miles among the bushes and never meet a soul'.[45] Porter's method was to graft onto a figure derived from historical accounts his own perceptions, theories and hypotheses. Although an historical purist or classical realist might argue with this kind of disruption of verisimilitude or the 'coherence' of a character, anyone sympathetic to postmodern subjectivism and fragmentation would not argue with an artist's right to do so. What Porter focussed on in this poem, though, was an exploratory and accepting view of the world, which the late twentieth century makes almost impossible: Marianne North submits her accumulated views on life to its conditions. In doing so, she raises some recognizably Porterian worries at this time, such as the human construction of God, Heaven and Hell, fear as a source of human understanding, and the involuntary nature of life. 'I cannot remember when I consented/To be born', she says disarmingly, and 'If there is a Heaven/It's because there must be something/We have moved away from'. Late in her monologue she confides a secret life behind her air of blithe Victorian optimism:

> ...To any watcher,
> Including God, I am a middle-aged lady
> Of the dauntless British sort feeling not
> Very well in a Burmese swamp — there she goes,
> You say, escaping the Victorian Sick Room
> But not evading fear. Do you know
> My furthest journey has been away from suicide?

Porter's imaginative delineation of a similarly evasive journey of his allegorical persona in work ranging from the unpublished play *The Losing Chance* to 'The Werther Level' (*ES, CP* 296−7), might seem oddly transposed in this apparently intrepid British Victorian lady. But her buoyant optimism and dark subterranean currents were projected from a sense of similar contradictions in the author himself. Above all she had her art, which could contain and even transform despair by a concentration of vision upon its 'objective correlatives' — in her case, the 'unreflective strugglers,/Vines, stamens, steamed-open lilies'.

The generally perfunctory critical treatment given to such poems in *Fast Forward* was probably a result of its appearance less than a year after the widely reviewed and acclaimed *Collected Poems*. The author's self-conscious search for new forms, different ways of expressing nuances of thought and feeling appropriate to his current concerns may also explain this. In this respect, *Fast Forward* — and to a greater extent *The Automatic Oracle* — represent a rush of experiment and change not unlike that which had occurred in *A Porter Folio* (1969) after the early successes of the 1960s, and the establishment in the poetry-reading public's mind of a 'representative' Porter. Porter had again failed to stand still for the memorialists. But continuities were evident. Evan Jones, who had schooled himself in Porter's work, saw a development in 'more firmly shaped' poems than in earlier work, less personal obscurity and arbitrariness of cultural reference.[46] Lachlan Mackinnon preferred the more straightforward poems to those which strove for 'the unintelligibility of dreams'.[47] Lawrence Lerner preferred a 'deeply personal' poem such as 'Clipboard' (*FF* 21) to what he dubbed 'poststructuralist poetry' which refused to 'come out with connections' and was 'unwilling to occupy the rational territory of the coherent subject'.[48] Charles Boyle, on the other hand, welcomed Porter's growing interest in postmodernist practice and theory. He discerned the 'liberating effect' of John Ashbery on Porter's work, with its demonstration that poetry 'does not have to make complete sense'.[49]

Certainly, a number of poems in *Fast Forward* flout accepted forms of logic and resist reduction to fixed referents. In 'Little Harmonic Labyrinth' (*FF* 48–9), as the title suggests, a verbal equivalent to a musical exercise is sought, in which Porter's speaker flashes a miscellany of images and ideas on the screen and asserts, teasingly, that 'the flight from meaning is our magic'. Chris Wallace-Crabbe applauded the direction taken in this poem and in 'Jumping to Conclusions' (*FF* 16–17), which have 'neither a demonstrable subject, a place, nor even an event to record'. In such poems, Wallace-Crabbe remarked, 'the speaking voice carries on in existential space, piling phrases together or linking them in chains, often in apposition with one another. His syntax is lovely — he orchestrates it like Ravel — but it is not governed by logic, or not by any traditional logic'.[50] While social reformers and even psychoanalytic critics would find little to engage them in these poems, such works have a place in Porter's output as *jeux d'esprit* which reveal a pleasurable randomness in the operations of the human spirit. Furthermore, they show that this 'serious' poet's reflections are seldom earnest, and that some 'accidents' of imagination may be more consequential than what is planned. In 'Dis Manibus' (*FF* 51−2) Porter's autobiographical persona remarks ironically: 'The theory is that what we write/mimics what we feel/and that's worth a prisoner's laugh'. From this stance no simple relationship should be theorized by readers between the words on the page and the feelings of their author. In this phase of self-conscious questioning of theory and practice, Porter was preparing for more audacious forays in his next volume into questions such as the relationship between language and the writer and their combined capacity for truth, deception and imaginative play.

◆

Although most critics had been caught napping by new turns of thought and developments in the exploration of religion, language and communities of belief in *Fast Forward*, the more perceptive of them did observe significant changes in Porter's eleventh volume, *The Automatic Oracle* (1987). Dedicated to Christine Berg, this collection was influenced in part by discussions with her of 'radical' literary, social and psychoanalytic theory, and by conversations with academic colleagues and friends in Australia and Britain. Despite evidence of such influences, in the new volume Porter expressed scepticism of the increasing specialization and abstraction of university studies of English, and reiterated his view of the value of a pluralistic metropolitan critical practice and a poetry free of domination by university professionals.

The *Automatic Oracle* won the Whitbread Award for Poetry. On 9 November 1988 the *Times* observed in soccer parlance that Porter was a 'First Division poet', whose latest collection was 'pre-occupied with our times, and their politics and dangers'.

The critics found their feet again. Sean O'Brien saw Porter 'accelerating away from the potential millstone of his *Collected Poems*', with poems which matched the best of earlier periods.[51] John Lucas found 'a new — or at the very least, renewed — playfulness' in *The Automatic Oracle*.[52] Tim Dooley remarked that, while embracing the contradictions and uncertainties of what he had called 'late modernity', Porter had 'gradually but steadily rehabilitated for himself the poet's roles as seer and spokesman'.[53] All three observations are valid, but the latter is especially interesting and contentious, and deserves closer investigation. Can a poet who has rejected the bardic and populist claims of Ted Hughes or Les A. Murray be considered a 'seer' or 'spokesperson'? In spite of (and even as a result of) his habit of self-deprecation and steadfast scepticism about inflated claims for humanity, Porter had achieved this status, in his unique way.

In *The Automatic Oracle*, Porter emerged as one of the great resistance fighters against false prophecies. His targets included prime ministers, the aristocracy, bureaucrats, businesspeople, academics, lawyers, poets and disc jockeys. Oracular utterances from such people should be treated with suspicion, he stated provocatively. In some respects, this may seem a reversion to his role as critic of admass, commercialism and class discrimination in the 1960s; and something of the energetic, resistant spirit of that decade certainly recurs in Porter's work of this period. However, an altogether more assured and worldly vision had been achieved in the interim, which was also disarmingly self-deprecatory. The change is evident in the contrast between an early poem, 'The Historians Call Up Pain' (*OBTB, CP* 30), and one twenty-six years later, 'Paradise Park' (*TAO* 20). In the earlier piece, the brittle desperation of the young expatriate in London emerges as a somewhat querulous self-pity: his ultimate appeal is to 'our loneliness/And our regret with which to build an eschatology'. A quarter-century on, a more expansive, worldly and less self-pitying tone emerges in a resonant evocation of a world 'beyond Apocalypse':

First Worlds and Third Worlds are equally fatigued
Importing and exporting barely describable butterflies
And alloys of the perfect miracle: trade goes on
Beyond Apocalypse and the dying and the hopeless,
Hurdles of misrule leapt lightly by the Press.

In spite of its gestures towards historicism, the first of these poems was plaintively personal; in the second, a world view which somehow incorporated both the metaphors of 'trade' and 'the perfect miracle' had been attained. No longer could the writer indulge in a simple 'regret', and from that construct an eschatology. The ways of the modern world were tough and ruthless, and to be confronted. International capitalism was the basis of world trade and popular understanding was dominated by a mass media which demonstrated a massive disregard for those with whom Porter identified and pitied, 'the dying and the hopeless'. (In another poem he wrote, 'We fume and Rupert Murdoch's none the worse'.) Yet his habit now was not to retreat into subjectivity, history or myth; nor was he inclined to condemn capitalism and the modern industrial state outright as a general evil against nature. Something of the Brisbane warehouseman's son remained. But so too did the idealist who dreamed of 'barely describable butterflies' and 'the perfect miracle'. The world was unutterably mixed and the thinking person's poet was bound to convey this.

The role of the marginalized thinker (intellectual is too grandiose a term for Porter), who understands the *realpolitik* of his times and makes occasional forays into that territory to sting its actors into an awareness of their folly or stupidity appealed to Porter. Exiles may be passive or active. Porter's spirited response to his inherent sense of exile was allegorized in the poem 'Spiderwise' (*TAO* 56–9), dedicated to fellow Australian expatriate Clive James, whose work as a writer of verse and prose (quite apart from his television work) Porter admired.

'Spiderwise' expands on, and revises, folkloric wisdom arising from the tall tales of British visitors or settlers in Australia. Michael Wilding, an English expatriate in Australia, wrote in his story 'As Boys to Wanton Flies' of spiders 'as large as a dinner plate' which haunt the nightmares of an English boy in his new country.[54] This gigantism is paralleled in Clive James's humorous reinventions of encounters with funnel-web and red-back spiders in his *Unreliable Memoirs*.[55] Peter Porter's poem 'Redback' also plays on such expectations. Less concerned with referential meanings, events or places, however, Porter brings to the fore the process of a poet's search for appropriate metaphors: the redback is not a 'natural' signifier for anything; it must be constructed by choosing from an enormous range of conventions and possibilities. Porter's choice is to capitalize his Redback, as a figure in allegory, and give it a narrative frame and argument. In the rhetoric of history the Redback becomes Porter's image of the terrorist who subverts conventional images of the 'normality' of Sydney Cove, where Australia's European history is said to have begun. In praising his 'naturally malevolent activist' the poet presents himself as a revisionist historian who overturns superficial and optimistic notions of Australia as untrammelled Arcadia:

> I'd like to think that he was waiting there
> When that initial cargo of strong rum
> And human refuse hit Australia's shore.
> To flog, to dream, to endlessly explore
> Might make Arcadia of a rural slum:
> The Redback gave the hedonists a scare.

For all its comic treatment, Porter's 'Redback' presented a quite savage view of Australian origins, akin to Robert Hughes's imaginative history of Australian convictism, *The Fatal Shore* (1987). In Porter's view, it was preferable to understand the white Australians' past — an outlook which was also implicit in the early poem, 'Sydney Cove, 1788' (*PAAM, CP* 50–1). An understanding of this horrific past would provide a foundation for national self-knowledge; better, certainly, than inflated hopes and idealism. By recognizing the dark side of their society's nature, Australians could commence a process of amelioration and restoration. The underlying moral tale here was similar to Porter's personal allegory as it has developed through his work: that the worst should be known, laid bare and judgement made, in a long slow haul towards restoration and selfhood. Rejection of the facile optimism of a Pangloss is an essential first step. The concluding lines of 'Redback' expose the sting in Porter's alternative spider-philosophy: his antipodean Red Cross Knight will ensure that 'Relieving nature won't be comfortable'. Typically, the pun has serious intent: the path towards civilization, sanity, peace and justice will be fraught with discomforts and difficulties. One may have to bite, sting, even poison parts of the body politic to return it to a proper liberal democratic course. This 'seer' or 'spokesperson' would retain the spider's prerogative to hide in the darkness and sting his opponents in unexpected places.

An important aspect of Porter's preferred body politic (he is not a utopianist) is its pluralism and diversity: a monolithic, conformist state is anathema. Hence his early interest in Australia's multicultural dimensions — the Chinese market gardeners and European refugees, and music from Germany and Italy in Brisbane's monocultural climate of the early post-war years. In the 1980s this pleasure in diversity revealed itself not only in the subjects of his poetry but also in its forms and styles. Porter's experiments in a variety of verse forms — a project which he had most explicitly widened in *A Porter Folio* (1969) — show a refusal to be hemmed in by current fashions. In *The Automatic Oracle* he exhibited a burst of enthusiasm for widening his repertoire, with new verse forms such as the triolet and the pantoum.

With a perky postmodernist self-consciousness, allied to a wry historical self-awareness unfortunately not common among critics, Porter invited readers to join him in 'Try a Triolet' (*TAO* 44–5): 'The nearly mindless triolet,/the only form that Auden shirked'. Porter gave some 'mind' to his triolet in the second of his four experiments, when he ironically scrutinized Sartre's contention that 'Hell is other people': 'The gaping O of self becomes a nought,/yet Hell is other people, Sartre thought'. The arrogance of the existential externalization of hell is thus exposed. By contrast, Porter's hell begins with himself. But this reader's pleasure lies also in enjoying the poet's negotiation of the French fixed form of eight lines, with only two rhymes and a thrice-repeated opening line. Porter meets the challenges of the form: the intricate repetition seems natural and inevitable, and he successfully negotiates shifts of emphasis within it. An additional pleasure lies in observing the tables turned on Sartre in a poetic form from his own language. All of this occurs with a deft touch which manages to avoid glibness, though it is difficult to avoid a hint of smugness, as in the successful negotiation of steps in an intricate dance. The form is particularly successful in the third triolet, which uses the repetitions to dramatize a preoccupation: 'Heaney, Hughes and Hill and Harrison — top poets' names begin with H'. Porter's sense of separation from success on the poetry scene in Britain is stylized into humour.

The pantoum was another challenge which Porter took up in *The Automatic Oracle*, and again he demonstrated formidable skill and an appearance of ease, if not grace. The 'Pantoum of the Opera' (*TAO* 64) is a wonderfully inventive fantasia around Andrew Lloyd Webber's musical 'The Phantom of the Opera', whose sentimental emotionalism is mocked by contrast with other operas, and rejected in the final stanza: 'Say No to Heavenly Takeaway'. A major source of enjoyment in the poem is the negotiation of the pantoum form, which derived from the Malayan *pantun* and was established in French by Victor Hugo. The pantoum is composed of quatrains in which the second and fourth lines of each stanza serve as the first and third lines of the next. Different themes must be developed concurrently, one in the first two lines and the other in the last two lines of each quatrain, the two pairs of lines being connected only by their sound.[56] Porter's poem builds a vivid picture of an emotionally disfigured opera lover who fills out his life by vicarious involvement in famous operatic dramas. The final stanza in a pantoum must end with the first line of the poem and, in some versions, the third line of the poem recurs as the second line of the concluding quatrain. Porter adopts both conventions in his concluding stanza:

> Say No to Heavenly Takeaway; it's clear
> Susannahs he has watched cannot replace
> the mother-love that fled from him in fear.
> Life has thrown its acid in his face.

For all its skilful deployment of poetic form, Porter's personal signature is heavily inscribed in this poem, reverting as it does, self-mockingly, to the chief source of his autobiographical persona's emotional deprivation and need for completion, 'the mother-love that fled from him in fear'. The poem gives another perspective on the powerful underlying emotionalism of the personal allegory in Porter's work, with its story of loss and compensatory creative activity.

Porter's response to the claims of postmodernist theory and practice was both responsive and resistant in *The Automatic Oracle*. He responded with zestful interest to aspects of the theory, but resisted takeovers. A wry comfort was found in notions of 'the death of the author', that the words on the page and not biography or history should be a poet's legacy. As Christopher Pollnitz pointed out in one of the most perceptive studies of Porter's poetry in the 1980s, Porter has listened thoughtfully to the fashionable academic theories of the decade, especially to the deconstructionists' tenets that language writes the poet and that the reader creates the poem.[57] Richard Machin and Christopher Norris forwarded the claims of poststructuralists more generally: 'The source of meaning always used to be an author. But it might be the reader (just another author), language (the medium as the message), or ideology (a mixture of all three)'.[58]

Through all of this, Porter discerned the major issue as the puzzle of subjectivity. In 'Sticking to the Text' (*TAO* 6–7), he posed the problem:

> Keeping ahead of death and Deconstruction
> We have the text we need to play the game,

But what should we do to make it personal —
Your text, my text — are they the same?

Although this poem appears to celebrate the circularity of language in itself, it ends by questioning this, as well as the endless deferrals of meaning and the ambiguities which poststructuralist readings are designed to perpetrate. In the end, Porter seems to support the ordinary reader's wish for a person-in-the-text, an identifiable author behind the various 'author-effects'. Yet this is no simplistic reversion to the notion of the poet speaking directly through the text to an audience. For one thing, the writer may adapt the poem according to his or her sense of what seems 'relevant' to a particular audience. In 'Tipp-Ex for the Oscar' (*TAO* 23—4), the poet who writes according to audience expectations is mocked: 'O guide me God/and make me choose the words which flatter best/the passeggiata of the relevant'.

The individual ego is the direct source of art, Porter suggests in 'Paradise Park' (*TAO* 20), the artist may choose to abandon his or her art, but that would not be to escape egotism. Furthermore, although the author may 'camouflage' himself, as Porter indicates in 'A Tribute to My Enemies' (*TAO* 15—6), he remains part of a war-game: the battle is between his own sense of truthfulness and those critics who urge him to preach positive messages, 'the towering/Yea-preachers whose laser scorn could track/Me down through all my dark of triteness/With the cry, "You, especially you, don't matter"'. But this author will not renege on taking responsibility for his words, what he has written he will stand by. Mysteries of authorship remain nevertheless, as he indicates in 'Throw the Book at Them' (*TAO* 8). The relationship of words to experience is often indirect and full of surprises: 'Proust could get ten thousand lines from/one night at a party and Robert Browning/knew he was in love only when he found he'd/said so on the page'. In the knowledge that writing can be pro-active, Porter is still not prepared to surround it, or himself as poet, in a cloudy mystique (the 'Claud of Unknowing', as he puts it elsewhere). The puzzling relationship of reader, writer and the world is cleverly summed up in 'Who Needs It?' (*TAO* 29—31):

We can't write well unless we think
Of something unapproached by ink
    And yet we're caught in rules
    Of Writing/Reading schools.

The outlook here is typically inclusive, recognizing the need for imaginative freedom and its actual limits. The multiplicity of voices and rhetorical modes in *The Automatic Oracle* perform what some deconstructionists have laboured to argue — in Christopher Norris's words, 'the endless displacement of meaning which governs language and places it forever beyond the reach of a stable, self-authenticating knowledge'.[59] This view is not anathema to Porter: as his poetry continually demonstrates, the old certainty of a single authoritative voice (in church, state or the institution of literary studies) is lost, and good riddance; pretence, betrayal and dissimulation abound and are part of any adult view of the world. Plainness is a linguistic device as much as sophistication. Porter's own baroque experiments with language, with their sudden changes of register, accommodate a cosmopolis of voices and styles. Yet by practice and persuasion, he enters a stubborn plea for the personal voice in poetry.

In 'Disc Horse' (*TAO* 49–50), the spinning record of sound parodically simulates the circulation of 'discourse', of postmodernist jargon and the obfuscations of the eighties:

> . . .Round and round
> go galaxies of talk and everyone knows
> not to intrude on anyone else's space
> for fear that no one then will have the power
> to recuperate the narrative.

Porter accepts, temporarily, for the purposes of this poem, that the modern technological and material world has, as Malcolm Bradbury put it, 'driven the subject into system, and the self into silence', leaving parody as the only appropriate art form for the age.[60]

When the individual, sentient poet is forced by the technologies, politics and fashions of the day into 'automatic pilot', what happens? The sound poetry imitated in 'Disc Horse' is one possibility: 'Zgtf-pxxlrjkdxrhgggkk ... ooof'. Another is parody, conscious or unconscious, as in Hamlet's famous soliloquy misremembered, or reinvented: 'What a work of peace is a man,/how golden in season,/how fond of the infinitive ...'. The claims of postmodernist and other theorists are pushed towards absurdity: 'These texts mean nothing, even Shakespeare's,/it's the shape the theory makes that counts'. While this is a potentially lethal shaft aimed at some literary theorists, Porter does not dismiss the diagnosis of the intellectual and literary environment which the modernist writer/critics and their successors, the postmodernists, have expressed in their arts. Bradbury encapsulated this attitude as 'a rage against this world, shaped by a haunting sense of fragmentation and loss, of waste and exposure and alienation', in which artists seem 'troubled by silences and absences, a consciousness of lost meanings and lost coherences, feelings of absurdity and nothingness'.[61] Against these tendencies, all of which may be felt in his poetry, Porter posits the fragile continuities of personal history, and of a voice which articulates it.

If any kind of theorizing takes precedence for Porter, it is that which derives from Freudian psychoanalysis. In addition to *Civilization and its Discontents*, which influenced Porter's outlook since his first book, *Once Bitten, Twice Bitten*, he had read *The Ego and the Id* and *The Interpretation of Dreams*, as well as articles and books by Charles Rycroft and other analysts. While remaining sceptical of Freud's explanations of dreams, and other aspects of his work, the concepts of repression, guilt and the unconscious provided fertile territory for Porter's own imaginings and conjectures. In particular, Porter increasingly recognized the importance of dreams in any deep account of human behaviour. This interest was revivified during Jannice Porter's last months in visits to a psychoanalyst (recalled in the poems 'That Depression is an Abstract', *LIACC*, *CP* 215–16 and 'Disc Horse' *TAO* 49–50), and after her death. The interest in dreams and their meanings was again renewed in discussion with Christine Berg and others in the 1980s and became a major feature of *The Automatic Oracle*, in which modes of accessing the oracles through dream-states was explored. But through all of this, a stubborn rationality resisted determinism and insisted on the human subject's right and obligation to shape his words, images and theories. Adam Thorpe effectively summarized Porter's stance on these matters: 'Ever aware of changing fashions ... he bravely outmanoeuvres the critical theorists by, in a sense, playing their own game, glossing his text in mid-poem, tragicomically certain of his life as process'.[62]

The signs of this particular life recur throughout Porter's work, building a composite picture in the reader's mind of a versatile figure 'Peter Porter'. In 'The Story of U' (*TAO* 62–3), his U-turn finds him reconstructing the childhood house from the dreams, memories and images of previous poems:

> This is the house they made for you
> With water-steps and angled palms,
> A cellar where your tears came true
> And terrors took you in their arms:
> Down by the water a boatshed
> Collected the dynastic dead
> Who heard cicadas keeping on
> Their etching of a single song.

The memories recalled here do not comprise autobiography in the traditional sense, but the 'single song' which persists through them like the non-stop sound of Australian cicadas, satisfies Tolstoy's criterion that art infects the reader with a writer's deepest emotions. Also built into the story is the subsequent transformation of the Lane Cove boatshed to myth, a place of the 'dynastic dead', whose fates would be grafted onto Porter's own in poem after poem. In 'Essay on Clouds' (*TAO* 36–7), this sense of self is fancifully noticed by clouds which speak out as they pass: 'we have marked you, flying over,/the last one of the dynasty of self'. While the politics of an outer world engage him intermittently, it is the changing rulers of this internal dynasty which involve him most fully. His interest in clouds is significantly different from Wordsworthian pantheism, or the elemental concerns of Hughes and Redgrove, in its insistence on human origins and experience:

> Night awaits the upper wind.
> I decide I should not like to live
> in a universe kept up by love
> yet unequipped to tell a joke
> or contemplate the sources of its fear.

The proper voice of authority, as it emerges in *The Automatic Oracle*, is a quieter, more humorous, irresolute and human-sounding voice than those issuing from prophets of the public culture, which confidently predict paradise or perdition. According to European classical legend, an oracle was an utterance, often ambiguous or obscure, spoken through a priest at a shrine as the response of a god to an inquiry. The priest or medium would sleep in the holy precincts and, using the method of incubation, receive an answer in a dream. The title poem (*TAO* 25−6) provides an example of modern, debased wisdom which is handed down by the state:

> . . .A world of innocence
> Is back, but do not look too closely at
> Its rich empleachment. Eden filters through
> The public parks, the weekend wilderness,
> As stiff as astroturf with fallen blood.

This warning is not 'automatic' like the state's pronouncements, but emanates resonantly from a source of sane disbelief in political promises.

What then is the quality of private dreaming? What kind of wisdom, if any, can come through dreams? Porter's poem 'Essay on Dreams' (*TAO* 38−9) begins with the difficulty (not solved by Freud or Jung) of even writing about dreams:

> To set the impossible as homework!
> The only thing more boring than
> someone else's dream is being told
> the plots of films. . .

The guiding principle of dreams, he asserts − in order to start writing his impossible essay − is arbitrariness. But as he enters the topic, he demonstrates, as he puts it later in the poem, that 'Arbitrary is arbitration'. The analogy between dreams and Italian landscapes turns out to be a statement of love and desire, the goddess Athena a displaced image of the mother, the Agatha Christie mystery an accusation turned upon its reader. Rather than divine intervention, a human self is declared: 'dreams are no one's coded messages,/merely the second life our flesh is fed'.

The reader is drawn further into reconstructing the scattered pieces of these dreams. The obsessive school-examination scenario is a classic piece of self-flagellation: 'They are coming round collecting papers,/you must hand yours in. You have covered it/with nonsense, or left it white with fear'. But the trope is again that of returning, the search for 'home'. Italian landscapes are one manifestation: like Totila — the returning Ostrogoth king in the sixth century A.D. who recovered most of Italy from the Goths — Porter's dreaming persona recognizes Italy as 'A country you know is yours/but totally indifferent to you'. If Italy presents itself as a surrogate homeland, the *mise-en-scene* from an Australian childhood bursts upon the dreamscape like a Christmas cake. The scene is recalled in the rhythms of a boy-man's wonder at these gifts from a departed childhood:

> Round the promontory, bursting like a cake,
> the ferry comes and soon its wave runs up
> the slimy ramp. Perhaps this is the spot
> where everything began; it could be we
> can leave the toys out on the lawn and start
> for home...

For 'the homeless mind', Porter implies, dreams are like this, circling and returning again and again to the most vulnerable aspects of the psyche, its sources of love and fear. A fully human response would learn to connect with such primal sources of feeling, from which the poet feels ostracized.

A co-operative medium, language provides many oracles for consideration in our times. Porter jokes about some of these, in a variety of poems through *The Automatic Oracle*, celebrating the revealing accidents of everyday discourse and showing his own willingness to subvert the messages of words. This enthusiasm for the vagaries of language, and his penchant for 'mixing it' with the rough argots of the streets, as well as the more abstruse jargons of the higher specializations, makes Porter a peculiarly versatile poet, rapidly shifting registers, and dreams. He has Babel at his command. But the voice which stands out, with the authority of experience, is that of the tentative survivor, who in the muted lyricism of his interior monologues, has achieved in his eleventh volume a certain equanimity. This figure is nowhere better represented than in 'The Melbourne General Cemetery' (*TAO* 53), in which a 'cautious man-amphibian' discovers that he can now breathe

the elements of both sea and land and 'take his chance of creeds and kinds'. Observing the tombstones, a 'homecoming' is felt, not unlike that experienced in the Highgate Cemetery in 'The Last of England' seventeen years earlier. A similarly calm penetration of tone is complemented in the Melbourne poem by a stronger sense of quiet authority, as if, having had many deaths imposed on him, he can now consider his own as a gift. This may be a temporary reconciliation only, but it stands as a point of balance in his whirling world of contemporary signs and events.

◆

Porter's artistic liaison with Arthur Boyd during the 1970s continued into the next decade with the publication of two more books, *Narcissus* (1984) and *Mars* (1988). The two Australian-born artists also demonstrated their growing friendship and interactive talents when paintings by Boyd appeared on the cover of the second edition of Porter's *Collected Poems* (1986) and on *A Porter Selected* (1989). The painting *Colonial Poet Under Orange-Tree*, 1979−80, which Porter selected for his *Collected Poems*, has been described by Ursula Hoff in these terms:

> Dressed in tight early nineteenth century clothing, so unsuitable for the Australian climate, the poet sits under an orange-tree, like himself of 'foreign' origin. Absorbed in his book and unaware of his surroundings, he has the cloven hoof of the false Jehovah in Blake's Book of Job: his alter ego, the beast, stares at its reflection in the water. Across the empty stretch of white-hot sand sits the log-cutter, from Streeton's *Whelan on the Log*, responsible for the denuding of the land.[63]

In so far as 'Colonial Poet' bears any reference to Porter himself, it would seem to be a self-mocking image of the poet as misfit, or exile. This state was reconstructed by Porter in mythopoeic terms as being 'locked out of paradise', and writing poetry was its chief reparation. The beast as alter ego is a reminder of poems ranging from 'Beast and the Beauty' to 'Paradis Artificiel', but Boyd's signification of an environmental consciousness was a relatively late development in Porter's work, dating from his return to Australia in 1974.

Narcissus was some nine years in the making, and was Boyd's idea. When Porter visited the painter in 1975, Boyd had shown him paintings on the Narcissus theme and they had agreed on a working title for a book, 'Narcissus at Nowra'. (The Australian location was subsequently dropped from the title, in the interests of international appeal.) Unlike previous collaborations, Porter had no clear idea of what he wanted to write about, other than that 'the poems were to be about the prevailing solipsism of self-awareness, and were to use the

Shoalhaven area as their playground'.[64] It is not surprising that in the period following his wife's death, Porter should have puzzled deeply over whether a self as dubious as his own could be lovable. The story of the beautiful youth who fell in love with his own reflection, pined away and was metamorphozed into the flower that bears his name, could have been taken up in a number of ways. What appealed to Porter was 'the sense that the self is somehow hateful and lovable at once', a notion in Pascal's *Pensées* to which he felt particularly drawn.[65] The intention to bend the legend towards 'an Australian cautionary tale'[66] without abandoning its European qualities was more problematic: the didactic intention seems subsumed in a fascination with the duality of a hateful/lovable self, although it recurs in occasional parodic or satiric instances.

Porter wrote the first poem for *Narcissus* in Sydney on a visit in 1976. Entitled 'The Painters' Banquet', this exuberant celebration of painters and painting first appeared in *The Cost of Seriousness* (1978) before finding its place in *Narcissus*. Other poems not included in *Narcissus* but which share some of the book's preoccupations include 'Piero di Cosimo on the Shoalhaven' in 'Three Transportations' (*TCOS, CP* 256−60), 'Roman Incident' (*TCOS, CP* 269−71), 'Landscape with Orpheus' (*ES, CP* 309−11), 'Paradis Artificiel' (*TAO* 19) and 'Pontormo's Sister' (*TAO* 68). 'Paradis Artificiel' is one of Porter's liveliest monologues, spoken as if by an appetite-ridden monster who has entered one of Claude Lorrain's ideal-landscape paintings: 'And into Arcadia, bump,/comes me. As Terry Southern said,/girls go for ghouls, hump! hump!' The monster is antipathetic to the enclosed, paradisal world in which narcissists like to believe, disrupting Claude's idyllic image of landscape. This monster had its progenitors further back, as early as 'Beast and the Beauty' (*OBTB, CP* 12), and is a recurrent metaphor of self in Porter's work.

While the idea of metamorphosis underpinned Porter's understanding of literary transformations, reflective art was not entirely dismissed and was explored outside the covers of *Narcissus* as well as within them. In 'Pontormo's Sister', a Browningesque dramatic monologue, Porter revealed through his speaker − the sixteenth-century Florentine painter, Jacopo Pontormo − how repetitions occur in art through the obsessive tendency of certain artists to paint the same face over and again. Pontormo reflects: 'All my people are the same person,/ as every artist shows'. Technique is found by 'bending Nature to the line/of my depression'. In Pontormo's case, the obsession and the face came together in the image of his sister, who died young and with whose fate the artist-brother identifies. This is what might be termed the acceptable face of narcissism, and is rendered in Pontormo's case in the Mannerist style.[67] Porter extends consideration of these matters through his melancholic, introspective persona, (whose mood recalls Browning's Andrea del Sarto), when he comments: 'This is the message/of the mannered style: God looks like/anyone who ever lived, but more so'. The thought is characteristically witty and subversive: an impression of 'universality' may be achieved by an infinite regression of faces like our own, or those in whom we see ourselves.

How appropriate was Narcissus to an Australian setting? At certain times and in certain cultural contexts, this question need not have been asked. Precedents for the use of European classical myths in Australian contexts certainly existed, in Sydney Long's paintings or Hugh McCrae's verse, to name two obvious examples. However, during a national revival of the arts in Australia in the 1970s and 1980s, the claims of indigenous Australian stories and experience were understandably being pressed by certain artists and critics, following their neglect through a protracted history of Australian cultural subservience. The new work of Boyd and Porter was therefore destined to swim against some prevailing currents. This was a situation to which Porter warmed, enjoying the adversarial role. He entered this part of the project with ebullient humour and, in a spirit of multiculturalism, ridiculed the 'exclusion' by some Australians of an allegedly unhealthy migrant:

> The native gods want to send me home.
> Go back where you come from,
> half-arse Greek! At their wombat
> supper and their cassowary tea,
> will they ask the Women's Circle
> to embroider them some scenes
> on a doily; a euchre game, a shower tea?

This is a jocular yet serious send-up of the exclusivist culture of Australian country towns, with their Country Women's Associations and sentimental regard for local recipes and customs; another shot in the cultural battle between Porter and Les A. Murray, based on their different socio-cultural assumptions. From Porter's metropolitan viewpoint, it would do Australians no harm whatsoever to ponder the story of Narcissus; indeed, it might reveal some rude truths.

A more serious challenge to the tone and tenor of *Narcissus* had been posed earlier, however, by Judith Wright's two poems, 'The Lake' and 'Interplay'.[68] At the end of 'The Lake', the anonymous Narcissus figure, seeing his image on the lake's surface, recognizes 'the terrible face of man'. Shirley Walker has pointed out that this poem demonstrates Wright's preoccupation with 'the danger of a complete detachment from nature and imprisonment within man-made concepts'.[69] The perception here of Narcissus as an entirely negative figure derives from a Romantic ideal of the unity of mind and world. By contrast, the companion piece, 'Interplay', contemplates the healthy interactions of separate and distinct lovers. But the contrast represents too rapid a dismissal of self-love. The multiple perspectives of Porter's *Narcissus* give more scope to the fairly widely accepted view, among Freudian psychoanalysts and others, that apart from extreme cases which express themselves in obsessive or perverse behaviour, 'the amount of love devoted to the self was reciprocal to that devoted to other objects or persons'.[70] Their different views on narcissism represented one of the main points upon which Freud and Jung parted company. Unlike Judith Wright's more metaphysical (and Jungian) approach, neither Porter nor Boyd avoided the implications of sexual narcissism. Images of erotic excitement or gratification through extreme admiration of the subjects's own attributes recur throughout *Narcissus*, and are given special intensity by the spellbinding etchings of Boyd on their intense black background.

The formal problems posed by *Narcissus* were in some respects greater for the poet than for the etcher. Clearly, a strong narrative or dramatic element would work against the theme. How then to give shape to the book? Porter has described the outcome: 'It is a series of deflected portraits of the self, as though one had entered a fun-fair alley of distorting mirrors'.[71] Childhood memories, details from his returns to Australia, and the setting of the Shoalhaven River preside. Like Boyd, Porter saw the river as a place of visual beauty and a source of symbolism: the river provided the element of 'reflection and self-knowledge', but also suggested 'laziness ... the other side of the coin of its life-giving function'.[72] In this sense, the river runs through the whole book. Against this agent of continuity, the self is shown to reproduce itself endlessly; even sunset becomes a 'playback' of the perceiver's eye:

Sunset, when the world runs blood —
a bird darts through the waterfall
and makes a rainbow
for a second. These colours
are a playback of my eye.

Here, as elsewhere, pantheism is held at bay by a strong sense of the creating self; yet magic and mystery remain, for the self's eye, heart, brain and hand are themselves a mysterious concatenation.

The chief temptation of the Narcissus story for the poet is to allow a mood of quiescent, melancholic reflection to dominate. However, Porter's lyric gift is the stronger for its masks of irony. In *Narcissus*, the poet's exuberance and desire for a crowded canvas (evident in 'The Painters' Banquet') is checked, giving greater scope for lyric reflection. Humour and parody cannot be excluded, however, jokes and ironies are necessary in Porter's world, and they continually ripple across the river's calm surface. In Part XIII, 'Echo's Moon-Calf', a barbecue with Les Murray on another river, the McDonald, is humorously recalled in the style of a casual Australian anecdote:

> His barbecue
> was across two rocks and he was throwing
> gum-balls on the fire to make it flame.
> To cut a long story short, it was
> Narcissus — not considering his own face
> but loving his body by stuffing it
> with food. He gave me half a glance
> and hid his uncooked steak.

Porter's versatility with verse forms was also extended to more intensely felt autobiographical images. One such employed the short 'emblem', originally a didactic device in Quarles's seventeenth-century *Emblemes*. The 'message' in Porter's recreated emblem from childhood is less evident than its underlying feeling:

> The little boy is watching his mother.
> Is it a sand-cake or a sponge?
> She beats the eggs and sugar for more
> than twenty minutes. He will meet
> his own face often in those thoughts
> that she must love him, so intent
> on pleasing him and long since dead.

Other images of the lonely child recur: 'I took the path/of double-looking and saw,/stretching into hell and childhood,/the old saw mill, the quarry/on the horizon, the disused railway-line,/cabbages asleep in the sun/and the creek polishing quartz/to paralyse the light'. These bright epiphanies of a childhood world before the complications of sex are counterpointed by a darkly self-involved sense of sexual disturbance, which contrasts radically with the innocence of Judith Wright's 'Interplay'. In a parenthetical comment (sex is a bracketed world for the persona) the problems of intimacy are revealed:

> (Among the girls I have been to bed with, there was an enormous pool of good will. Yet my intransigence was never more than a measure of impatience. Getting closer keeps us always a surface apart.)

The paradoxes of intimacy between men and women are revealed with an oblique assurance, diffracted in the broken images of pools and surfaces.

Is such solipsism a disqualification from social theory and action? Christopher Lasch has argued that the plight of Narcissus, as it applies to contemporary western societies, is based on a confusion of self and not-self, not just 'egoism' in the usual sense: 'The minimal or narcissistic self is, above all, a self uncertain of its own outlines, longing either to remake the world in its own image or to merge into its environment in blissful union'.[73] To the extent that Porter's *Narcissus* is a cautionary tale, it warns against both these forms of narcissism. A blissful merging into the environment had never been possible for Porter himself, nor did his socio-political inclinations translate readily into programmatic activism. In one of 'Narcissus's Revolutionary Sayings' (Part XVII), Porter exposed the adolescent unreality of his own Fabian Socialist thinking in the years before he left Brisbane in 1951:

> In the years when I called myself a Socialist, I did not believe that the world around me was real in any sense. How easy then to alter it, to submit it to improvements which I found comforting to imagine and probably much more concrete than existing systems. The idealism of impotence needed very little work.

This recollection of an earlier self represents neither a journey into the interior nor a flight from selfhood; rather, it indicates a blurring of boundaries between the self and the world, wherein an unassimilated theorizing is imposed upon the act of observation. As we have seen, 'Paradise Park' (*TAO* 20) and other poems of the 1980s represent a significant shift away from such self-involvement towards a maturer vision in which world, text and author are more interfused.

◆

Porter's fourth book with Arthur Boyd, *Mars* (1988), was a complete transformation from the images of lassitude and voluptuous self-reflection in *Narcissus*. Here was a minefield of explosive energy. The legend of Mars was rendered very freely by Porter and Boyd: the Roman god was closely associated with agriculture and the season of growth, ripening and harvests. Porter and Boyd transposed the myth into modern industrial society, where weather and seasons were less threatening than the appetite for power demonstrated by politicians and technocrats.

The theme of war, its origins and consequences, was not new to Boyd or Porter. Boyd had previously worked on a series of visual responses to the holocaust and the concentration camps of Europe. And one of Porter's most quoted poems was his translation of holocaust images into contemporary urban life, 'Annotations of Auschwitz' (*OBTB*, *CP* 30−2) and 'Somme and Flanders' (*1962/3*, *CP* 40) had summoned up images of three uncles killed in the First World War.

Sentimental patriotism and didactic denunciation detract from any art dealing with war. Porter and Boyd were concerned not to 'drown in good intentions'[74] and avoided a one dimensional treatment through the exercise of humour and ingenuity, qualities which the main subject of this book, Mars, is lacking: 'Mars has no sense of humour/and his seriousness turns living to a joke', quips Porter. The alliance between verse and visual art freed Porter to exercise a form of anachronism which his persona in 'The Prince of Anachronisms' (*TAO* 5) had proposed: 'History's perspective is a lie./Art is the same age eternally'. While this 'late modernist' view of history poses problems for an historiographer, it allowed the poet to present Mars after his birth in the following terms: 'He lies in a basinette, a seed of time./Like all babies, he resembles Winston Churchill/or some Japanese War Lord'.

In Porter's verse, as in Boyd's images, war is principally a masculine exercise of power, and sexual prowess and domination are its co-respondents. Thus Mars's offer to Venus (via Homer's *Odyssey*): 'He offers her a hungry war, a body count among/politicoes, the world as strategy where pre-lunch/Martinis are Lives of the Saints, and a mist of semen/hangs over battlefields and the dowsing of the dead'. In one of his many metamorphoses in this book, Mars is a grizzled general looking back to his training in a British-style public school, resembling the ethos of Australian public schools recalled in early Porter poems such as 'Mr Roberts' and 'Eat Early Earthapples'. In *Mars*, the satire suggests a direct sequential line from such schools to the big war games:

Life is constant service,
first head down in the scrum
or bellowing in chapel,
then buried deep in nerves and mud
hearing howitzers glissade,
finally the tiring house of women.

The stupidity of war is mimicked in doggerel: 'An average boy of average guilt,/he often used to wank/but now the knob his hand is on/is on a Chieftain tank'.

The principal milieu for modern Mars is America. The verse contains allusions to Nixon, impeachment, Exocet missiles and Side-winders, Vietnam, microchips, IVF clinics, baseball: 'At blast-off the whole caboodle seemed in order;/a searing imitation of the sun/poured from its tail and every dialled recorder/pointed where it should; each nervous pun/by the controllers at the console assured a/ready laugh: they'd done it, they'd hit a home run'. But Boyd's motif of the digger's hat is a reminder, too, of Australia's part in this. At one point Boyd's imagery leaps beyond Porter's verse evocation of 'The Field of Mars', when an erect Mars, trousers off and vest rolled up, is shown copulating with a donkey. Later, it's a goat — the metamorphoses are multiple, but the iconography is consistent: martial and sexual lust are from the same stable.

This is strong material, showing the endless recurrence of war despite the best intentions of peacemakers. The warlords change, but the perversions of lust and power remain. In one of Boyd's images, while Mars molests a naked woman above planet earth, her tears of blood rain upon Australia. Australians are part of this world of power-play, violence, anger and lust, attracted, like others, to a 'heedlessness of life'.

Porter's verse also serves to remind us of the metaphors of war which pervade everyday discourse. How then to 'attack' it? (So many of our verbs are aggressive.) Porter offers an approach on several fronts. The seemingly harmless, television-induced fantasies of a child may themselves be an insidious beginning: 'I wish I were a little bird/ that on the sea is seen,/an Exocet, I'd skim the waves/below the radar screen'. Porter is aware of the temptation to trade rhetorical blasts with counter-blasts, and steps out of the line of cross-fire. But how to avoid it without seeming feeble? Against the arrogance of a Vulcan, producer of military hardware, the artist-as-humanist can set only the 'worthless tropes of a Humanist culture'. We meet again 'Old Stuffy, signaller of sickness'. Aware of the vulnerability, and ultimate ineffectuality of humane values in a world of violence and anger, their proponent is left acknowledging weakness. Weakness is strength. Or is it?

The question persists: how can intelligence, humour, generosity and love counter the forces of destruction? 'It's human nature anyway/ to kill and to be slain.' Porter offers some perspectives on this problem, nothing as organized as battle plans or 'strategies', but what might be summed up as sites of resistance. First, while acknowledging the 'Wound' of birth, he mocks the superficial belief in a Freudian-style determinism which insists, for instance, on the necessity to 'kill the father'. Second, a new angle on history, and a motive for resistance, is offered in a 'Soliloquy of the Unborn', wherein the subjects tell us they are listening for the 'melody' from the human species which makes 'one sound cling to another', which cancels out perpetual dissonance. Death's certainty, even off the battlefield, should also be a chastening influence: in spite of the frustrations and *angst* of living, nothing we do to hasten death will make us happier. Finally, Porter questions the stories we tell each other: 'The gods keep mortal men on tenterhooks/ With stories of their love affairs and wars/And nurture them and kill them unawares'. Throughout the welter of images and mini-narratives in a variety of voices and styles offered to us in *Mars*, this note of a resistant sanity counters the intoxicants of power and aggression.

Nevertheless, the book presents problems for the reader. Its multi-plicity of voices, its tumult, contrast radically with the silence and spaces of *Narcissus*, calling for a different kind of response from the reader. The sense a reader makes of a text is prescribed to an extent by its rhythm and circumstances.[75] Porter's text implies an uncommon reader, who can find enjoyment both in the jargon of modern journalism and in the myths of ancient Greece. The profligacy of images, stories and allusions in *Mars* will sometimes bewilder the reader. Nevertheless, taken as a whole, *Mars* is a highly successful extended treatment of war — its motives, manifestations and consequences — equivalent in its genre to Joseph Heller's novel, *Catch-22*. It presents a sophisticated socio-political commentator alongside the excavator of self.

POSSIBLE WORLDS

◆

Poised on the cusp of the 1990s, Porter's twelfth volume, *Possible Worlds* (1989), was unusual in not taking its title from a poem included in the book, inventing instead a title which implied a generic significance. The book was dedicated to fellow poet and friend, Gavin Ewart, who shared a similar talent for comic and satiric poems across a wide range of verse forms. *Possible Worlds* was appropriately dedicated; one of its characteristics was its revival of the 1960s 'social poet' in Porter, whom Ewart had recalled in the 1980s as 'the serious satirical Colonial-in-London.../lighting Latimer candles to Culture':

> So we were into satire. Our London was brash, immoral
>   surprising —
> 'What a city to sack! — It was sacked by advertising.
>
> . . .
>
> That satire rings truer now, in the money-mad world of a
>   Thatcher,
> and in the rye, alas, we're left without any catcher;
>
> but writers, wrote Wystan (to Christopher?) are ironic points
>   of light.
> And I think you've certainly been one, before you go, and I go,
>   and we all
> go into that not-so-good night.[1]

The poem by Ewart, seen in its entirety, contains some nice touches, and shrewd insights. The 'Latimer candles to Culture', for instance, suggest the Protestant vigour of the young Porter's veneration of high culture and his desire for a (largely undefined) Reformation. Also implied is Porter's modulation towards irony rather than satire which, as in Auden, ironically reveals the 'points of light' in a civilization. Ewart's reference to the materialism of Britain in the 1980s indicates one of the major targets of the irony and satire in *Possible Worlds*, which builds on insights in *Fast Forward* and *The Automatic Oracle* in showing 'the world' as a global network of rapid interconnections, from which no country can remain excluded.

The title of Porter's twelfth volume seems to derive principally from the chief of cautionary tales, Voltaire's *Candide*, on which, in his mid-twenties, he had composed fragments toward an opera.[2] Thereafter, at different points in his poetry, Porter had alluded to the false expectations engendered by Candide's absurdly optimistic tutor, Dr Pangloss; namely that, in spite of overwhelming evidence to the contrary, we live in 'the best of all possible worlds'. Porter would probably agree with the adage about the difference between optimists and pessimists: the optimist believes that we live in the best of all possible worlds; the pessimist fears this is right. He might, on the other hand, be heartened by recent psychological evidence that, contrary to popular opinion, the depressed are more accurate in evaluating life, that it is the optimist who distorts the evidence.[3] Whatever the validity of such psycho-social findings, Porter's allusion to *Candide* raises a further intriguing question. Could he be hinting at an acceptance of Candide's diminished world of the possible at the end of Voltaire's satiric tale, the necessity to 'cultivate our garden'?

On the contrary, the epigraph to *Possible Worlds* offers a direction which acknowledges pragmatics but surmounts them. Taken from Wallace Stevens's poem 'Credences of Summer', it points to certain operations of the human spirit which will not be circumscribed by the practical requirements of living:

> The trumpet cries
> This is the successor of the invisible.
>
> This is its substitute in stratagems
> Of the spirit. This, in sight and memory,
> Must take its place, as what is possible
> Replaces what is not.

Against the restrictions of everyday life, such trumpets will sound out, transforming actual places, objects or events into surrogates of the invisible world. In a lecture in Florence in 1989, Porter glossed his epigraph:

> We all start with that invisible. As Thomas Mann puts it, 'We come out of the dark and go back into the dark again.' But we find in daily life, as we live it, the revelations which Stevens calls 'stratagems of the spirit.' Thereafter we live in possible worlds and they may be centred anywhere.[4]

Stevens's phrase 'stratagems of the spirit' was taken up again by Porter as the title of a key poem in *Possible Worlds*. 'Stratagems of the Spirit' (*PW* 55–6) is a verse essay framed by an introductory sentence, 'You've reckoned without the world' and a concluding declaration which invokes Stevens's 'invisibles': 'we must pace our footfall on/The temple steps and listen for the sun'. The essay's argument oscillates between these poles, reminding us both of the sun's heat and light and the technical sense of 'stratagem' as a plan or scheme designed to trick an enemy. Here, as elsewhere in Porter's work, the twin enemies are despair and death: the stratagems of the artful imagination are to create a sense of the numinous in everyday life. As in Pope's *Moral Essays*, Porter fits poetry to exegesis.

Characteristically, such stratagems are elucidated best in the visual arts and music. Porter's chief example in 'Stratagems of the Spirit' is Piero della Francesca's painting *Madonna del Parto*, which is reproduced on the cover of *Possible Worlds*. (The poem refers also in part to the fresco panels of the life of St Benedict by Luca Signorelli and Sodoma in the Abbazia di Monte Olivetto Maggiore in Italy.) Whereas Piero's fifteenth-century cohorts saw the world sententiously, with every allegory pointing to God, twentieth-century observers read sexuality into the painting, 'a marvellous smile which runs/Through flesh *del parto*, headlong on to love'. Both are of the spirit, just as Linnean-style scientific classifiers of data may also, 'star-faced', hear 'musical/Communiqués no erudition robs/Of freshness'. By various means, in different historical circumstances, the spirit's strategists will outmanoeuvre the old enemies.

The 'real' material world of contemporary Britain is evoked in *Possible Worlds* as dystopia, in the manner of satirists such as Voltaire, Swift, or, in the long concluding poem, the pseudonymous Sir John Mandeville. 'The New Mandeville' (*PW* 63–8), set in three 'musical' movements, recalls *The Travels of Sir John Mandeville*, in which this fictional knight leaves England in 1322 on a pilgrimage to Jerusalem and then journeys over much of the known world. Porter had read extracts from an English translation of the Travels while at school, and the story of Mandeville's words freezing in the cold of northern latitudes and hanging everywhere on trees or blown onto sills, so that every word which had ever been spoken was recoverable, appealed to him. But more central than Mandeville's travels, in spite of the poem's title, are those of Gulliver — in the third and fourth parts. Porter writes here as a semi-Modernist, with strong backward-looking enthusiasms. He would be lost without the profuse present and its dreams of modernity: the 'ever-changing Expo/Of the newest redeployments of the Spirit/Where greedy fables are precisely nimble/Whichever laser-digits phrase the murk/Of history'. Against this, however, are set the 'true books': 'The true books came as liberators, they/Reproduced the best historic hells,/The Lagers, Genocides, Apostalates,/But nothing here would print such images'. The true horrors of twentieth-century history and society, in Britain as elsewhere, have been suppressed: the poet's task is to restore them to the commerce of the imagination.

Porter's London satires of the 1960s and the 1980s were linked by the poem 'He Would, Wouldn't He' (*PW* 21). The title, as its last stanza indicates, comes from a statement by one of the most famous 'good-time girls' of London's legendary sixties, Mandy Rice-Davies who, along with Christine Keeler, was a central figure in the much publicized scandal of the decade, the so-called 'Profumo Affair'. Porter wrote his poem before the appearance of the film *Scandal* in 1989, which recalled these events, as the *Guardian's* reviewer noted:

> *Scandal* takes as one of its central texts the surely immortal words of Mandy Rice-Davies, 'He would, wouldn't he'. And around that remark, made at the trial of Stephen Ward when the prosecuting counsel denied that Lord Astor had any knowledge of her, carnal or otherwise, a web is weaved, a web that catches more than one fly.[5]

In Porter's sardonic joke-poem, this often-repeated headline of the tabloid newspapers becomes an unconvincing public denial of guilt by the British ruling classes. The tone of mockery is cleverly sustained by the sing-song iambic repetitiveness of the lines and the single 'ee' rhyme throughout (reputedly the easiest rhyme in the English language, but craftily deployed by Porter here):

> He'd say, if pressed, the bomb has kept us free,
> We sing the future in an awkward key
> But, look, the orthodox revives in me.
>
> As well in acid rain command the tree
> To stop its leafing: the moving hands decree
> Time has not changed by even one degree.

The major target here is the continuing irresponsibility and hypocrisy of the political and social ruling classes; in the second stanza the register shifts to suggest just how pervasive this menace is: 'acid rain' works with chilling metonymic efficiency to suggest the corrosive influence of a society's governors.

Linguistic habits or unconscious slips of the tongue are fixed unerringly in Porter's best satiric portraits. Former British Prime Minister Margaret Thatcher, for instance, is alluded to in the third stanza of the poem, with her public use of the royal pronoun 'we', which Porter extends to a more generic point about the imperial ego. Another target is those right-wing media commentators who 'praise the Family/And say the Poor live irresponsibly'. By capitalizing the nouns, Porter highlights the folly of such generalizations, and their orotund condescension. The chief enemy identified is the kind of ambition which becomes a way of perceiving the world: 'The world made new by new ambition'. Those who lust after a CBE, or the Chat Show Host elevated to Superbard (even Clive James does not escape notice) are emblematic of those caught on a wheel of ambition which controls their view of the world. Porter gives his last words to humanity's mythical first man, the disappointed Adam, whose request to stay in the Garden of Eden was turned down:

> Adam, when told by angels that his plea
> Had been refused by God, just shrugged like Mandy
> And answered, 'Well, He would, wouldn't He.'

Adam is yet another manifestation of the emigré who bears most of the emotional force of Porter's poetry, and is also his chief carrier of wisdom. It is a role which Porter plays out through protean mask figures, and provides an emotional locus for the expression of both regret and a spirited relish of adversity.

'An Ingrate's England' (*PW* 33), presented a jaundiced perspective on England, but the irony was broad enough to include the speaker and his own English origins, which had enabled him to endure his early years in Australia: 'This is the England in your flesh,/A code enduring Summer while/Tasteless birds flap at the edge of/Civilizing concrete'. While implicated in this England, Porter's persona remains an outsider, still not using pronouns in the proper English way nor owning any property. Displacement stimulates a satiric energy directed towards the property owners and wealthy merchants of contemporary England; but by the late 1980s Porter had achieved a tone of authority rather than spite, envy, or (as in some of the London poems), despair. On the verge of the 1990s, Porter unreservedly condemned the attack by market-force Tories on the natural and built environment of Britain, a political policy with which some artists unfortunately seemed to agree, and which perhaps they profited from:

> The selling of the past to merchants
> Of the future is a duty pleasing to
> The snarling watercolourist. Prinny
> Used to ride by here, and still the smoke
> Of loyalist cottages drips acid rain on voices.

A dark, ironic tone is finely controlled in these lines, mocking while it characterizes the voices of monarchists who refer intimately to the Prince Regent as 'Prinny' while dripping 'acid rain' on accents other than their own. Porter's republican views are implicit, as is a conservationist outlook and a tendency, noted by some fellow English poets, to be more informed about Britain's past than they themselves are — a result perhaps of the emigré's enthusiasm for the culture of his adopted country.[6]

The dystopian vision of 'An Ingrate's England' is epitomized in the poem's concluding stanza, which rehearses an allegorical pilgrimage by train from Porter's nearest station, Paddington, to an old England being destroyed by the new vandals:

> The trains in their arched pavilions leave
> For restless destinations, their PA Systems
> Fastidious with crackle; nobody
> Will ask you to identify yourself
> But this will lead to hell, the route
> The pilgrims take — down the valleys
> Of concealed renewal to the pier-theatre,
> The crinkle-crankle wall, the graveyard up for sale.

Although Porter was never a proponent of Merrie England, he did believe that the country had provided him with an accommodating past, as it had for other Australian and Commonwealth expatriates. Now, not only the picturesque 'crinkle-crankle walls' of East Anglia would go, but — worst sacrilege of all — graveyards were up for private sale; nothing epitomized better the crass materialism of the 'privatization' boom of those years. Porter's allegorical pilgrimage becomes, then, a journey to hell. His ironic placement of himself as an 'ingrate', again positions him as an outsider, ungrateful for favours given, not 'one of us'.

Porter's discomforting intelligence does not allow him to devote restrictive love to any particular country. The abstractness of patriotic feeling, and its capacity for extension into war-mongering attitudes had been demonstrated in *Mars*. In his public joust with Les A. Murray and other Australian nationalists, 'On First Looking into Chapman's Hesiod' (*LIACC*, *CP* 210−13), Porter's speaker preferred an ideal notion of 'the permanently upright city' to 'a blunt patriotism,/A long-winded, emphatic, kelpie yapping/About our land'. More recently, he allowed a humorous 'misprint' to express his prejudice against officially fostered patriotism: 'Patriots always stand for the/National Anathema' ('A Bag of Pressmints', *TAO* 40). But his 'Essay on Patriotism' (*PW* 31−2) went beyond these previous bites at the topic and made a meal of patriotism, its motives and manifestations. Not so much a 'metaphor man' as a propositional poet, Porter adopted a dialectical structure marked by tendentious opening phrases in each of the poem's seven paragraphs, such as 'Compared to', 'So when they say', 'Let us therefore . . .'. Compared with an early, relatively undistinguished poem in dialectical form, 'A Moral Tale has a Moral End' (*OBTB*, *CP* 20), this poem (like the verse essays on clouds and dreams in *The Automatic Oracle*) successfully combines wit, erudition and feeling. The main difficulty with the 'Essay on Patriotism' is that it begins with an elaborate opening conceit on the superior sovereignty of self over nation. Although the notion of a 'wide Porterian peace', was probably intended to convey a tone of self-mockery and thereby guard against taking too seriously any pompous over-estimation of self, it has the opposite effect. Being a patriot for oneself smacks of an overblown hubris, however self-aware the speaker shows himself to be.

The 'Essay on Patriotism' strengthens greatly, however, when it considers instances of patriotism and finds them either absurd or criminal, or both. Britain's post-Empire reduction to 'a Third Class Power ... without a decent satellite to spy from' is of no concern to the speaker, because a diminished national status and power provide less scope for cruelty and injustice, associated with the grandiose rhetoric of, for example, the Boer War, when the poor, the sick and the fearful were duped to 'bray the victories of their rulers/on air they couldn't warm'. As in much of Porter's work, a principal focus, indeed the major actor, is language itself: here, the vocabulary and syntax of the proposers of patriotism are suspect because they are engaged in exercises of power and deception. Above all, Porter warns, the word 'great' must be scrutinized. It may be an appropriate word for Blake and Milton, but is 'much too big for Cromwell/and generally should watch itself in mirrors,/bearing down like Yeats's Nobel head'. The pun on 'Nobel' is compounded in this satiric image of Yeats proceeding at snail's pace down the steps of the Savile Club, vainly observing himself in the stairway mirrors. This 'King of the Cats' had expressed what for Porter was a suspect patriotism. Other wordsmiths, such as journalists, could be even more dangerous, because more widely read. Porter's persona digs beneath journalistic phrases such as 'the patriotic proletariat' and concocts in their place more truthful (and wittier) headlines for articles which reveal a pervasive middle-class propaganda. Like England's northerners abandoning their workmen's caps and acquiring special doors for their cats ('From flat-cap to cat-flap/in one generation'), and 'dual income no children' couples replacing 'yuppies' ('Dinkies/are not toys today'), the Mafia too has gone up-market ('Designer/Murder comes to Sicily'). Language does not simply reflect such attitudes and perceptions, it also constructs them. When specialist jargons invade the language (Porter gives examples from 'Hi-Fi' magazines) we are forced into islands of confusion. Similarly, when the head is filled with patriotic jargon, we can become automata for uncomprehended causes, like the macho protagonist of the American *Rambo* films of the 1980s: 'no wonder/Rambo gobbled up the gooks/if he had such voices in his head'.

By fusing together statements from three opponents of restrictive national feeling, Porter moves towards the apex of his case: 'Patriotism is not enough/of a scoundrel's last refuge/even if you love/your neighbour as yourself'. In this collage, British nurse Edith Cavell (who was shot by the Germans for her resistance work in Brussels in the First World War), Dr Johnson and Jesus Christ can be identified as authors of the

statements Porter has combined. But the figure behind them is clearly
Porter — not magisterially in control, but aware of the uncertainties
and vicissitudes of life. As he goes on to suggest, his birth in Australia
at a certain time was a chance event which had imposed upon it
certain possessive pronouns, the chief of which were '*ours*' and '*mine*':
'Where I landed I named *ours*/though it was never *mine*'. The attitude
here is informed by a migrant's sense of provisionality, both in relation
to a particular country and to life on earth. But Porter's verse essay
ends with a reaffirmation of the 'dynasty of self' with which he
imagined the passing clouds accusing him in 'Essay on Clouds'. His
closing paragraph recalls Australian beginnings and compares his fate
with that of other contributors to the dubious concept of a 'national
interest', encountered during return visits to Australia:

> True patriots all,
> the still-swimming lobster in the tank,
> the lambs that face the ocean through steel bars,
> the opals in the open-cut —
> I left my mother's and my father's house
> and stepped on to a road beneath the stars.

'True patriots all' recalls an anonymous poem in couplets, allegedly
the first poem composed at Sydney Cove, which prologued the convict
performance of Farquhar's *The Recruiting Officer* there in 1789.
Although doubt has been cast on the provenance of the poem, this
does not affect Porter's employment of the phrase, which suggests the
bizarre uses to which patriotism can be put. Porter had observed
Australian crayfish in tanks, which would become lobsters for export
'in the national interest', and at Fremantle in 1987 he had seen the
infamous 'sheep ships' taking live lambs to the Middle East for slaughter.
A swift change of gear in the final lines returns to the personal,
individual fate, which is central to the poem: this exile has *chosen* his
fate, which is not restricted to a single country; its lyrical universality
is held in check only by the restrictive factors enumerated earlier.

For Porter, the self was able to realize dreams but its achievements
were always limited by the contexts of possibility. After visiting the
jarrah forests of south-western Australia in 1987 and seeing the
evidence of the disease called 'die-back' there, Porter became intrigued
with the metaphor and gave it a wider application. 'Die-Back' (*PW* 15–
16) includes the observation that 'A beetle, a fungus,/a vagrant virus,
evolution's other part of the forest,/has changed the whole map of the
bush. The cities soak/up soft artesian poisons, the quokkas change
their diet/to gleanings from McDonald's, and the desert/seen at night
on speculators' charts/ scintillates with little spots of death'. A chill of
excitement is evident in this revelation, which is not the conventional
ecologist's condemnation. Porter has observed humans poisoning

themselves and their environment before. What can be done? His satiric images of the business people of Enterprise Economy reveal their self-perception as inviolate: 'This unkillable species kills/and goes on being nice to parents, strong on law and order,/its system of despair decked out in hope'. Artists, however, 'map at least the edges/ of the possible, the unendurable'. Though they hold a key, their limits lie in the inability of writing to cope with the intensity of feeling or the complexity of the world: the human spirit is unequal to the task of conveying either 'the chaos of the brain' or 'the blaze of dying'. Porter's lines in 'At Lake Massaciuccoli' (*ES*, *CP* 306–7) are recalled here: 'No one produces the art he wants to,/Everything that he makes is code,/To be read for its immaculate intention'. The suffering artists partake of these limitations, they are not prophets. Their task is to bring the 'juggernaut' self back to a recognition of the human life-cycle, to realize that it too is involved in the 'die-back' of the species.

As Porter grappled with the concept and application of 'possible worlds', his capacity to incorporate social thinking into his verse increased. This development was especially noticeable in two poems, 'Decus et Tutamen' (*PW* 3–4) and 'Civilization and its Disney Contents' (*PW* 35–6). The first was concerned with money, materialism and power, the second with the capacity of theory to explain the sources of discontent in society. The title, 'Decus et Tutamen', was taken from the rim of certain English coins; roughly translated, it means 'This coin is guaranteed'. But in Porter's burlesque ode to money, the whole question of measures of value in contemporary society is raised. Satiric relish is taken in the curious objects which are held to signify value: 'Turn the small disc in your hand,/The coin, the CD of desire; it proves its plastic cousins/Men of substance . . ./It sings across all latitudes'. The strength of Porter's satire lies in its capacity to 'sing' such dreams, to evoke their allure rather than merely pronouncing judgement upon them.

Porter's ironic scrutiny of capitalism — its assumptions, uses and abuses — is international and free-ranging. Significantly, the America which began with Columbus's discovery is highlighted by Porter as the home of capitalism. The emblematic capital is Los Angeles which produces the crassest of spokespeople: 'Ecology is shit, says the white-shooed/Magnate in his barbican of glass; the universe is stable,/The sewers of LA can take a furlough, what is/A littoral of winding tides upon a fret of mud/If not a model sewage farm?' Is there an answer to such gross materialism? Porter reveals its pervasiveness, from the Tyrian traders to media magnate Rupert Murdoch, a son of the manse who now heads an empire:

> So beautiful is money, it is the nation in low carving,
> The miracles by St Democracy among
> The bargain car-lots, the blue-rinse hyper-rise,
> Selling short the wimps and setting up sons of the Manse
> As Neo-Emperors.

The Brisbane skyline, with its modish display of blue-tinted plateglass windows may be suggested here, but the point is that such skylines are replicable around the world. The way citizens use money provides a symbolic *bas-relief* of a nation. Implicit is Porter's preference for a more caring, equitable society, not based on communist collectives, but comprising a version of social welfarism which gives individuals the best chance while not encouraging huge disparities of wealth.

A positive outcome of this kind — the impossible dream come true — would require desires stronger than for money. The concluding section of 'Decus et Tutamen' offers a visionary alternative, which (like the earlier 'Cities of Light', *FF* 60–2 and 'Next to Nothing', *PW* 1) incorporates the poet's gathering awareness of Australia's Aboriginal past as a counter to the arrogant myths of modern materialism. While the poem acknowledges the fact that money 'sings across all latitudes', its lyrically expansive conclusion suggests another scenario, incorporating present and past realities:

> . . . Now, on sunny afternoons,
> Our yachts at anchor, their cordless phones asleep,
> We see ourselves the ancient tribesmen of the land
> Telling snake legends to the rainbow trees,
> Reverted like the glasshouse hills and mortared gullies
> To pure landscape till our very death becomes
> A record in a rock, perfection of the will to live.

Though lyrically expressed, this vision is qualified by its hedonistic yachtsperson's setting (Moreton Bay and the Glasshouse Mountains south of Brisbane are suggested), and lacks the sense of personal urgency and integrity of another recurrent image in Porter's work — of a boy, his father and a dinghy on this same bay. Furthermore, the distortion of Aboriginal legends of waugyl, the sacred rainbow serpent, into snake legends being told to rainbow trees suggest a somewhat superficial surrealism, rather than a deeply imagined alternative. So although the dystopia of a western society dominated by money is countered to some extent by this longer and more generous view which puts humanity's fortunes in a broader context, the consolatory end seems not entirely justified. Paradise lost still seems to be more imaginatively realizable than any paradise regained.

Porter's internationalist view envisaged a world of liberal democratic societies unhampered by nation-state considerations. But he was too aware of problems, confusions and the generally fallen state of humankind to adopt the mantle of a utopianist, except ironically. In 'Civilization and its Disney Contents' (*PW* 35–6), as the title implied, he was at least as aware of the comic cartoon possibilities of contemporary society as he was of the explanatory power of Freud's theories in *Civilization and its Discontents*, or indeed of other generalizations. Nevertheless, the use of Freud as the putative speaker in the poem indicates that *Civilization and its Discontents* had indeed remained alive in Porter's thinking since he had first encountered it in the mid−1950s. Whereas Freud's late work had been a significant formative influence on Porter's first volume, in the late 1980s it was set aside in a questioning of the validity of all broad theories of human behaviour.

Although a poet of propositions, aphorisms and paradoxes, Porter resisted totalizing theories which claimed to subsume the oddities of individuals. While he was drawn at times to great generalizers such as Freud or Marx, he worried about the convergent nature of their intellects, feeling more drawn to the divergent imagination exemplified in Empson's *Seven Types of Ambiguity*. Poems such as 'Throw the Book at Them' and 'Sticking to the Text' in *The Automatic Oracle*, which ridicule pompous, dogmatic or opaque theorizing, were followed in *Possible Worlds* by 'The Poem to End Poems' (*PW* 37), in which the poet complains that 'Today the theorists are the avant-garde,/The artists make it in the Supplements'. Porter's poem 'Civilization and its Disney Contents' widened the base of his discontent with the idea that criticism and theory were the pre-eminent forms of contemporary discourse.

The speaker in 'Civilization and its Disney Contents' is a Freud/ Porter composite, demonstrating the poet's concept of the individual not as monolithic but as protean and flexible, capable of housing a considerable number of contending pressures, desires and ideas at once. The 'Freudian slip' of the poem's title both ridicules and reinforces the Ur-psychoanalyst's need for systems and theories. The case for systematic thought is put succinctly by Freud/Porter: 'systems are what separate us from the animals;/are the sublimity of our reasoning, the jolt/to eye and brain of the façade of San Miniato'. (Mention of Florence's most harmoniously beautiful church places Porter firmly in control of his Freudian alter ego here.) While admitting that other observers have noted that all people want is 'food or sex or comfort', Porter's Freud claims a 'duty' to find organizing systems behind such mundanity. An unconscious Moses-complex is revealed in this Freud in a humorously ironic collocation of the speaker's religious and sexual needs: 'And so you have before you my Hellenic-Hebraic laws,/the tables of which I brought down from the attic,/the universality of family trunks and secrets'. However, half-way through the poem, this Freud (as if in psychoanalysis) begins to reveal the real confusion in which people live, much messier than his theories. All around us we see the contradictions between what people esteem and what they do; where business and aesthetics intersect, for example, we have the spectacle of 'reverent aesthetes' whose 'incomes derive from Mail Order jugs of Charles and Di'. Porter challenges the reader to construct a theory from such paradoxes.

The level of seriousness in 'Civilization and its Disney Contents' rises discernibly when it turns towards death. Freud's essay on death is brought to mind here, in which he observed that, following the death of people close to us, we behave 'as if we belonged to the tribe of Asra, we must die too when those die whom we love'. In such circumstances, Freud's essay continued, we become disinclined to court danger and the will may be paralyzed; other 'renunciations and exclusions' may follow.[7] For Porter, death is the great obsession: 'And all the Mickey, Dopey, Bambi bits/are to keep your eyes from wandering to the dancing dials'. An audacious and curiously affecting conclusion combines comic and tragic modes with ironic sententiousness in a peculiarly Porterian mix:

> Bind up the sticks for strength. We are not Fascists.
> What will they dig up afterwards of us? Donald Duck is quacking
> his charges off to school. He will not tell them he has cancer.

Can jocularity and seriousness live together like this? How much discordance can a reader take? Disjunctions and contradictions recur: this Freud exhorts his readers/listeners to get their individual lives together, then denies any fascism on his part, though his theorizing is authoritarian. We are faced with the difficulty of resolving or accepting these contradictions. At the end, Freud is legitimized not in the act of theorizing but in reporting on a case study. What the archeologist of the future might dig up, then, is not just dirt, or a model as beautifully harmonious as San Miniato, but an everyday misery, whereby a parent feels bound to conceal from his charges the fact that he is dying. With some finesse, Porter has introduced a sudden pathos to the comic book world, indicating the often unwitting proximity of comic and tragic conditions. The effect is chilling, at first, but an aftertaste of stoical suburban heroism follows, suggesting another angle on Freud's theory of repression.

The worlds which Porter sought to inhabit imaginatively at this time were physical as well as psychic and social. Australian settings were uppermost among these. In 'A Headland Near Adelaide' (*PW* 14) and 'The Wind at Bundanon' (*PW* 12–13), for instance, place and experience were located precisely, though physical and topographical detail was subsumed by wider reflections, in these poems on evolution and romantic explanations of nature, respectively. 'The Ecstasy of Estuaries' (*PW* 10–11) was more engagingly personal. Arising from a visit to the south-west coast of Western Australia near Augusta in 1987, Porter's use of setting was important psychologically as a stimulus to memories of a Brisbane childhood on the other side of the country. Although not a topographical poet, Porter admired the beaches, estuaries and high karri forests of Australia's south-west. He was also impressed with the annual rescue of beached whales by local inhabitants, who would turn them around and take them back out to sea. This generous

act — reversing the torque towards suicide of the whales — appealed
to Porter, and his evocation of an adjacent estuary, of the Blackwood
River, is lyrical and celebratory:

> What might rock sleep is breaking out at sea
> On reefs which bear the Southern Ocean;
> Up-river pelicans on posts applaud
> Such widening to island esplanades,
> A shallow onus of the tide, the whiting
> Sketching on the bottom their own shapes.

Behind the estuary are the tall karri forests with trees 'as various
as signatures'. The speaker seems relaxed, and receptive to a spirit of
place. He allows his persona to respond to nature (as he seldom does
elsewhere in his poetry), to indulge pleasurably in a fantasy, shared
with others, of being 'Slaves to the ancient brightness of the sea'.

Such pleasurable ecstasy in stillness can only be temporary for
Porter however. His mind is quickly at work again, prompted by 'a
magistracy of memory'. These waters and sandbanks of south-western
Australia recall the Moreton Bay of childhood and a familiar melodic
refrain re-enters the verse:

> To scatter toast crumbs to the gulping gulls
> And let the dinghy flutter on the tide,
> To be reliving what was hardly lived
> When years ago the boat came back at dusk,
> A father and a son, strange strangers, home,
> This is the storytelling of the blood.

Nowhere in Porter's work is the lyrical temper more finely tuned, yet
this is not simply nostalgia; within it lies a recognition that the need to
revive the memory arises from a sense of incomplete experience: 'to be
reliving what was hardly lived'. In the poem's final stanza a dialectic
between 'movement' and 'permanency' is suggested. An abiding quality
resides in these estuaries, and the tableland behind, for which there is
no equivalent in life. They offer consolatory possibilities, a stimulus to
that personal memory which is a guarantee of humanity: 'This ecstasy
of estuaries prepares/A tableland for time to wander in'. The lyrical
force of these images is related to the writer's sense of exile from the
country he has left, but towards which he feels magnetically drawn,
through childhood memories.

Two other Australian 'place' poems enlarge Porter's concept of
'possible worlds'. 'Woop Woop' (*PW* 5–6) and 'Porter's Retreat' (*PW*
50–1) take up Australian place names (the first mythical, the
second actual) and filter them through a remarkable process, whereby
they become testaments to the pressures and ingenuities of the verbal
imagination.

'Woop Woop' is a jocular Australian term denoting any remote or supposedly backward town or district. A story in the *Bulletin* in the early 1970s recalled its use in another place: 'A London taxi ground up to the front door and a sun-tanned, behatted head emerged from its window. "That airport must have been in bloody Woop-Woop", it said in unmistakeably Australian cadences'.[8] Porter savours the humour and local colour in the term, but also reveals its origins; in this, he writes not as an etymologist but as a poet. His point of view is unmistakably metropolitan, demonstrating the need among city-dwellers to name, and thereby claim, places to which they may feel humorously superior; but he also reveals, ironically, their need to name and localize the idea of simplicity. In Porter's hands, 'Woop Woop' enters an international vocabulary. It is 'The backtrack Trebizond of everyone' — referring to the exotic port on the Black Sea in north Turkey, popularized for the English by Rose Macaulay in *The Towers of Trebizond*. The Australian version of such exotica is a 'backtrack' to a mythical place beyond civilization, which is used by writers, film makers and intellectuals to show how far they have come. Porter seems to have had in mind Sydney poet John Tranter, who came from Cooma, as the type of 'open-shirted young sophisticate' who comes to 'tame the capital', but fiction writer Frank Moorhouse's journey from Nowra to Sydney would also fit: both are versions of the same myth, the liberations of urban living are made the more 'intelligently decadent' by the distance such figures seem to have come. The film industry would be lost without such myths. One of the reasons for the international success of Australian films in the 1970s and 1980s was their evocation of a Victorian or Edwardian outback, exemplified in the film of Miles Franklin's novel *My Brilliant Career*, set in Possum Gully. It could be Porter's Woop Woop, give or take the odd detail:

The movie industry could not exist without it:
wasp-waisted girls are seen riding after Schumann
to the soup-tin letter-box to hear that London
wants their novels, and following riots in Europe,
amuse their company just naming its odd name.

For all its local colour, Porter's poem suggests that Woop Woop is archetypal, a mythical construction of those who, in the midst of turmoil and complication, find it necessary to entertain themselves with an idea of simplicity and innocence. Australian literature is spiced with stories of the inland journey, ranging from the explorers' journals to Patrick White's *Voss*. Porter is aware of this, and of similar journeys by painters such as Nolan and Boyd: 'every journey to simplicity/is inland and the parrots dress in ever-brighter/greens and scarlets the emptier the lakes they lap'.

While acknowledging the needs which drive our mythmakers towards such inland journeys — some pure, some commercial —

Porter also records the threats to the safeguards of distance and isolation which once allowed Woop Woop to exist. Is a regional consciousness still possible? Is such a 'site' (Porter uses the word with some of its literary-theoretical associations) still feasible?

> Perhaps it has no future; we know already,
> despite remoteness and the different sorts of fly,
> it has suburban aspects: nobody here must wait
> a day to hire his favourite video, and one of its sons
> read 'The Death of Virgil' through his Sunday School.

Improved transport and video-technology have brought suburban 'comforts' to such places, but their mythical simplicity was already undermined by the complex minds which occasionally inhabited them, like worms in the bud. (Porter's fellow Australian expatriate, Oxford academic Peter Conrad, is alleged to have asked at his local Tasmanian library when he was eight for a copy of James Joyce's *Ulysses*. Herman Broch's *The Death of Virgil*, Porter's example, is considered more difficult than *Ulysses*.) Porter delights in such contradictions, which resist the reducibility of the human intellect and imagination. The outback is not, for Porter, a place of simpletons, as his recurrently expressed admiration for Les Murray's 'bush baroque' demonstrates. With the fellow-feeling of a Brisbane suburban boy who knew more about Mozart than trams, Porter's parting words on Woop Woop are characterized by a friendly irony:

> It is full of details we agree to love —
> the cat called Fortunata, the minestrone
> made in milk-churns, an aunt who mounted 'Tosca'
> in a shearing shed: outside town, it offers you
> the peace inside your mother's mind, the need to get away.

For Porter, the country is not unsophisticated, but nor is it a repository of the virtue with which Les Murray and others have endowed it. What it offers is a duality: a certain security ('the peace inside your mother's mind'), and the need to flee that security as soon as possible.

'Porter's Retreat' (*PW* 50–1) is in some respects a companion piece to 'Woop Woop'. The place named 'Porter's Retreat' differs from Woop Woop in being in fact a small town, some 150 kilometres east of Sydney across the Great Dividing Range. This piece contributes to a breaking-down of distinctions between actual and imaginary. The metaphor of inland explorations was employed again in 'Porter's Retreat', but this time in the service of personal allegory. Porter's persona guides his readers to certain vantage points from which they may survey the prospect. The high expectations of early life are rendered in terms of the challenge of crossing 'the Divide', urged on by the pro-

vision of 'endless-seeming vistas ... whose common epithet was
Felix,/appointed place of all felicity'. A typical phase in the human
life-cycle is cleverly linked with a phase of Australian history, in
the nineteenth-century gold rushes, when the country was known as
Australia Felix — a prefigurement of the Lucky Country myth.

The second phase of Porter's allegory has moved beyond the Divide.
It begins with criticism of cultural historians who theorize that naming
places is an attempt to 'straighten out the land with names,/to fit a grid
of the accounting gods/on plastic otherness'. Porter was contending
here with Paul Carter's book *The Road to Botany Bay*, which he had
been critical of in a review in the *Times Literary Supplement*.[9] One of
Porter's main criticisms of the book was that it 'luxuriated' in a theoretical vocabulary which 'dispensed with the nimbus of common experience'. As with other aspects of contemporary theory and criticism, the
more antipathetic he was, the more challenged he felt to assimilate its
message. 'Porter's Retreat' is therefore one of the best tributes to a book
that could be hoped for, a poet's acceptance of the challenge to hypo-
thesize alternative strategies of naming in his own practice, to give
them 'the nimbus of common experience':

> Here by Disappointment Bluff, gaze
> across the Vale of Sixty to Uncanniness;
> beyond the rock-strewn creek
> is an escarpment called Incalculable
> and the fields are more elided than Elysian —
> mark this tree the point of going back
> and set it on the map — Porter's Retreat —
> the place at which all further progress
> ceases to have consequence.

This sense of a cessation of 'consequence' might be a liberation for
a poet/commentator, freeing the imagination from unnecessary constraints. The third and concluding stanza reveals Porter's persona
further inland. He is beyond 'the people's fort/where families collaborate
with the sun/to make home-movies of divinity'. The allegorical
explorer has now aged and is 'digging for nurture under balding trees,/
only too willing to fold up the map/and start the evening's diary entry'.
The next day will be 'a scorcher'. In the meantime, he will name a
nearby hill 'Mt Misery/and the muddy tank-full where the river/dips
into the underworld will make/a just impression as Lake Longevity'. In
authentic Porterian tones, these lines make jokes and music of the
sadness of an unchosen life.

As important as the myths of pioneering are those of memory and
the past. This is one of the central insights of *Possible Worlds*. It is
dramatized strikingly in the poem 'They Come Back More' (*PW* 46–8),
in which the pronoun 'they' refers to ghosts of a familial past — father,

mother, friends, an Australian house, a wife who died. These actors in the poet's memory convincingly replace the overblown Lowellian portraits of forebears in the early poem 'Five Generations'. 'They Come Back More' is arranged in five parts, each of which starts *in medias res* with the conjunction 'and', the simplest of all narrative devices, suggesting that each ghostly story is in mid-flow. The first part commences: 'And I thought they all had gone'. The fifth part, summing up the other four, opens with 'And they come back more, the more to kill'. Memory can be destructive as well as creative and life-enhancing. No resolution is offered to the crises of living. Porter suggests that we are on an eternal loop of memory, destined to keep returning to the past: this is one of the chief guarantees of suffering, and of our humanity. The revenants thus gain new lives of their own. In accusing himself of an unfulfilled life, Porter has always imagined richer experience elsewhere: 'Where I should have packed/My sandshoes, taken a striped towel and zinc cream/To the sun, I sulked instead on Genoa velvet/For old Vienna and old manses'. The sense of exclusion is often a front for such rich ante-chambers of possibility.

By the late 1980s, Porter was also exploring actual and possible worlds in his newspaper reviews of contemporary poetry, fiction, television and cinema. One of the films reviewed at this time was the much-vaunted *Sammy and Rosie Get Laid*,[10] in which Porter observed that filmmakers Kureisha and Frears were right to show Britain as 'a vicious and self-divided society'. Moreover, their 'picture of nastiness, if only sketched in, is concentrated in the right places — property developers, the police, a long-held-back national vindictiveness'. Porter's criticism of the film was on other grounds: that its main figures were 'a troupe of sleepwalkers hardly seeming to notice the misery around them'. With an unusual capacity to place cinema in relation to the other arts, Porter observed that 'you have to work on a scale like Delacroix's to turn Brixton riots and Ladbroke Grove evictions into properly demonic tableaux', and that Frances Barber's Rosie 'wears one smirk throughout, a pedal-note of self-concern'. Artistic and moral concerns were intertwined here, as they were in Porter's comment that the film was full of sex but no feeling. Whereas the previous film by Kureishi and Frears, *My Beautiful Laundrette*, had been 'psychologically intimate and personal — a piece of cinematic chamber music', *Sammy and Rosie Get Laid* seemed 'inflated and hectoring, a small-budget *Sardanapalus*'. While Porter's abhorrence of racism was evident in this review, so too was his preference for art which gives scope to feeling — a reminder of this important feature of his own work.

Through his reviews in a variety of newspapers and magazines, mainly in Britain and Australia, Porter explored a range of utopian and dystopian visions. The catchment zone for *Possible Worlds*, as with previous volumes of poetry, was showered upon by books, films, paintings, music and television, which his agile and retentive mind stored for their insights into his poetic concerns. In the *Guardian*, for instance, Porter praised Primo Levi's *The Drowned and the Saved* for its refusal to 'indulge in Right Wing Realpolitik which asserts that Auschwitz is nothing beside Gulag'. Furthermore, Levi had convinced him that 'dictatorships end up poisoning their own tongues'.[11] Porter's comments on Levi's representations of hell in the concentration camps were no mere surface scratchings: he was searching for the language, motives and feelings of the experience, and some of the book's insights parallel those in Porter's own 'Essay on Patriotism' (*PW* 31−2).

An increasing recognition of local and regional differences in Australian society and culture was evident in Porter's reviews in the late 1980s, thus extending his view of possible worlds. Fellow expatriate Peter Conrad's return to Tasmania, recorded in his book *Down Home* (1988), was praised by Porter for the richness of its prose and its strong images: 'evocations of the Huon Valley, the horrors of Macquarie Harbour, the iconic world of hydro-electricity, and the efforts of early artists to cope with Tasmania's landscape'.[12] In this return to his roots, Conrad had thrown away his conjuror's gear, reined in his cleverness and was writing like a 'Ruskin of the South Pacific, illuminating the moral map as well as the physical one'. Porter's was of course a very personal reading of Conrad's project, confirming the impression of his suite of Australian poems in *Possible Worlds*, that once a talented writer has achieved sufficient distance from 'home', the place and its associations can engage the intellect and emotions with an almost preternatural force.

Significantly, Porter's imaginative and emotional affiliations with Australia were most fully tested in his reviews of Australian poetry. Lengthy reviews of recent books by Les A. Murray and Chris Wallace-Crabbe indicated some directions in his thinking about Australia as a renewed source of imaginative possibility. In 1981, after publication of Murray's *The Boys Who Stole the Funeral*, Porter observed how different Murray was from the bush poets he had struggled to read in his youth. Murray was 'a true virtuoso, who protects his chosen people — the small farmers, the outback experts, the beleaguered women — with rituals and spells couched in Pentecostal language'.[13] In 1988, after publication of *The Daylight Moon*, Porter's praise was no less stinting, Murray was 'the finest Australian poet alive and one of the best in the English-language world'.[14] Having said this, Porter was the better able to dispute Murray's ideology, which was, in Porter's view, 'to make Australia into the land he believes it should be (and, more alarmingly, is intended to be by some shaping spirit — God or the National Will)'. When writing about the country people around Bunyah in New South Wales, Murray 'delights in idiosyncrasy, ritual, prejudice and even brutality, since the people are in possession of the spirit'. When Murray switches from dramatizing to polemics however, he becomes a 'talkative priest' preaching '*Das Volk*'. The deliberate overstatement relates to Porter's own counter-ideology expressed in his 'Essay on Patriotism', 'On First Looking into Chapman's Hesiod' and elsewhere, which favours metropolitan secular pluralism over any monistic view of 'the people' or of 'God'. Porter shrewdly observed that Murray 'both is and yet is not a religious poet: his Catholic universality is always running aground on the residual reefs of Presbyterian tribalism'. An agrarian identity for Australians is insufficient, in Porter's view, but more significantly, it may reduce Murray's prodigious poetic talents: 'Retreat, even inside your tribe, may not be the best way to cultivate your voice'. With affectionate humour, Porter exhorts Murray to keep his road open from the farm where he lives to the city, for 'his country and its literature need him at the centre of things'.

Porter was also reading his affinities and views of Australia at this time into Chris Wallace-Crabbe's poetry. Reviewing Wallace-Crabbe's *The Emotions are not Skilled Workers* in 1981,[15] Porter had observed that this writer was clearly on the Melbourne side of 'the great Melbourne/Sydney divide in Australian literature': 'His poetry moves without difficulty through the England-rinsed scenes of Melbourne to reflections of a world facing economic and political breakdown. He writes about banana trees and yams as happily as about Herrick, Galileo and the Emperor Frederick, Stupor Mundi'. Such versatility clearly appealed to Porter. In fact, Wallace-Crabbe seemed to be heading in some similar directions to Porter himself:

> We start a bit grey and elderly: only later, after much experience,
> do we throw off ponderousness, embrace wit and light-
> spiritedness and appear verdant to the public gaze.[16]

If, as Porter asserts, Wallace-Crabbe was writing, 'after thirty years on the job, with twice the élan [he] had at the beginning', the same could be said about Porter's work at least since *The Automatic Oracle* (1987). A significant formal development in Wallace-Crabbe's work (as in Porter's own) was what Porter termed 'the extended discourse ... a relaxed structure, with its obligatory conversational tone and an easy habit of flaring into purple'. The capacity of such poetry to be 'lyrical while disputatious' — the written speech of a liberal humanist who 'wants to modulate to beauty without interrupting [the] speech' — reflects Porter's tendencies as much as Wallace-Crabbe's. The ode-like structure, conversational, yet capable of lyrical flights as of aphoristic injunctions, and often evoking a state of wakeful dreaming, had been used effectively by Porter since his 'Ode to Afternoon' (*LIACC, CP* 200—1). It was now being used to good effect in Australia by Wallace-Crabbe, Murray and others such as David Malouf, Philip Salom and Peter Rose. The 1960s campaign for the lyric *pur sang* as Australia's only 'natural' form, accompanied by attacks on A. D. Hope's 'discursive mode', appeared to have been laid to rest by the late 1980s. Porter's unique mix of disputation, humour and lyricism was being taken up in different ways by an Australian company of scribes.

In some respects then, Porter's prose reviews may be seen as a running commentary on developments in his verse at a particular time. While the poetry has a clear priority, the literary journalism has kept the poet alert to the *dailiness* of experience, the just-possible world of deadlines and anger. But the anxious strivings of literary journalism could also provide encounters with eternal issues. Porter's extensive review of Colleen McDannell and Bernhard Lang's book *Heaven: A History*,[17] for example, raised important issues for the 'hungry soul' of 'Stratagems of the Spirit' (*PW* 55—6). The history of ideas and images of heaven (and the more human-centred notion of paradise) clearly fascinated Porter, though he was not dislodged from a sceptical humanist stance: 'What you dress your paradise in must derive from what you have known in life, however much you expect it to transcend its earthly model'. And later: 'It could be that we are not judged by God but are contributing to inventing him'. With these premises, it is not surprising to find that fanaticism and fundamentalism are equally repugnant to this author. Porter makes some shrewd observations of his own on artistic representations of heaven and hell: 'Hell has done better at the hands of the painters: its iconography gives them something to work with. However, God continues to get the best tunes'. Preferring the Mahlerian Heaven to Swedenborg's 'ultimate welfare state', Porter ultimately reverts to the best approximation to heaven he knows: 'Schubert, wavering between A minor and A major, gives us all we can ever really know about Heaven until we get there: "Schöne Welt, wo bist du?"'

The music of other worlds may be revealed in stranger ways. In 'The Blazing Birds' (*PW* 9), a wonderfully discordant song of praise seems to emerge from Australia's birds which were called 'songless' by Kendall and other early colonial writers. Written in Western Australia in 1987, the poem celebrates through a marvellous display of punning, alliteration, onomatopaeia and syntactic gymnastics the cacophonous energy which is its own song of praise:

> On a mat of pier, Australia's noisy birds
> are sucking anthems. So much suck comes out
> with lumps of sun it spells Magnificat.

The poet, caught up in the game, admits, 'I almost lose my way among similitudes'. Like Mahler, he knows that 'Bird cries may seam into our symphonies'. For Porter, a host of celebrants are at work:

> The 'twenty-eight', the spangled drongo, kooka
> with its caco-credo, magpie mutts,
> what messages they drag across the sky!

Kings to fish for, larks to scent the air,
a parlement of fowles refuelling,
and Bib and Bub expelled from Paradise.

The innocent anthropomorphisms in which romantic namers of
nature like to indulge will be tolerated by these birds only up to a
point: they live in a much tougher world. May Gibbs's Australian bush
paradise and its creatures in her books for children (such as *Bib and
Bub in Gumnut Town*) is not the one Porter imagines these birds
inhabiting. Like earlier magpies and kookaburras in Porter's doxology,
these birds are predators, however exotic their names, and share their
murderous impulses with humans. This, more than their conventional
romantic qualities, makes them worthy of our attention, even praise.
Like Chaucer's allegorical fantasy, *The Parlement of Fowles*, to which
Porter alludes, his poem is a dream-vision. It proposes no belief in
'Nature, the vicaire of the almyghty Lord'; rather, it is human-centred
and mixed in tone:

All at once and always changing gear
the Sistine servers shrike. Perennial
the praise and every liturgy a laugh.

Such promiscuous rapidity of praise, phrase, laughter and discord co-
exist in Porter's dazzling imaginative cosmopolis of birds.

Love, the source towards which the poems in *Possible Worlds*
move, will be found, if at all, as a harmonic key emerging from the
welter of experience. There are cautionary tales. 'Open-Air Theatre,
Regent's Park' (*PW* 44–5), indicates the dependency and humiliation
which sensual and sexual love may impose on people, as it recalls a
visit to the park with Jannice Porter, not long before her death, to see
Shakespeare's *The Two Noble Kinsmen*. 'The Camera Loves Us' (*PW*
34) offers the false god of self-love, previously explored in *Narcissus*.
A more sustaining form of contemplative love is suggested in 'Little
Buddha' (*PW* 41–2), in which the household figure's 'unchanging
look' draws the observer to 'The lure of impersonal/Truth, a silence of
the stars'. This is one, but not the only stratagem. Alternatively, one
may seek love in a variety of partners. This 'lure' is tested in 'Hand in
Hand' (*PW* 57–8), where the implicit search is for a Platonic wholeness, a joining of the hemispheres. But the speaker here perceives love
as a quality recognized chiefly in retrospect:

That hand's in this hand, so no one sees
Their partnership is change, that love is like
A signal passing which may startle air
Only by its afterglow ...

This 'afterglow' is the aura in which Porter's most profound poetic romances occur.

The chief consolation of a spirit which sees itself as warped, and exiled from happiness, is music. One of the key poems in *Possible Worlds*, 'The Orchard in E-Flat' (*PW* 53—4) recalls Porter's early poem 'Walking Home on St Cecilia's Day' (*OBTB, CP* 13), where the speaker found himself 'Saddled with Eden's gift, living in the reins/Of music's huge light irresponsibility'. The reins are lighter in 'The Orchard in E-Flat', a relaxed and fanciful ode which cogitates on the ways music may fit into human lives. The world here is perceived as a 'chord of limitless additions' beyond the central chord which controls life. That central chord is the E-Flat major, the opening key of Wagner's *Rhinegold* Prelude:

> Behind us is the deep note of the universe,
> The E-Flat pedal on which time is built,
> Spreading and changing, both a subtle
> Growth of difference and a minimalist
> Phrase, with bridges crossing it and staves
> Of traffic on its tide, a broad bloodstream
> To carry to the delta full mythologies.

The poem confirms its writer's view of music's greater self-sufficiency than literature, yet it manages the difficult accommodation of rhythm and sound to dream-like details within a verbal context suggestive of music. Porter has correctly observed that poetry is an art of signs, to which meanings are attached, however much the poet may attempt their detachment: musical equivalences such as onomatopaeia work only when 'the mimetic noises are fixed firmly in their sphere of operation by overt meaning'.[18] Music is 'an abstract system of sounds which is totally real at the same time'; nor does the fact that we associate these sounds with emotions and parts of the concrete world detract from music's 'self-sufficiency'.[19] Porter has stressed that poetry can only ever be analogous to music, as he has attempted to demonstrate in distinguishing between degrees of 'musical' control in poetic 'Deep Form, Middle Form and Light Form'.[20] Not only has Porter encouraged and inspired contemporary composers such as Geoffrey Burgon,[21] he has also assimilated musical ideas and practices into his own work, ranging from the early 'Lieder' of 'The Porter Song Book' (*APF* 110—15) to the fantasia of 'The Orchard in E-Flat'. His appreciation of music, as of painting, is a source of his virtuoso approach to language. In this light, the opening and closing phrases of 'The Orchard in E-Flat' deserve attention. The opening phrase 'The waves are weeping vaguely' is taken up in the last words of the poem, 'the weeping waves'. The macaronic pun of the first line (the French word for wave is 'vague') indicates the kind of playful manipulation of words which is analogous to a musical composer's 'limitless additions' to his central chord.

'The Orchard in E-Flat' is emblematic of Porter's work to date in that it takes up into its 'everlasting anthem' fragments of the author's remembered or imagined life. Some have a personal plangency: 'A mother and a boy come to the orchard/To turn a cow back to its field'. Other phrases emerge as a free association of memory and sound. In the last stanza, Caliban is quoted. In *The Tempest*, Prospero tells Caliban to bring in some more wood. Caliban replies, as in Porter's poem, 'There's wood enough within'. The pun has a serious intent, suggesting that the fuel we require is inside ourselves: the human imagination may in itself sustain us, keep us alive and warm. If this seems a somewhat limited outlook, even bleak, it must be set in the context of a poem which recognizes a music of the spheres created by human hands and minds. Porter's vision does not deny the misery or helplessness of people, but it suggests that art (for which music stands as the purest form) can alert humans to a 'chord of limitless additions', offering a space 'Where the bruise of exile turns to timeless rose'. But this will not be the final word. Such an imaginative and inventive poet will surely compose many further variations on his major theme.

NOTES

◆

INTRODUCTION

1. Clive James, *The Review*, No. 24 (December 1970), 53.
2. Stephen Spender, 'Wakeful Dreaming', the *Observer*, 27 March 1983, 33.
3. See 'Peter Porter in Profile', *Westerly*, 27, 1 (1982), 45−57.
4. Anthony Thwaite, *Poetry Today: A Critical Guide to British Poetry 1960−1984* (Longman, Harlow, 1985), 67.
5. Peter Porter, 'In Exile', Part One, ABC Script, Radio Helicon (15 November 1987), unpublished typescript.
6. Ibid.
7. Stephen J. Greenblatt, (ed.) *Allegory and Representation* (The Johns Hopkins University Press, Baltimore, 1981), viii.
8. Quoted in John Keane, 'The rebel with a democratic cause', the *Weekend Australian*, 23−4 December 1989, 1.
9. James Olney, *Metaphors of Self: The Meaning of Autobiography* (Princeton University Press, Princeton N.J.), 1972.
10. *Quadrant*, 37, 9, (September 1983), 30.
11. See Ursula Hoff, *The Art of Arthur Boyd* (André Deutsch, London, 1986), 76.

1

THE BRIGHT LOCKED WORLD

1. 'A Philosopher of Captions' in *English Subtitles*, *Collected Poems*, 279.
2. See Bruce Bennett, 'Peter Porter in Profile', *Westerly*, 27, 1 (March 1982), 49.
3. Peter Porter, 'Locked Out of Paradise', *The New Review*, 3, 36 (March 1977), 15−20.
4. Peter Porter, 'In Exile: An Autobiographical Anti-Biography', Part One of a three-part script for ABC Radio Helicon, unpublished typescript (Sydney, 1987).
5. See Peter Porter, 'Locked Out of Paradise', 15.
6. 'Peter Porter in Profile', 46.
7. 'In Exile', 4.
8. Peter Porter, 'Brisbane Comes Back', *Quadrant*, 98, XIX, 6 (September 1975), 53−8.
9. Ibid.

10. See Peter Porter, *Sydney* (Time-Life, Amsterdam, 1980).
11. 'In Exile', 5.
12. Ibid., 6.
13. Porter has commented that the biblical quotation reached him characteristically through its musical setting and not through reading the Bible — via Handel's *Messiah* and Gibbons's anthem 'This is the record of John'.
14. 'Locked Out of Paradise', 19.
15. Ibid., 18.
16. Ibid.
17. 'Peter Porter in Profile', 46.
18. Ibid., 46–7.
19. 'In Exile', Part One: Inventing the Past, 12.
20. Ibid.
21. See Bruce Dawe, *Sometimes Gladness: Collected Poems, 1954–1982*, second edition (Longman Cheshire, Melbourne, 1983), 39.
22. 'Peter Porter in Profile', 42.
23. Ibid.
24. Ibid.
25. Manning Clark, *A Short History of Australia* (Heinemann, London, 1969), Ch. 1 and *Occasional Writings and Speeches* (Fontana, Melbourne, 1980), Ch. 1.
26. Interview with Peter Spearitt for the 1938 volume of the Bicentennial History Project, unpublished typescript (London, 18 July 1979).
27. See Henry Handel Richardson, *Myself When Young* (Heinemann, London, 1964), 178.

2

EACH UNSHINING HOUR

1. Interview with Peter Spearritt, London, 18 July 1979 (unpublished typescript), 6.
2. Ibid. 'My mother, I think, was what you could call an "aspiring atheist". Neither of them had the slightest religious conviction that I was ever aware of'.
3. Ibid.
4. Ibid., 7.
5. Peter Porter, 'Brisbane Comes Back', *Quadrant*, 98, XIX, 6 (September 1975), 53–8.
6. Letter from P. D. Edwards, 27 August 1987. Edwards was two years Porter's junior. At the time of writing he was a Professor of English and Pro-Vice-Chancellor at the University of Queensland.
7. 'Brisbane Comes Back', 33–4.
8. Letter to Bruce Bennett, 13 February 1986. The quotes which follow are also from this source.
9. Letter from P. D. Edwards, 27 August 1987.

10. Letter from Peter Porter to Bruce Bennett, 13 February 1986.

11. Ibid.

12. The most striking of these is Martin Bell's 'Headmaster: Modern Style' which, however, lacks the compact force of 'Mr Roberts'. See Peter Porter (ed.) *Martin Bell: Complete Poems* (Bloodaxe, Newcastle upon Tyne, 1988), 48−51.

13. Peter Edwards recalls that the boy in 'Mr Roberts' who shot himself not long after leaving TGS was named Boadle. Letter, 27 August 1987.

14. Dorothy Green, 'Consumer's Report', *Hemisphere*, 28, 5 (March-April 1984), 297−8.

15. It seems that Porter here, as elsewhere, took liberty with the facts. The train journey from Toowoomba to Brisbane took approximately three hours in 1945 and did not involve travelling through the night. The literal accuracy of details is unimportant beside the poem's central thematic concerns.

16. See Eric W. White (ed.) *15 Poems for William Shakespeare*, The Trustees and Guardians of Shakespeare's Birthplace, Stratford-upon-Avon, 1964. The other contributors were Edmund Blunden, Dom Moraes, Charles Causley, Roy Fuller, W. J. Snodgrass, Thom Gunn, Stephen Spender, Randall Jarrell, Derek Walcott, Thomas Kinsella, Vernon Watkins, Laurie Lee, David Wright and Hugh McDiarmid.

17. Interview with Bruce Bennett.

18. See 'Three Transportations', *TCOS, CP* 256−8.

19. Michael Hulse, 'Love and Death: Nine Points of Peter Porter', *Quadrant* (September 1983), 32.

20. For a discussion of Furphy in this respect, see R. S. White, *Furphy's Shakespeare*, The Centre for Studies in Australian Literature, University of Western Australia, 1989.

21. 'Peter Porter in Profile', *Westerly* 27, 1 (March 1982), 47.

22. Bruce Bennett, 'Australian Literature and the Universities', in Stephen Murray-Smith (ed.) *Melbourne Studies in Education 1976* (Melbourne University Press, Melbourne, 1976), 117.

23. A. A. Phillips, 'The Cultural Cringe', *Meanjin*, 9, 4 (1950).

24. Unpublished address by P. D. Edwards at Toowoomba Grammar School, 1983.

25. P. R. Stephensen, *The Foundations of Culture in Australia: An Essay Towards National Self Respect* (W. J. Miles, Gordon, NSW, 1936).

26. Letter to Bruce Bennett, 25 July 1989.

27. *The Toowoomba Grammar School Magazine and Old Boys' Register*, 1945 and 1946. Essay on 'Richelieu', 23; 'Santa Lucia', A One-Act Play in Five Scenes, 23−5 (Alan Dunn Memorial Prizes), 1945. 'Great Men I Should Have Liked to Have Met', 19 (Eric Partridge Essay Prize); 'The Secret House', An Interlude in One Scene, 22−3; and 'Napoleon', A Short (Verse) Trilogy, 1946, 23−4.

28. This extract is from the *Toowoomba Grammar School Magazine*, 1946. The author is 'Todd, VIA'.

29. Interview with Peter Spearritt, 10.

3

SUMMER HERMIT

1. David Malouf, 'Escaping the Circle of Hell: the Poet's Journey', the *Weekend Australian*, 15—16 October 1988, 13.
2. Brian Penton, *Advance Australia — Where?* (Cassell, London, 1943), 234.
3. David Malouf, *12 Edmonstone Street* (Penguin, Harmondsworth, 1986), 10.
4. David Malouf, *Johnno* (Penguin, Harmondsworth, 1976), 4.
5. Peter Porter, 'Brisbane Comes Back', *Quadrant*, 98, XIX, 6 (September 1975), 53—8.
6. See Dorothy Jones and Barry Andrews, 'Australian Humour', in Laurie Hergenhan (ed.) *The Penguin New Literary History of Australia* (Penguin Books, Ringwood, 1988), Ch. 4.
7. Peter Carey, 'A Love Affair with Losers', the *West Australian* (26 September 1987), 45.
8. Peter Porter, 'Locked Out of Paradise', *The New Review*, 3, 36 (March 1977), 18.
9. Peter Porter, 'Brisbane Comes Back', 57.
10. 'Locked Out of Paradise', 19.
11. Interview with Peter Spearritt, 18 July 1979, unpublished typescript, 4.
12. See Interview with Germaine Greer in Clyde Packer (ed.), *No Return Ticket* (Angus and Robertson, Sydney, 1984), 86.
13. 'Brisbane Comes Back', 53.
14. Interview with Peter Spearritt, 14.
15. Hugh Lunn, *Joh*, 37, quoted in Ross Fitzgerald, *From 1915 to the Early 1980s: A History of Queensland* (University of Queensland Press, St Lucia, 1984), 245.
16. Ross Fitzgerald, *From 1915 to the Early 1980s*, 93—5.
17. Charles Osborne, *Giving It Away: The Memoirs of an Uncivil Servant* (Secker and Warburg, London, 1986).
18. 'Brisbane Comes Back', 57.
19. Ibid.
20. 'Peter Porter in Profile', 49.
21. Letter from P. D. Edwards, 28 August 1987.
22. Ibid.
23. Robert Gray, 'Peter Porter and Australia', *Poetry Review*, 73, 1 (March 1983), 18
24. 'Brisbane Comes Back', 58.
25. Ibid., 55
26. John La Nauze, *Walter Murdoch: A Biographical Memoir* (Melbourne University Press, Melbourne, 1977), 122.
27. Walter Murdoch, *Speaking Personally* (Angus and Robertson, Sydney, 1930), Preface.
28. See Bruce Bennett, *Place, Region and Community* (Foundation for Australian Literary Studies, Townsville, Monograph Series No.11, 1985), 41—60.
29. A. A. Phillips, *The Australian Tradition* (Longman Cheshire, Melbourne, 1985), 82.

30. Alan Moorehead, *Rum Jungle* (London, 1953), 9–12, quoted in G. Serle, *From Deserts the Prophets Come* (Heinemann, Melbourne, 1973), 125.
31. C. J. Koch, 'Maybe It's Because I'm a Londoner', *Kunapipi*, VIII, 1 (1986), 7.
32. 'Locked Out of Paradise', 18.
33. See 'Peter Porter in Profile', 56; and '"In the New World Happiness Is Allowed"' (*TCOS*, *CP* 254–5).
34. 'Brisbane Comes Back', 54.
35. 'In Exile', 68.
36. 'Locked Out of Paradise', 17.
37. 'In Exile', 9.
38. Ibid.

4

AT HOME, AWAY

1. Jill Neville, *Fall-Girl* (Panther Books, London, 1965).
2. Ibid., 29.
3. Ibid., 27.
4. Ibid., 114.
5. Peter Porter, 'Locked Out of Paradise', *The New Review* 3, 36 (March 1977), 17.
6. Peter Porter, 'Living in London', *London Magazine*, 13, 2 (June/July 1973), 62.
7. Ibid., 63.
8. Jill Neville, interview, London 21 February 1989.
9. Ibid.
10. *Fall-Girl*, 30.
11. Ibid., 50–5.
12. Jill Neville, interview, 21 February 1989. A prelude to this incident for Peter Porter had been almost disastrous. Distraught with rage and a confusion of feelings he was run down by a car and ended up in hospital, though without serious injury. (Interview with Roger Covell, Sydney, 4 May 1989.)
13. Philip Toynbee, 'From Virgil to Kingsley Amis', the *Observer*, 22 April 1962.
14. E.g. 'Peter Porter in Profile', *Westerly*, 27, 1 (March 1982), 49.
15. Clive James, 'The Boy from Brisbane', *Poetry Review*, 73, 1 (March 1983), 25–7.
16. 'Living in London', 63.
17. Ibid., 64.
18. Ibid.
19. See Roger Garfitt, 'The Group' in M. Schmidt and G. Lindop (eds.), *British Poetry Since 1960: A Critical Survey* (Carcanet Press, Oxford, 1972), 13–69; and Edward Lucie-Smith and Philip Hobsbaum (eds.) *A Group Anthology* (Oxford University Press, London, 1963).
20. See G. S. Fraser, 'The 1950s, The Movement and the Group' in *The Modern Writer and His World* (André Deutsch, London, 1964), 346–55.

21. *A Group Anthology*, vi.
22. Blake Morrison, 'A Philosopher of Captions', the *Times Literary Supplement*, 1 April 1983, 321—2
23. Philip Hobsbaum, Letter to the *Times Literary Supplement*, 15 April 1983, 379.
24. 'The Group', 13.
25. 'Peter Porter in Profile', 49—50.
26. BBC Radio Script, 'Poetry Up to Now: the Movement', 24 November 1980.
27. Ibid.
28. See Blake Morrison, *The Movement: English Poetry and Fiction of the 1950s* (Oxford University Press, Oxford, 1980), 248—53. 'Fear of the left became dominant in the Movement ideology in the 1960s, with Amis, Conquest, Davie, Larkin and Wain in particular being affected by it'.
29. *The Modern Writer and his World*, 352.
30. Peter Porter (ed.) *Martin Bell: Complete Poems* (Bloodaxe Books, Newcastle upon Tyne, 1988), 15.
31. Edward Lucie-Smith's foreword to *A Group Anthology* gives details of the usual format of these meetings.
32. Ibid., vii.
33. *A Group Anthology*, Epilogue.
34. Peter Porter, BBC script, 'Poetry in the Sixties', 1980.
35. Peter Porter, 'Grub Street versus Academe', *The New Review*, 1, 4 (1974), 47—52.
36. Roger Garfitt, 'The Group', 26.
37. *Martin Bell: The Complete Poems*, Introduction, 15.
38. Ibid., 17.
39. Ibid., 20.
40. Ibid., 19.
41. Peter Porter, 'Peter Redgrove: A Brief Memoir', *Poetry Review*, 71 (2/3 September 1981), 9—14.
42. Ibid.
43. Ibid.
44. Anthony Thwaite, 'Porter: Man and Poet', *Poetry Review*, 73, 1 (1983), 21—2.
45. Ibid.
46. Ibid.
47. The poem was 'An Anthropologist's Confession'. The exhibition was held at the British Museum in 1975.
48. Patrick White, *Flaws in the Glass* (Jonathan Cape, London, 1981), 13.
49. Margaret Owen, 'Peter Porter and the Group', *Poetry Review*, 73, 1 (1983), 12.
50. Ibid., 13.
51. Ibid.
52. Ibid.

5

CARNIVAL WINTER

1. See Brian Masters, *The Swinging Sixties* (Constable, London, 1985), 13.
2. Ibid. The *Time* cover story appeared on 16 April, 1966.
3. Ibid., 14.
4. John Press, the *Sunday Times*, 25 November 1962. The poems in *New Poems 1962* (Hutchinson, 1962) were selected by Patricia Beer, Ted Hughes and Vernon Scannell. Other poets in the volume include Larkin ('Ambulances'), Kinsella, MacBeth, Thomas, Thwaite and Tomlinson.
5. The *Observer*, 19 July 1963.
6. John Rolph, 'A Note on the Scorpion Press', *Poetry Review* 73, 1 (1983), 13.
7. Ibid. See also John R. Kaiser, *Peter Porter: A Bibliography 1954–1986* (Mansell Publishing, London and New York, 1990), 2.
8. Of the eight poems published before 1961, five had appeared in *Delta* and one each in the *Times Literary Supplement* ('Metamorphosis'), *Outposts* ('Tobias and the Angel') and the *Observer* ('Mr Roberts').
9. Richard Kell, the *Manchester Guardian*, 17 March, 1961.
10. Ibid.
11. Anthony Cronin, the *Daily Telegraph*, March 1961.
12. Donald Davie, the *Spectator*, 24 March 1961.
13. A. Alvarez, 'Poetry Round-Up: Ginsberg and the Herd-Instinct', the *Observer*, 14 May 1961.
14. Ibid.
15. 'Stylish Rhymes', the *Times Literary Supplement*, 9 June 1961.
16. Ray Mathew, 'From Brisbane to Kensington', the *Bulletin* (17 June 1961).
17. Sylvia Lawson, 'Elizabethan Smoothies', *Nation*, 85 (Sydney, 13 January 1962), 21–2.
18. Ibid., 21.
19. Dorothy Green, 'Consumer's Report: Peter Porter, *Collected Poems*', *Hemisphere*, 28, 5 (March-April 1984), 297–8.
20. 'Elizabethan Smoothies', 21.
21. Michael Hulse, 'Love and Death: Nine Points of Peter Porter', *Quadrant* (September 1983), 31–2.
22. Anthony Thwaite, 'Porter: Man and Poet', *Poetry Review*, 73, 1 (1983), 21.
23. Peter Porter interviewed by Alan Brownjohn, *The Literary Review*, 59 (May 1983), 22.
24. Telephone conversation with Richard Neville, 5 July 1989.
25. Interview with Ian Hamilton in London, 3 January 1986.
26. John Fuller, 'Of their Age', *Time and Tide* (30 March 1961), 525.
27. Letter from Peter Porter, 13 February 1986.
28. Interview with Diana Phillips at Kingston-upon-Thames, 15 December 1988.
29. Letter from Peter Porter, 13 February 1986.
30. Frederick Grubb, 'Exile, Vigour, and Affluence: Peter Porter', in *A Vision of Reality: A Study of Liberalism in Twentieth Century Verse* (Chatto and Windus, London, 1965), 191.
31. Ibid.

32. Ibid., 193.
33. Ibid., 199–200.
34. Interview with Diana Phillips, 15 December 1988.
35. 'Forward the Anti-Dressers', the *Evening Standard*, 19 July 1961.
36. See Edward Timms and David Kelley (eds.) *Unreal City: Urban Experience in Modern European Literature and Art* (St Martin's Press, New York, 1985), 3.
37. Ibid., 4.
38. For a further discussion of this issue, see Milton J. Bates, *Wallace Stevens: A Mythology of Self* (University of California Press, Berkeley, 1985), Ch. 4.
39. Interview with Alan Brownjohn, *The Literary Review*, 59 (May 1983), 22.
40. G. S. Fraser, *The Modern Writer and his World* (André Deutsch, London, 1964), 352.
41. Jonathan Raban, *The Society of the Poem* (Harrap, London, 1971).
42. Margaret Owen, 'Peter Porter and the Group', *Poetry Review*, 73, 1 (1983), 12.
43. Blake Morrison, 'A Philosopher of Captions', the *Times Literary Supplement*, 1 April 1983, 321.
44. Michael Hulse, 'Love and Death: Nine Points of Peter Porter', *Quadrant* (September 1983), 34.
45. Ibid.
46. 'A Philosopher of Captions', 321.
47. 'Peter Porter in Profile', *Westerly* 27, 1 (March 1982), 53.
48. Four of the fifty epigrams from Martial first appeared in *Poems Ancient and Modern* (1964) and another eleven were in *The Last of England* (1970).
49. Peter Porter, *After Martial* (Oxford University Press, London, 1972), xiv.
50. A. Alvarez, *The New Poetry*. 'The New Poetry or Beyond the Gentility Principle' (Penguin, Harmondsworth, 1966), 21–32.
51. See William H. Wilde, Joy Hooton and Barry Andrews, *The Oxford Companion to Australian Literature* (Oxford University Press, Melbourne, 1985), 554.
52. Leonie Kramer, *A. D. Hope* (Oxford University Press, Melbourne, 1979), 8.
53. Michael Hulse, 'Love and Death', 32.
54. A. Alvarez, 'The Poems of an Ad-Man', the *Observer*, 26 December 1964.
55. [Clive James], the *Times Literary Supplement*, 18 September 1969, 1021.
56. See Martin Seymour-Smith, *Guide to Modern World Literature* (Wolfe, London, 1973).
57. Peter Porter interviewed by Peter Ryan, London, 21 September 1976 (unpublished typescript).
58. Interview with Trevor Cox, London, 24 February 1989.
59. Quoted in *The Most Beautiful Lies*, (ed.) Brian Kiernan (Angus and Robertson, Sydney, 1977), 49.
60. Michael Levey, *The National Gallery Collection* (The National Gallery, London, 1987), 34.
61. Porter interviewed by Peter Ryan, unpublished typescript (London, 21 September 1976).
62. See Hal Porter's Introduction to *Coast to Coast* (Angus and Robertson, Sydney, 1963); and Michael Wilding, 'Tabloid Story' in *Cross Currents: Magazines and Newspapers in Australian Literature*, (ed.) Bruce Bennett (Longman Cheshire, Melbourne, 1981), 232.

63. The *Listener* (11 October 1962).
64. Brian Adams, *Sidney Nolan: Such is Life* (Hutchinson, London, 1987), 154.
65. The *Listener* (10 January 1963).

6

THE STRETCH MARKS ON HISTORY

1. The *Listener* (10 January 1963).
2. Martin Seymour-Smith, 'The True and the False', the *Daily Mail*, January 1965.
3. Christopher Ricks, 'Desperate Hours', the *New Statesman*, 15 January 1965, 79.
4. Letter from Peter Porter, 13 February 1986.
5. 'Desperate Hours', 79.
6. Cyril Connolly, 'Martial Splendours', the *Sunday Times*, December 1964.
7. Blake Morrison, 'A Philosopher of Captions', the *Times Literary Supplement*, 1 April 1983, 321.
8. Thomas Dilworth, Windsor University, unpublished typescript.
9. Cyril Connolly, 'Modern Musings', the *Sunday Times*, 17 August 1969.
10. Brian Jones, 'Black Feathers', *London Magazine*, 9, 7 (October 1969), 97.
11. The *Times Literary Supplement*, 18 September 1969, 1021
12. Carl Harrison-Ford, Review in *Poetry Magazine*, 18 (June 1970), 35–40.
13. Ian Hamilton, 'In the Public Service', the *Observer*, 13 July 1969.
14. Anthony Thwaite, 'Ingenious Chameleon', the *New Statesman*, 11 July 1969, 53.
15. See discussion of this poem in Chapter 1.
16. Candida Baker, Interview with Peter Porter, *Yacker 3: Australian Writers Talk About Their Work* (Picador, Sydney, 1989), 261–83.
17. Interview with Blake Morrison, London, 21 February 1989.
18. 'Peter Porter: A Profile', *Westerly*, 27, 1 (March 1982), 50.
19. Cyril Connolly, 'Modern Musings', the *Sunday Times*, 17 August 1969.
20. Interview with Trevor Cox, London, 24 February 1989.
21. 'Modern Musings', the *Sunday Times*, 17 August 1969.
22. Interview with Peter Porter, London, 26 October 1988.
23. Peter Porter, Discussion of *The Last of England*, the *Poetry Book Society Bulletin*, 66 (Autumn 1970), 1–2.
24. See, for example, 'Huts of Words', *London Magazine* (July/August 1969), 194–200.
25. R. A. Simpson, 'Two Worldly Poets', the *Age*, 11 September 1971, 18.
26. Carl Harrison-Ford, 'Between the Two Voices', *Nation* (26 June 1971), 21.
27. Peter Joseph, Review of *The Last of England*, in *New Poetry*, 19, 4 (August 1971), 36.
28. K. L. Goodwin, 'Poetry Chronicle, 1969–70', *Meanjin* 30, 3 (September 1971), 369.

29. Philip Roberts, 'In Two Voices', the *Sydney Morning Herald*, 3 July 1971, 19.

30. Edgar Dent, 'The Loneliness of the Absentee Poet', the *Sunday Review* (Melbourne), 22 January 1972, 392.

31. See, for example, Fay Zwicky, Review of *English Subtitles*, *Westerly*, 27, 2 (June 1982), 108. Zwicky refers to 'the dampening pressures of (Porter's) adopted country's mores'.

32. See the *Times Literary Supplement*, 30 October 1970, 1245 and Alan Brownjohn, 'Post-War', the *New Statesman*, 25 September 1970, 384.

33. Michael Schmidt, 'A Defence', *Poetry* 120 (June 1972), 177.

34. John Fuller, 'Porter's Complaint', the *Listener*, 84, 2178 (24 December 1970), 888.

35. Jonathan Raban, *The Society of the Poem* (Harrap, London, 1971), 140.

36. Alexander Craig (ed.), *12 Poets* (Jacaranda Press, Brisbane, 1971), 164

37. See John Tranter, 'Having Completed My Fortieth Year', *Under Berlin: New Poems 1988* (University of Queensland Press, St Lucia, 1988), 43.

7

FUEL FOR THE DARK

1. Peter Porter, 'Grub Street *versus* Academe', *The New Review*, 1, 4 (1974), 47–52.

2. Peter Porter, 'John Woodby, 1910–1970' (unpublished BBC typescript).

3. R. M., *Classical Outlook*, April 1974.

4. Gavin Ewart, 'Cat, Tib, Prop and Mart', *Encounter* 40, 3 (March 1973), 64. '*The Kid from Spain*' was the title of a 1930s film starring Eddie Cantor.

5. Terry Eagleton, *Exiles and Emigrés: Studies in Modern Literature* (Chatto and Windus, London, 1970).

6. Ibid., 9.

7. Terry Eagleton, 'New Poetry', *Stand* 14, 2 (1972), 77.

8. Andrew Taylor, 'The Outsider', *Australian Book Review*, 11 (April 1973), 70–1.

9. Clive James, 'A Slough of Despond', the *Observer*, 26 November 1972.

10. Ibid.

11. David Selzer, 'Porter Major', *Phoenix: A Poetry Magazine*, 10 (July 1973), 83.

12. Ibid.

13. 'New Poetry', 78.

14. 'The Poet in the Sixties: Vices and Virtues — a recorded conversation with Peter Porter', Michael Schmidt and Grevel Lindop (eds.), *British Poetry Since 1960: A Critical Survey* (Carcanet Press, Oxford, 1972), 203.

15. Charles Osborne, *Giving it Away: Memoirs of an Uncivil Servant* (Secker and Warburg, London, 1986), 116.

16. D. J. Enright, *The Alluring Problem: An Essay on Irony* (Oxford University Press, Oxford, 1986).

17. See 'La Déploration sur La Mort d'Igor Stravinsky' (*PTTC, CP* 190) and 'The Tomb of Scarlatti' (*PTTC, CP* 189).

18. Damian Grant, 'Verbal Events', *Critical Quarterly* (Spring 1974), 84.

19. Seamus Heaney, 'North', in *North* (Faber and Faber, London, 1975).

20. 'Verbal Events', 84.

21. Edward Neill, 'Peter Porter's Poetry: An Appreciation', *Basa Magazine* 2, 2 (Autumn 1985), 10.

22. Peter Porter, 'In Exile: An Autobiographical Anti-Biography', Part One: 'Inventing the Past', ABC Radio, 1987, unpublished typescript, 4.

23. Terry Eagleton, 'New Poetry', *Stand*, 14, 2 (1972), 77.

24. Peter Porter, 'Working with Arthur Boyd', *Westerly*, 32, 1 (March 1987), 69.

25. Ibid., 71.

26. See Ursula Hoff, *The Art of Arthur Boyd*, Introduction by T. G. Rosenthal (André Deutsch, London, 1986), 23.

27. Letter from Arthur Boyd, Ramsholt, England, 24 August 1989.

28. See 'Working with Arthur Boyd', *Westerly*, 32, 1 (March 1987), 71.

29. *The Art of Arthur Boyd*, 24.

30. 'Working with Arthur Boyd', 69.

31. Ibid., 71.

32. Ibid.

33. 'De Profundis', the *Times Literary Supplement*, 2 November 1973, 1348.

34. Roger Garfitt, 'Whales, Women, Weather', *London Magazine*, 13, 6 (1974), 102.

35. Robert Adamson, 'Extravagant Gesturings But Without Any Passion', the *Australian*, 5 January 1974, 16.

36. *The Art of Arthur Boyd*, 70.

37. Ibid., 71.

38. 'Working with Arthur Boyd', 75.

39. Ibid., 73.

40. Elizabeth Riddell, 'Now Swarm Many Versifiers', the *Australian*, 15 July 1973.

41. Ibid.

42. Peter Porter, 'Return to Arcadian Australia', the *Listener*, 92, 2372 (12 September 1974), 333–4.

43. See Peter Porter, 'Byron and the Moral North', *Encounter*, XLIII, 2 (August 1974), 65–72.

44. Ibid., 66.

45. Peter Porter, 'Huts of Words', *London Magazine*, 9, 4–5 (July-August 1969), 194–200.

46. Ibid.

47. 'Huts of Words', 198.

48. Jonathan Raban, *The Society of the Poem* (Harrap, London, 1971), 140.

49. Peter Porter, Review of *The Boy in the Bush* and M. L. Skinner, *The Fifth Sparrow* in the *New Statesman*, 86, 226, 16 November 1973, 741–2.

50. Seamus Heaney, *Wintering Out* (Faber and Faber, London, 1972).

51. Porter says that the anecdote came from Dr Vivian Smith of Sydney University: its truth mattered less to Porter than its hypothetical possibility. But the story does have some basis in fact. When he was considering a future outside Vienna in the 1880s, where he was having difficulty establishing a practice, Freud wrote: 'We shall see whether I can go on living in Vienna ... I really no longer care where this will be, whether here or in America,

Australia or anywhere else'. Ronald W. Clark, *Freud: The Man and the Cause* (Paladin/Collins, London, 1982), 85. Vivian Smith told Porter the story after reading it in Ernest Jones's Life. (Letter from Vivian Smith, 6 October 1989).

52. Douglas Dunn, 'Quotidian Tasks', *Encounter*, 46, 2 (February 1976), 75.
53. Chris Wallace-Crabbe, 'Making sense of the Imaginary Museum', the *Australian*, 22 May 1976.
54. Les A. Murray, 'My New Country Enjoys Good Relations With Death', the *Sydney Morning Herald*, 14 February 1976, 15.
55. 'My New Country Enjoys Good Relations With Death', 15.
56. Bruce Bennett, 'Peter Porter in Profile', *Westerly*, 27, 1 (March 1982), 56.
57. See *Australian Poems in Perspective*, P. K. Elkin (ed.), (University of Queensland Press, St Lucia, 1978), 172–84. This essay was republished as 'On Sitting Back and Thinking About Porter's Boeotia', in *The Peasant Mandarin: Prose Pieces by Les A. Murray* (University of Queensland Press, St Lucia, 1978), 172–84.
58. Peter Porter, 'Country Poetry and Town Poetry: A Debate with Les A. Murray', *Australian Literary Studies* 9, 1 (May 1979), 39–48.
59. E.g. Andrew Taylor, 'Past Imperfect: The Sense of the Past in Les A. Murray' and Bruce Bennett, 'Versions of the Past in the Poetry of Les A. Murray and Peter Porter', in Kirpal Singh (ed.), *The Writer's Sense of the Past* (Singapore University Press, Singapore, 1987), 189–97 and 177–88.
60. Peter Porter, 'Barding it up in Bunyah', *Scripsi* 5, 1 (1988), 192.
61. 'Country Poetry and Town Poetry', 39.
62. Russell Davies, 'West of Eden', the *New Statesman*, 24 October 1975, 516.
63. James McAuley, 'Terra Australis', in Leonie Kramer (ed.), *James McAuley* (University of Queensland Press, St Lucia, 1988).
64. Michael Hulse, 'Love and Death: Nine Points of Peter Porter', *Quadrant* (September 1983), 32.

8

THE EASIEST ROOM IN HELL

1. See Peter Porter manuscript, Australian National Library, Canberra, MS 6640.
2. Interview with Susanna Roper, London, 11 March 1989. The following quote also comes from this interview.
3. Interview with Peter Porter, London, 10 December 1985.
4. A. Alvarez, *The Savage God: A Study of Suicide* (Penguin, Harmondsworth, 1974, first published 1971).
5. Interview with Peter Porter, London, 3 November 1988.
6. See Jeffrey Meyers, *Manic Power: Robert Lowell and his Circle* (Macmillan, London, 1987); and John Haffenden, *The Life of John Berryman* (Routledge and Kegan Paul, London, 1982).
7. Interview with Roger Covell, Sydney, 4 May 1989.
8. Interview with Ian Hamilton, London, 3 January 1986.

9. Interview with Trevor Cox, London, 24 February 1989.

10. Interview with Jill Neville, 21 February 1989.

11. Letter from Peter Porter, 23 October 1989.

12. Interview with Porter on London-Oxford train, 10 December 1985.

13. Patricia Beer, 'The Notation of Pain', the *Observer*, 9 April 1978.

14. Herbert Lomas, 'Taproots', *London Magazine* (November-December 1980), 126–30.

15. Ibid., 129.

16. Michael Millgate, *Thomas Hardy* (Oxford University Press, London, 1982), 487.

17. Desmond Graham, 'The Lying Art', *Stand*, 20, 1 (1978), 69.

18. Bruce Beaver, 'Virtuosity and Sophistication', the *Age*, 16 December 1978, 27.

19. Peter Porter, 'In a Trance Through Paradise', unpublished paper delivered at a conference on Australian-Tuscan cultural relations at the University of Florence, January 1989.

20. 'Taproots', 129.

21. James Olney, *Metaphors of Self: The Meaning of Autobiography* (Princeton University Press, Princeton, 1973), x.

22. Stephen J. Greenblatt, (ed.) *Allegory and Representation* (Johns Hopkins University Press, Baltimore, 1981), viii.

23. See Peter Porter, 'Melting Moments: Puccini and his Critics', *Encounter*, LIX, 5 (November 1982), 46–52.

24. Ibid., 46.

25. 'In a Trance Through Paradise'.

26. Interview with Peter Porter, London, 3 November 1988.

27. Pamela Law, 'This Gift of Despair', *Quadrant* 22, 8 (August 1978), 77.

28. Interview with Jacqueline Simms, Oxford University Press poetry editor, London, 20 February 1989.

29. Interview with Trevor Cox, London, 25 February 1989.

30. Charles Tomlinson, *Notes from New York and Other Poems* (Oxford University Press, Oxford, 1984), 32.

31. 'In a Trance Through Paradise'.

32. Interview with Peter Porter, London, 3 November 1988.

33. 'In a Trance Through Paradise'.

34. 'Peter Porter on *The Cost of Seriousness*', *Thirty Years of the Poetry Book Society 1956–1986*, (ed.) Jonathan Barker (Hutchinson, London, 1988), 144.

35. Interview with Ian Hamilton, London, 3 January 1986.

36. Interview with Susanna Roper, London, 11 March 1989.

37. Les A. Murray, *Poems Against Economics* (Angus and Robertson, Sydney, 1972), 3–9.

38. Peter Porter, 'Working with Arthur Boyd', *Westerly* 32, 1 (March 1987), 75.

39. See Fiona Giles (ed.) *From the Verandah* (McPhee Gribble, Melbourne, 1988).

40. Peter Porter, Review of Robert Duncan's *The Fast Decade* in *London Magazine*, 148 (1969).

41. Clive James, 'To Peter Porter: a letter to Sydney', *Other Passports: Poems 1958–1985* (Jonathan Cape, London, 1986), 103. First published in the *New Statesman*, 90, 2317 (15 August 1975).

9

THE LYING ART

1. Anne Stevenson, 'Night-time Tongue', the *Listener*, 100, 2568 (13 July 1978), 62.
2. Herbert Lomas, 'Taproots', *London Magazine* (November–December 1980), 128.
3. John Lucas, 'A Claim to Modesty', the *Times Literary Supplement*, 17 April 1981, 429.
4. John Lucas, *Modern English Poetry — From Hardy to Hughes: A Critical Survey* (B. T. Batsford, London, 1986).
5. 'A Claim to Modesty', 429.
6. Ibid.
7. Claude Rawson, 'Moving Pictures', *London Review of Books* (16 July–5 August 1981), 14–15.
8. Ibid.
9. 'Peter Porter in Profile', *Westerly*, 27, 1 (March 1982), 46.
10. See Laurence Lerner, 'Wrestling with the Difficult', *Encounter*, 57, 3 (September 1981), 62–3.
11. Ian Hamilton, 'At home with the Muse', the *Sunday Times*, 12 April 1981.
12. James Lasdun, the *Spectator*, 18 April 1981, 22.
13. Quoted in Deborah Mitchell, 'Evangelical Duty', the *Literary Review*, July 1981, 16.
14. Ibid.
15. John Fuller, 'The age of disillusion', the *Observer*, 5 April 1981.
16. John Lucas, 'A Claim to Modesty', 429.
17. Elizabeth Riddell, 'A welcome, authentic voice', the *National Times*, 6–12 December 1981, 48.
18. John Tranter, 'Porter carries a cultural passport', the *Sydney Morning Herald*, 17 October 1981.
19. See John Tranter (ed.) *The New Australian Poetry* (Makar Press, Brisbane, 1979). The twenty-four poets represented in this anthology include Nigel Roberts, Michael Dransfield, Robert Adamson, John Forbes, Laurie Duggan, John A. Scott and Tranter himself.
20. Fay Zwicky, Review of *English Subtitles*, *Westerly* 27, 1 (March 1982), 106–109.
21. See Joan Kirkby (ed.) *The American Model: Influence and Independence in Australian Poetry* (Hale and Iremonger, Sydney, 1982).
22. 'Peter Porter in Profile', *Westerly* 27, 1 (March 1982), 47.
23. Robert Gray, 'Humane Sceptic', the *Age Monthly Review*, March 1982.
24. Jamie Grant, the *Age*, 19 December 1981, 21.
25. Peter Porter papers, Australian National Library, MS 6640.
26. See Juliet Clutton-Brock, *The British Museum Book of Cats: Ancient and Modern* (British Museum Publications, London, 1988).
27. Paul Harvey, *The Oxford Companion to Classical Literature* (Oxford University Press, Oxford, 1984), 5, 17.
28. Samuel Beckett, *Molloy, Malone Dies, The Unnameable* (Calder and Boyars, London, 1959).

29. Ibid., 74.
30. Dorothy Green, 'Consumer's Report', *Hemisphere*, 28, 5 (March/April 1984), 297—8.
31. Peter Porter, 'In a Trance Through Paradise'.
32. Evan Jones, 'Peter Porter: Poet into Person', *Scripsi*, 1, 2 (October 1981), 10.
33. *Lempriere's Classical Dictionary*, 430.

10

BEYOND APOCALYPSE

1. Peter Porter, *Collected Poems* (Oxford University Press, Oxford, 1983, reprinted 1988), Preface.
2. Ibid.
3. Stephen Spender, 'Wakeful Dreaming', the *Observer*, 27 March 1983, 33.
4. Ibid.
5. Alan Brownjohn, 'From the Eighth Floor of the Tower: The Collected Peter Porter', *Encounter* 358 (September-October 1983), 80—4.
6. Ibid., 80
7. Les A. Murray, 'Porter: an infinitely more serious Bertie Wooster', the *Sydney Morning Herald*, 18 June 1983.
8. Vincent Buckley, 'The Collected Porter', *Australian Book Review*, 63 (August 1984), 21—2.
9. Robert Gray, 'Peter Porter and Australia', *Poetry Review: Peter Porter — Special Issue*, 73, 1 (March 1983), 16—20.
10. Ibid., 17.
11. Julian Croft, 'Responses to Modernism, 1915—1965', in L. T. Hergenhan (ed.) *The Penguin New Literary History of Australia* (Penguin, Ringwood, 1988), 426—7.
12. See Clive James, 'The Boy from Brisbane', *Poetry Review*, 73, 1 (1983), 26.
13. Thomas D'Evelyn, 'Think Poetry is in Decline? Try Reading Peter Porter', *The Christian Science Monitor* (16 January 1985), 19.
14. Patrick Hanrahan, 'Peter Porter: *Collected Poems* for delightful reading', the *Daily American*, 12 May 1983, 6.
15. Peter Levi, 'Funny Poet', the *Spectator*, 23 April 1983, 24.
16. Ibid.
17. Dick Davis, 'An absent vision', the *Listener* (18 August 1983).
18. Christopher Reid, 'Rhetoric as rich as wedding cake', the *Sunday Times*, 27 March 1983.
19. Blake Morrison, 'A philosopher of captions', the *Times Literary Supplement*, 1 April 1983, 321.
20. Evan Jones, 'The Poetry of Peter Porter', *Scripsi* 2, 4 (1984), 65—75.
21. Ibid., 68.
22. Ibid., 74.
23. Jamie Grant, 'The Essential Porter', the *Age Monthly Review*, September 1983, 13—5.

24. Chris Wallace-Crabbe, 'Porter's strengths on show', the *Age*, 28 May 1983, 8.
25. Dorothy Green, 'Consumer's Report', *Hemisphere*, 28, 5 (March-April 1984), 297–8.
26. Douglas Dunn, 'A Piece of Real', *London Magazine* (June 1983), 74–8.
27. Ibid., 75.
28. John Lucas, 'A new Daks suit', the *New Statesman*, 1 April 1983, 21.
29. Martin Booth, *British Poetry 1964–84* (Routledge and Kegan Paul, London, 1985), 34.
30. Alan Bold, 'Grace in a Grotesque World', the *Scotsman* (Edinburgh), 26 March 1983.
31. Peter Porter, 'In Exile', Radio Helicon, ABC Radio typescript (15 November 1987).
32. Interview with Christine Berg, London, 7 March 1989.
33. 'Peter Porter interviewed by Martin Harrison', *Australian Literary Studies*, XI, 4 (October 1984), 458–67 (459).
34. Interview with Christine Berg, London, 7 March 1989.
35. Bruce Bennett, 'Peter Porter in Profile', *Westerly* 27, 1 (March 1982), 54–5.
36. Chris Wallace-Crabbe, 'From plain diction to fast forward', the *Age*, 27 April 1985.
37. 'Peter Porter in Profile', 54–5.
38. Ibid.
39. Ibid., 55.
40. Ibid.
41. Ibid.
42. Evan Jones, *Australian Book Review*, (December 1985–January 1986), 42.
43. 'Doll's House' was first published in *The Animal Programme: Four Poems*, (Anvil Press Poetry, London, 1982).
44. Mrs John Addington Symonds (ed.) *Recollections of a Happy Life: Being the Autobiography of Marianne North*, 2 vols, (Macmillan, London, 1892).
45. Ibid., vol. 2, 149.
46. Evan Jones, 'Two Literate Poets', *Australian Book Review* (December 1985–January 1986), 41–2.
47. Lachlan Mackinnon, 'A high-tech primitive', the *Times Literary Supplement*, 11 January 1985, 54.
48. Laurence Lerner, 'Forms of Difficulty', *The Sewanee Review*, (Spring 1988), 312–22, (318).
49. Charles Boyle, 'Leaking Categories', *London Magazine* (March 1985), 85–9 (85–6).
50. Chris Wallace-Crabbe, 'From plain diction to fast forward', the *Age*, Saturday Extra, 27 April 1985, 14.
51. Sean O'Brien, 'Abrasive Sensuality', *London Magazine* (April-May 1988), 118–22 (120).
52. John Lucas, 'Best rhymes with zest' the *New Statesman*, 8 April 1988, 28.
53. Tim Dooley, 'From late modernity', the *Times Literary Supplement*, 15–21 January 1988.
54. Michael Wilding, 'Like Boys to Wanton Flies', in B. Bennett, P. Cowan and J. Hay (eds.), *Perspectives One: Short Stories* (Longman Cheshire, Melbourne, 1985), 142–51.
55. Clive James, *Unreliable Memoirs* (Jonathan Cape, London, 1980).

56. Alex Preminger, (ed.) *Princeton Encyclopedia of Poetry and Poetics*, (Princeton University Press, Princeton, 1974), 597.

57. Christopher Pollnitz, 'Peter Porter: Whether "the World is But a Word"', *Scripsi*, 5, 1, (1988), 197–208.

58. Richard Makin and Christopher Norris, *Post-Structuralist Readings of English Poetry* (Cambridge University Press, Cambridge, 1987), 3.

59. Christopher Norris, *Deconstruction: Theory and Practice* (Methuen, London, 1982), 28.

60. Malcolm Bradbury, 'An Age of Parody', in *No, Not Bloomsbury* (André Deutsch, London, 1987), 48.

61. Ibid.

62. Adam Thorpe, 'All Tangle and Terror', *The Literary Review*, 116 (February 1988), 56.

63. Ursula Hoff, *The Art of Arthur Boyd* (André Deutsch, London, 1986), 76.

64. Peter Porter, 'Working with Arthur Boyd', *Westerly*, 32, 1 (March 1987), 75.

65. Ibid., 76.

66. Ibid., 75.

67. S. J. Freedberg, *Painting in Italy 1500 to 1600* (Penguin, Harmondsworth, 1971), 119.

68. Judith Wright, *Collected Poems 1942–1970* (Angus and Robertson, Sydney, 1971), 192.

69. Shirley Walker, 'The Philosophical Basis of Judith Wright's Poetry', in Chris Tiffin (ed.), *South Pacific Images* (SPACLALS, Department of English, University of Queensland, St Lucia, 1978), 164.

70. Ronald W. Clark, *Freud: The Man and The Cause* (Paladin, London, 1982), 336.

71. 'Working with Arthur Boyd', 76.

72. Ibid.

73. Christopher Lasch, *The Minimal Self: Psychic Survival in Troubled Times* (Pan/Picador, London, 1985), 19.

74. 'Working with Arthur Boyd', 78.

75. See David Trotter, *The Making of the Reader: Language and Subjectivity in Modern American, English and Irish Poetry* (Macmillan, London, 1984), Ch. 1.

11

POSSIBLE WORLDS

1. Gavin Ewart, 'The Peter Porter Poem of '83', *Poetry Review* 73, 1 (March 1983), 30.

2. Michael Hulse, 'Stratagems of the Spirit', *Quadrant*, 33, 12 (December 1989), 48. Porter wrote a scenario and the first Act of his would-be opera, 'The Best of all Possible Worlds'. The music was to have been written by the Sydney composer David Lumsdaine, who wrote no more than the Prologue and some bits of the first scene.

3. Shirley Fisher, a psychologist at Strathclyde University, reported these findings at a meeting of the British Psychological Association in Sheffield: 'Recent laboratory tests suggested that the depressed judged evidence wisely, especially when judging between what they did and what happened. It was the non-depressed who appeared to distort the evidence in the direction of optimism'. The *Guardian Weekly*, 141, 12, 24 September 1989, 23.

4. Peter Porter, 'In a Trance Through Paradise' (University of Florence, January 1989, unpublished typescript).

5. Derek Malcolm, 'Funny dirty business', the *Guardian*, Thursday, 2 March 1989, 21.

6. E.g. Peter Reading, 'Endurance, elegies and a cosmopolitan collection', the *Sunday Times*, 7 January 1990, 14.

7. Sigmund Freud, *Civilization, War and Death*, John Rickman (ed.) (Hogarth Press, London, 1953), 16.

8. The *Bulletin* (Sydney), 5 February 1972. Quoted in *The Australian National Dictionary*, W. S. Ramson (ed.) (Oxford University Press, Melbourne, 1988), 748.

9. See Peter Porter, 'A local habitation and a name: The Road to Botany Bay', the *Times Literary Supplement*, 27 November 1987.

10. Peter Porter, 'Polemical Pairings', the *Times Literary Supplement*, 22—8 January 1988, 87.

11. Peter Porter, 'A dialect of death', the *Guardian*, 15 April 1988, 26.

12. Peter Porter, 'Back to his Roots', the *Observer*, 30 October 1988, 44.

13. Peter Porter, 'The Muse in the Outback', the *Observer*, 17 May 1981.

14. Peter Porter, 'Barding it up in Bunyah', *Scripsi* 5, 1 (1988), 191—2.

15. Peter Porter, 'High Season: Chris Wallace-Crabbe's Melbourne', *Australian Book Review* (May 1986), 9—10.

16. Ibid.

17. Peter Porter, 'Imagining the After-Life', the *Times Literary Supplement*, 18—24 November 1988, 1270.

18. Peter Porter, 'The Shape of Music and the Shape of Poetry', *Quadrant*, 24 (June 1980), 5.

19. Ibid.

20. Ibid., 11.

21. Interview with Geoffrey Burgon, London, 22 February 1989.

BIBLIOGRAPHY

◆

PRINCIPAL WORKS BY PETER PORTER

BIBLIOGRAPHY

A full-scale comprehensive bibliography of the work of Peter Porter
has been compiled by John R. Kaiser, librarian at the Pennsylvania
State University. Kaiser's *Peter Porter: A Bibliography 1954–1986*
(Mansell Publishing, London and New York) was published in 1990.

MANUSCRIPTS AND NOTEBOOKS

Since the 1950s, Porter has composed poems in notebooks, sometimes
having three or four at the same time. He generally works from both
ends of a notebook towards the middle.

The majority of Porter's notebooks and looseleaf manuscripts are
held at the Australian National Library, Canberra. These include manuscript notebooks containing poems written c1955–60, 1971–7 and
1973–80 (MSS series 6640). More recent acquisitions include manuscript notebooks covering the period 1978–89, together with general
notebooks and assorted papers, manuscripts and letters relating to
publications.

Looseleaf manuscripts from the 1950s are held at the Lockwood
Memorial Library, Buffalo. The British Museum in London holds notebook manuscripts containing poems written between 1958 and Christmas 1963. Indiana University holds some loose manuscripts and notebooks relating to *A Porter Folio* and poems for a cantata with David
Lumsdaine. Reading University Library holds some notebooks, including
translations from Martial's epigrams and poems which appeared in
The Last of England and *Preaching to the Converted*.

BOOKS OF POETRY

Once Bitten, Twice Bitten, Scorpion Press, Northwood, Middlesex,
1961.
Poems Ancient and Modern, Scorpion Press, Lowestoft, Suffolk, 1964.
A Porter Folio, Scorpion Press, Lowestoft, Suffolk, 1969.
The Last of England, Oxford University Press, London, 1970.
After Martial, Oxford University Press, London, 1972.
Preaching to the Converted, Oxford University Press, London, 1972.
Jonah by Arthur Boyd and Peter Porter. Martin Secker and Warburg,
London, 1973.

The Lady and the Unicorn by Arthur Boyd and Peter Porter. Martin Secker and Warburg, London, 1975.

Living in a Calm Country, Oxford University Press, London, 1975.

The Cost of Seriousness, Oxford University Press, Oxford, 1978.

English Subtitles, Oxford University Press, Oxford, 1981.

Collected Poems, Oxford University Press, Oxford, 1983.

Fast Forward, Oxford University Press, Oxford, 1984.

Narcissus by Arthur Boyd and Peter Porter. Secker and Warburg, London, 1984.

The Automatic Oracle, Oxford University Press, Oxford, 1987.

Mars by Arthur Boyd and Peter Porter. André Deutsch, London, 1988.

Possible Worlds, Oxford University Press, Oxford, 1989.

A Porter Selected, Oxford University Press, Oxford, 1989.

BOOKS WITH CONTRIBUTIONS BY PETER PORTER

Penguin Modern Poets 2: Kingsley Amis, Dom Moraes, Peter Porter. Penguin Books, Harmondsworth, 1962.

A Group Anthology, Edward Lucie-Smith & Philip Hobsbaum (eds.). Oxford University Press, London, 1963.

15 Poems for William Shakespeare, Eric W. White (ed.). The Trustees and Guardians of Shakespeare's Birthplace, Stratford-upon-Avon, 1964.

London Magazine Poems 1961–66, selected by Hugo Williams, introduced by Allan Ross. London, 1966.

The Voice of Love: Song Cycle. Nicholas Maw. Poems by Peter Porter. Boosey and Hawkes, London, 1968.

Australian Poetry 1969, selected by Vivian Smith. Angus and Robertson, Sydney, 1969.

Breakthrough: Poetry in Britain during the 1960s, R. B. Heath (ed.). Hamish Hamilton, London, 1970.

Contemporary Poets of the English Language, Rosalie Murphy (ed.). St James Press, Chicago and London, 1970.

A Choice of Pope's Verse, selected with introduction by Peter Porter. Faber and Faber, London, 1971.

Twelve Poets 1950–1970, Alexander Craig (ed.). Jacaranda Press, Milton (Qld.), 1971.

Peter Porter Reads from his Own Work. Poets on Record 12, University of Queensland Press, St Lucia, 1974.

The English Poets from Chaucer to Edward Thomas, Peter Porter and Anthony Thwaite. Secker and Warburg, London, 1974.

Roloff Beny in Italy, designed and photographed by Roloff Beny with an epilogue by Gore Vidal. Text and anthology by Anthony Thwaite and Peter Porter. Thames and Hudson, London, 1974.

New Poetry 1, Peter Porter and Charles Osborne (eds.). Arts Council of Great Britain, London, 1975.

Poetry Book Society: The First Twenty-Five Years, Eric W. White (ed.).

Poetry Book Society, London, 1979.

Sydney, by Peter Porter and the Editors of Time-Life Books, Photographs by Brian Brake, The Great Cities. Time-Life Books, Amsterdam, 1980.

Landscape Poets: Thomas Hardy, selected and introduced by Peter Porter. Weidenfeld and Nicolson, London, 1981.

Spirit of Wit: Reconsiderations of Rochester, Jeremy Treglown (ed.). Basil Blackwell, Oxford, 1982.

Martin Bell: Complete Poems, Peter Porter (ed.). Bloodaxe Books, Newcastle upon Tyne, 1988.

INTERVIEWS

'Peter Porter' in Orr, Peter (ed.). *The Poet Speaks*. Routledge and Kegan Paul, London, 1969, 179—84.

Douglas, Denis. 'Conversation with Peter Porter', *Overland*, 44, Winter 1970, 33—4.

'The Poet in the Sixties: Vices and Virtues — a recorded conversation with Peter Porter' in Schmidt, Michael and Grevel Lindop, *British Poetry Since 1960: A Critical Survey*, Carcanet Press, South Hinksey (Oxford), 1972, 202—12.

Martin, Philip. 'Interview with Peter Porter', *Quadrant*, 87, XVIII, No. 1 (Jan-Feb. 1974), 9—19.

Bennett, Bruce. 'Peter Porter in Profile', *Westerly*, 27, 1 (March 1982), 45—56.

Brownjohn, Alan. 'Alan Brownjohn talks to Peter Porter', *The Literary Review* No. 59 (May 1983), 18—22.

'What I Have Written ...' *Hemisphere*, 28, 3 (Nov/Dec 1983), 150—6.

Harrison, Martin. 'Peter Porter. Interviewed by Martin Harrison', *Australian Literary Studies*, 11, 4 (October 1984), 458—67.

Kavanagh, Paul. 'Little Harmonic Labyrinths: An Interview with Peter Porter', *Southerly*, 1 (March 1985), 12—22.

Peter Porter: What I Have Written, Writers Talking, Australian Film Institute video, 1988 (30 minutes).

Baker, Candida. *Yacker 3. Australian Writers Talk About Their Work*, Pan Books (Australia), Sydney, 1989, 260—83.

ARTICLES AND COMMENTARIES BY PETER PORTER

'Living in London', *London Magazine*, 13, 2, June-July 1973, 57—64.

'Grub Street *Versus* Academe', *The New Review*, 1, 4, 1974, 47—52.

'Byron and the Moral North', *Encounter*, August 1974, 65—73.

'Return to Arcadian Australia', the *Listener*, 92, 2372, 1975, 333.

'Poetry and Madness', *Southerly*, 36, 4, 1976, 385—405.

'Miracles of Metamorphosis: Gustav Mahler', *Encounter*, August 1976, 55—60.

'Brisbane Comes Back', *Quadrant*, 98, XIX, 6, September 1975, 52—8.

'Locked Out of Paradise', *The New Review*, 36, 3, March 1977, 15—20.

'The Achievement of Auden', *Sydney Studies in English*, 4, 1978–9, 73–113.

'Country Poetry and Town Poetry: A Debate with Les Murray', *Australian Literary Studies*, 9, 1, 1979, 39–48.

'The Shape of Music and the Shape of Poetry', *Quadrant* XXXIV, 6, June 1980, 4–12.

'Les Murray: An Appreciation', *The Journal of Commonwealth Literature*, XVII, 1, 1982, 45–52.

'The Professional Amateur', *Spirit of Wit: Reconsiderations of Rochester*, Jeremy Treglown (ed.), Blackwell, Oxford, 1982, 58–74.

'Melting Moments: Puccini and his Critics', *Encounter*, LIX, 5, November 1982, 46–52.

'The Mystery and the Music: On Mozart's Life', *Encounter*, LX, 6, June 1983, 53–8.

'Composer and Poet', *The Britten Companion*, Christopher Palmer (ed.) Faber and Faber, London, 1984, 271–85.

'An Expatriate's Reaction to his Condition', *Westerly*, 32, 2, December 1987, 43–7.

'Barding it up in Bunyah', *Scripsi*, 5, 1, 1988, 191–6.

'Incandescence of the Ordinary', *Overland*, 113, December 1988, 14–16.

'Australian Expatriate Writers in Britain', *A Passage to Somewhere Else*, Doireann MacDermott & Susan Ballyn (eds.), Promociones y Publicaciones Universitarias, Barcelona, 1988, 135–42.

ARTICLES AND COMMENTARIES ON PETER PORTER

Bennett, Bruce. 'Peter Porter's Expatriate Vision'. *Poetry of the Pacific Region*, Robert Sellick (ed.), CRNLE Essays and Monograph Series No. 2, Centre for Research in the New Literatures in English, Adelaide, 1984, 19–30; reprinted with revisions in *European Relations: Essays for Helen Watson-Williams*, The Centre for Studies in Australian Literature, University of Western Australia, Nedlands, 1985, 103–14.

Bennett, Bruce. 'Versions of the Past in the Poetry of Les Murray and Peter Porter', *The Writer's Sense of the Past*, Kirpal Singh (ed.). Singapore University Press, Singapore, 1987, 178–88.

Bennett, Bruce. 'Passports to Paradise: Peter Porter and Clive James', *A Sense of Exile: Essays in the Literature of the Asia-Pacific Region*, Bruce Bennett & Susan Miller (eds.). The Centre for Studies in Australian Literature, University of Western Australia, Nedlands, 1988, 67–80.

Brownjohn, Alan. 'From the Eighth Floor of the Tower: The Collected

Peter Porter'. *Encounter* 358, September — October 1983, 80–4.

Croft, Julian. 'Responses to Modernism, 1915–1965'. *The Penguin New Literary History of Australia*, L. T. Hergenhan (ed.). Penguin, Ringwood, 1988, 426–7.

Dooley, Tim. 'Acting Against Oblivion'. *Poetry Review*, 73, 1, March 1983, 27–9.

Dunn, Douglas. 'A Piece of Real'. *London Magazine*, 23, June 1983, 74–8.

Eagleton, Terry. 'New Poetry'. *Stand* 14, 2, 1972, 74–8.

Eagleton, Terry. 'Recent Poetry'. *Stand*, 23, 2, 1982, 62–8.

Ewart, Gavin. 'From Notions to Emotions'. The *Times Literary Supplement*, 19 May 1978, 550.

Fraser, G. S. 'The 1950s, the Movement and the Group', in *The Modern Writer and his World*. New and revised edition, André Deutsch, London, 1964, 346–55.

Garfitt, Roger. 'The Group', *British Poetry Since 1960: A Critical Survey*, M. Schmidt & G. Lindop (eds.). Carcanet Press, South Hinksey, Oxford, 1972, 15–59.

Grant, Damian. 'Verbal Events', *Critical Quarterly*, 16, Spring 1974, 81–6.

Grant, Jamie. 'The Essential Peter Porter'. The *Age Monthly Review*, September 1983, 13–15.

Gray, Robert. 'Humane Sceptic'. The *Age Monthly Review*, March 1982.

Gray, Robert. 'Peter Porter and Australia'. *Poetry Review*, 73, 1, March 1983, 16–20.

Green, Dorothy. 'Consumer's Report'. *Hemisphere* 28, 5, March-April 1984, 297–8.

Hulse, Michael. 'Love and Death: 'Nine Points of Peter Porter'. *Quadrant*, 193, XXVII, 9, September 1983, 31–8.

James, Clive. 'The Boy from Brisbane'. *Poetry Review*, 73, 1, March 1983, 25–7, reprinted in *Snakecharmers in Texas: Essays 1980–87*. Jonathan Cape, London, 1988, 44–8.

Jones, Evan. 'The Poetry of Peter Porter'. *Scripsi* 2, 4, 1984, 65–75.

Lomas, Herbert. 'Concubinages'. *London Magazine*, 21, June 1981, 73–9.

Lucas, John. 'A Claim to Modesty'. The *Times Literary Supplement*, 4072, 17 April 1981, 429. Reprinted in *Moderns and Contemporaries*, Harvester Press, 1985.

Morrison, Blake. 'A Philosopher of Captions'. The *Times Literary Supplement*, 4174, 1 April 1983, 321–2.

Murray, Les A. 'On Sitting Back and Thinking About Porter's Boeotia', in *The Peasant Mandarin: Prose Pieces by Les Murray*. University of Queensland Press, St Lucia, 1978, 172–84. First published in *Australian Poems in Perspective*, P. K. Elkin (ed.). University of Queensland Press, St Lucia, 1978.

Neill, Edward. 'Peter Porter's Poetry: An Appreciation'. *Basa Magazine*, II, 2, Autumn 1985, 9–14.

O'Neill, Michael. 'The Lying Art: An Aspect of the Poetry of Peter Porter'. *Durham University Journal*, 48, June 1987, 367–72.

Owen, Margaret. 'Peter Porter and the Group'. *Poetry Review* 73, 1, March 1983, 12.

Pollnitz, Christopher. 'Peter Porter: Whether "The World is But a Word"'. *Scripsi*, 5, 1, 1988, 197–208.

Rawson, Claude. 'Moving Pictures'. *London Review of Books*, 16 July - 5 August 1981, 14–15.

Richards, Max. 'The Citizenship of Peter Porter'. *Australian Literary Studies*, 8, 3, May 1978, 351–359.

Selzer, David. 'Porter Major'. *Phoenix*, 10, July 1973, 83–5.

Steele, Peter. 'The Radiations of Peter Porter'. *Westerly*, 29, 3, October 1984, 65–74.

Szirtes, George. 'Humane Astringencies'. *Poetry Review*, 73, 1, March 1983, 34–6.

Taylor, Andrew. 'The Outsider'. *Australian Book Review*, 11, April 1973, 70–1.

Thwaite, Anthony, 'Porter: Man and Poet', *Poetry Review*, 73, 1, March 1983, 21–2.

Thwaite, Anthony. '"The Group" and After', *Poetry Today: A Critical Guide to British Poetry 1960–1984*. Longman, London, 1985, 66–81.

Wallace-Crabbe, Chris. 'Porter's Strengths on Show'. The *Age*, Saturday, 28 May 1983, 8.

Williams, David, 'A Map of Loss: The Recent Poetry of Peter Porter'. *Critical Quarterly*, 25, 4, 1983, 55–62.

WIDER READING

Adams, Brian. *Sidney Nolan: Such is Life*. Hutchinson, London, 1987.

Alvarez, A. *The Savage God: A Study of Suicide*. Penguin, Harmondsworth, 1974.

Attridge, Derek. *The Rhythms of English Poetry*. Longman, London, 1982.

Axelrod, Steven Gould. *Robert Lowell: Art and Life*. Princeton University Press, Princeton, 1978.

Barr, Ann and York, Peter. *The Official Sloane Ranger Handbook*. Angus and Robertson, Sydney, 1982.

Booth, Martin. *British Poetry 1964 to 1984: Driving through the Barricades*. Routledge and Kegan Paul, London, 1985.

Bradbury, Malcolm. *No, Not Bloomsbury*. André Deutsch, London, 1987.

Breslin, James E. B. *From Modern to Contemporary: American Poetry 1945–1965*. University of Chicago Press, 1984.

Clutton-Brock, Juliet. *The British Museum Book of Cats, Ancient and Modern.* The British Museum, London, 1988.

Clark, Ronald W. *Freud: the Man and the Cause.* Paladin/Collins, London, 1982.

Curtis, Tony (ed.). *The Art of Seamus Heaney.* Poetry Wales Press, 1982.

Day, Roger. *Larkin.* Open University Press, Milton Keynes, 1987.

Eagleton, Terry. *Exiles and Emigrés: Studies in Modern Literature.* Chatto and Windus, London, 1970.

Edwards, Michael. *Poetry & Possibility.* Macmillan, London, 1988.

Enright, D. J. *The Alluring Problem: An Essay on Irony.* Oxford University Press, Oxford, 1986.

Fraser, G. S. *The Modern Writer and his World.* André Deutsch, London, revised edition 1964.

Fussell, Paul. *Abroad: British Literary Travelling Between the Wars.* Oxford University Press, Oxford, 1980.

Gay, Peter. *Freud: A Life for Our Time.* Papermac, Macmillan, 1989.

Goffman, Erving. *The Presentation of Self in Everyday Life.* Allen Lane, London, 1959.

Goodwin, Ken. *Adjacent Worlds: A Literary Life of Bruce Dawe.* Longman Cheshire, Melbourne, 1988.

Greenblatt, Stphen J. (ed.). *Allegory and Representation.* The Johns Hopkins University Press, Baltimore, 1981.

Greer, Germaine. *Daddy, We Hardly Knew You.* Hamish Hamilton, London, 1989.

Grubb, Frederick. *A Vision of Reality: A Study of Liberation in Twentieth Century Verse.* Chatto and Windus, London (and Barnes and Noble, New York), 1965.

Haffenden, John. *The Life of John Berryman.* Routledge and Kegan Paul, London, 1982. Prentice-Hall, Englewood Cliffs, N. J. 1971.

Hallberg, Robert von. *American Poetry and Culture 1945–1980.* Harvard University Press, Cambridge, 1985.

Hamburger, Michael. *The Truth of Poetry: Tensions in Modern Poetry from Baudelaire to the 1960s.* Methuen, London, 1969.

Hardy, Barbara. *The Advantage of Lyric: Essays on Feelings in Poetry.* The Athlone Press, London, 1977.

Hartman, Charles O. *Free Verse: An Essay on Prosody.* Princeton University Press, Princeton, 1980.

Hergenhan L. T. (ed.). *The Penguin New Literary History of Australia.* Penguin, Ringwood, 1988.

Hobsbaum, Philip. *Essentials of Literary Criticism.* Thames and Hudson, London, 1983.

Hoff, Ursula. *The Art of Arthur Boyd.* André Deutsch, London, 1986.

James, Clive. *The Metropolitan Critic.* Faber and Faber, London, 1974.

James, Clive. *Snakecharmers in Texas: Essays 1980–87.* Jonathan Cape, London, 1988.

Jones, Peter and Schmidt, Michael. *British Poetry Since 1980: A Critical Survey*. Carcanet Press, Manchester, 1980.

Lasch, Christopher. *The Culture of Narcissism*. W. W. Norton, New York, 1978.

Lasch, Christopher. *The Minimal Self: Psychic Survival in Troubled Times*. Pan Books, London, 1985.

Levey, Michael. *The National Gallery Collection*. National Gallery, London, 1987.

Makin, Richard and Norris, Christopher. *Post-Structuralist Readings of English Poetry*. Cambridge University Press, Cambridge, 1987.

Marsack, Robyn. *The Cave of Making: The Poetry of Louis MacNeice*. Clarendon Press, Oxford, 1982.

Martin, Graham and Furbank, P. N. (eds.) *Twentieth Century Poetry: Critical Essays and Documents*. Open University Press, Milton Keynes, 1975.

Meyers, Jeffrey. *Manic Power: Robert Lowell and his Circle*. Macmillan, London, 1987.

Millgate, Michael. *Thomas Hardy: A Biography*. Oxford University Press, Oxford, 1982.

Morrison, Blake. *The Movement: English Poetry and Fiction of the 1950s*. Oxford University Press, Oxford, 1980.

Morrison, Blake. *Seamus Heaney*. Methuen, London, 1982.

Motion, Andrew. *Philip Larkin*. Methuen, London, 1982

Niall, Brenda. *Martin Boyd: A Life*. Melbourne University Press, Melbourne, 1988.

Olney, James. *Metaphors of Self: The Meaning of Autobiography*. Princeton University Press, Princeton, N. J. 1972.

Osborne, Charles. *Giving it Away: The Memoirs of an Uncivil Servant*. Secker and Warburg, London, 1986.

Packer, Clyde. *No Return Ticket*. Angus and Robertson, Sydney, 1984.

Paulson, Ronald (ed.) *Satire: Modern Essays in Criticism*. Prentice-Hall, Englewood Cliffs, N. J., 1971.

Perloff, Marjorie. *The Dance of the Intellect: Studies in the Poetry of the Pound Tradition*. Cambridge University Press, Cambridge, 1985.

Petch, Simon. *The Art of Philip Larkin*. Sydney University Press, Sydney, 1981.

Pinsky, Robert. *The Situation of Poetry: Contemporary Poetry and its Traditions*. Princeton University Press, Princeton, 1976.

Preminger, Alex (ed.) *Princeton Encyclopedia of Poetry and Poetics*. Princeton University Press, Princeton, N. J. enlarged edition 1974.

Raban, Jonathan. *The Society of the Poem*. Harrap, London, 1971.

Reeves, James. *Commitment to Poetry*. Barnes and Noble, New York, 1969.

Ricks, Christopher. *T. S. Eliot and Prejudice*. Faber and Faber, London, 1988.

Robinson, Peter (ed.) *Geoffrey Hill: Essays on his Work*. Open University Press, Milton Keynes, 1985.

Ross, Andrew. *The Failure of Modernism: Symptoms of American Poetry.* Columbia University Press, New York, 1986.

Smith, Stan. *W. H. Auden.* Basil Blackwell, Oxford, 1985.

Stead, C. K. *The New Poetic: Yeats to Eliot.* Penguin, Harmondsworth, 1967.

Steele, Peter. *Expatriates: Reflections on Modern Poetry.* Melbourne University Press, Melbourne, 1985.

Thwaite, Anthony. *Poetry Today: A Critical Guide to British Poetry 1960–1984.* Longman, London, 1985.

Timms, Edward & Kelley, David. *Unreal City: Urban Experience in Modern European Literature and Art.* St Martin's Press, New York, 1985.

Tolley, A. T. *The Poetry of the Forties.* Manchester University Press, Manchester, 1985.

Trotter, David. *The Making of the Reader: Language and Subjectivity in Modern American, English and Irish Poetry.* Macmillan, London, 1984.

Vendler, Helen. *Wallace Stevens: Words Chosen Out of Desire.* Harvard University Press, Cambridge, 1986.

Wallace-Crabbe, Chris. *Toil and Spin: Two Directions in Modern Poetry.* Hutchinson, Richmond (Vic.), 1979.

West, Thomas. *Ted Hughes.* Methuen, London and New York, 1985.

Williams, John. *Twentieth-Century British Poetry: A Critical Introduction.* Edward Arnold, London, 1987.

INDEX

◆

Abse, Dannie 125
Adelaide 2
  Festival of Arts 118, 170
adolescence 6, 21–8, 69, 95
advertising, Porter's employment in
  73–4, 82, 86–8, 89, 109
'Affair of the Heart' 122–3
After Martial 82, 118, 121, 122
'Alcestis and the Poet' 149, 188–9,
  196
allegory xiv, xvi–xvii, 7, 88, 164, 219,
  220, 222, 242, 251–2
Alvarez, Al xi, 61, 64–5, 79, 83–4,
  86, 119, 151
Ambit 89
America
  and Australia 200–1
  poets of 110, 112, 120, 138
  politics 108, 234
  Porter's concept of 245
Amis, Kingsley 54, 61, 86
anachronism 83, 94, 106, 133, 134,
  212
'Angel in Blythburgh Church, An'
  166–7
anger 75, 103, 141
'Anger' 141, 143, 165
'angry young man/men' xi, 54, 61
'Annotations of Auschwitz' 77–9,
  84, 233
'Anson Jones' 18; *see also Losing*
  Chance, The
'Anthropologist's Confession, An' 58,
  95
'Applause for Death' 111
Arnold, Matthew 17, 26, 130, 187
art 29, 105, 113, 169, 177, 229
  moral purpose 79
  nature and 185
  poetry and 182–3
  relationship with pain xii
Ashbery, John xiii, 179, 182, 217

'At Lake Massaciuccoli' 165, 177,
  180, 195, 245
'At Ramsholt' 167
'At the Porta Humana' 208
'At Whitchurch Canonicorum' 113
Auden, W. H. xiii, 52, 56, 63, 65–6,
  92, 104, 112, 118, 120, 133, 137,
  138, 160, 168, 177, 188, 197, 210,
  214, 237
Australia 26, 31, 32, 66, 67
  and England 174–5, 178–9
  and Europe 168–9, 198–9
  as setting for poems 189, 248, 250
  British view 2, 47, 58–9, 64–5, 88,
  90, 109
  Porter's return to xii, 136–9,
  142–8, 170
  Porter's view 13, 23, 24, 51, 85,
  101, 103, 173
  reaction to Porter 181, 182–3
'Australian Garden, An' 8, 143,
  146–8, 171, 204
Automatic Oracle, The xiii, 197, 207,
  217, 218–28, 239, 243, 247, 255
'Automatic Oracle, The' 226
'Away, Musgrave, Away' 82

Bach, J. S. and W. F. 105, 107
'Bag of Pressmints, A' 242
'Barjai' group 35–6
baroque language and imagination
  22, 85, 125, 188, 189, 196
'Battle of Cannae, The' 82
'Beast and the Beauty' 60, 69, 73,
  74–5, 78, 84, 147, 228, 229
Beethoven, Ludwig van 96, 104, 105
Bell, Martin 18, 55, 56, 58, 72, 102
Berg, Christine 203, 205–7, 218, 225
Bible xiv, 7, 104, 132, 160; *see also*
  Christianity; *Jonah*; religion
Black Mountain Group 138, 181
'Blazing Birds, The' 256–7

Boccherini, Luigi 173
Boyd, Arthur xvii, 172, 173, 250
  collaboration with Porter xvi, 49,
  61, 90, 118, 132–6, 187, 197, 228,
  230–5 *passim*
Brisbane 1, 2–3, 5, 6, 7, 29, 30–42,
  43, 100, 115, 251
Brisbane Church of England
  Grammar School 14–15, 16
British Empire xvi, 21, 52, 54, 129,
  216
Brock, Edwin 62
Browning, Robert 26, 46, 48, 70, 104,
  137, 200, 229
Brownjohn, Alan 18, 53, 54, 61, 121,
  198–9
Buckley, Vincent 199, 200
bushman image 89; *see also*
  Australia
Byron, Lord xiv, 36–7, 46, 82,
  114–16, 117, 137–8, 200, 204

'Camera Loves Us, The' 257
capitalism 88, 219, 245
Carey, Peter 31, 87
Carne, Brian 29, 44
'Castle in Sight, The' 241
'Cat's Fugue' 187
'Cats of Campagnatico, The' 215
Chekhov, Anton 49, 76, 190–1
childhood 1, 2–11, 12, 100, 105–6,
  107, 108, 141, 153, 196, 227, 232,
  249, 250
Children's Crusade, The 118
Choice of Pope's Verse, A 118
Christianity xv, 1, 36, 77, 126, 208;
  see also Bible; religion
'Christmas Recalled, A' 4, 5, 63, 67
'Cities of Light' 246
'Civilization and its Disney Contents'
  245, 246–7
class 46, 64, 72, 73; *see also* social
  poet
classicism 41, 101, 121, 169
'Clipboard' 217
collaboration 61, 90, 104, 118,
  132–6, 187, 228, 230
Collected Poems xii, xiii, 11, 23, 24,
  99, 104, 125, 197–202, 208, 217,
  228
commercialism 73–4, 88, 128, 219

'Competition is Healthy' 104, 107–8
'Condolences of the Season' 9
Connolly, Cyril 95, 98, 106, 119
'Consumer's Report, A' 88, 111
'Conventions of Death' 69, 75–6
Cosimo, Piero di 173
Cost of Seriousness, The xii, 21, 135,
  149, 152, 154–76, 177, 180, 183,
  185, 212, 229
'Cost of Seriousness, The' 177
Courier Mail 29, 35
Covell, Roger 29, 35, 40, 44, 48, 50,
  51, 152, 153
Cox, Trevor 82, 87, 153, 167
'cultural cringe' 26
'Cyprus, Aeschylus, Inanition' 210

Davie, Donald 54, 63–4, 121, 177,
  183
Dawe, Bruce 9, 34, 85–6, 181, 200
death 3, 32, 92, 100–1, 107, 122,
  123, 125, 131, 158, 160, 164, 177,
  210, 248
  father's 10, 204
  mother's 1, 9, 29, 135–6, 149, 153
  wife's 153–4, 156, 157, 160, 180,
  194
'Death in the Pergola Tea-Rooms' 69,
  79–82, 112, 155, 162
'Decline of the North, The' 204–5
'Decus et Tutamen' 245–6
'Dejection: An Ode' 213
'Delegate, The' 149–50, 157–9, 160,
  165, 188
'Deliverance for Doctor Donne, A' 46
'Delphi' 124, 128, 129
despair 74, 75, 156, 166
diaspora 11, 179
'Die-Back' 244–5
'Dis Manibus' 100, 218
'Disc Horse' xv, 152, 224, 225
discontent 8, 117, 145
disillusionment 121, 129, 177
'Doctor's Story, The' 155
'Doll's House' 214
Donne, John 36, 103, 105, 120, 156,
  205
Down Cemetery Road' 142
dreams 4, 10, 188, 192, 225, 226–7
Drysdale, Russell 89–90
Duff Cooper Memorial Prize xiii, 197

Duggan, Laurie 181
'Dust, The' 131

'Easiest Room in Hell, The' 154−7,
    161
'Eat Early Earthapples' 21−3, 34, 46,
    67, 69, 89, 173, 234
'Ecstasy of Estuaries, The' 249
education xiv, 9, 10, 14−28, 29, 56
elegy xii, 157, 164
'Elegy and Fanfare' 214−15
Eliot, T. S. xvi, 41, 48, 71, 75, 98,
    110, 118, 138, 161, 185, 212
England
    compared with Australia 17
    influence on Porter 41, 111, 137,
    241
    perception of, in Australia 27−8
English literature 25−6, 44, 56, 104,
    110, 200
*English Poets, The: From Chaucer to
    Edward Thomas* 118
English Subtitles xiii, 169, 177−96,
    197, 207, 208
Enright, D. J. 54, 125, 126
'Essay on Clouds' 225−6, 244
'Essay on Dreams' 4, 42, 226
'Essay on Patriotism' 242−4, 254,
    255
'Euphoria Dies' 82
Europe xvi, 96, 97, 112, 122, 181,
    190, 191, 198, 200
'Europe' 112, 113, 138
'Evensong' 169
'Evolution' 101, 124, 125, 126
Ewart, Gavin 19, 82, 86, 87, 121, 239
'Exequy, An' 160−4, 165, 189
exile xiv, xvii, 1, 2, 10, 24, 27, 29, 41,
    44, 61, 66, 83, 107, 127, 137, 143,
    164, 172, 175, 189, 190, 195, 215,
    219, 228, 241, 250; *see also*
    expatriatism; home; outsider
'Exit, Pursued by a Bear' 23
expatriatism xii, 9, 39, 46, 47, 48, 66,
    85, 90, 107, 168, 200, 242

failure 12, 22, 31, 50
'Fair Go for Anglo-Saxons' 102
family, sense of 6, 11−14, 18, 32−3;
    see also historicism
'Fantasia on a Line of Stefan George'
    101
'Farewell to Theophrastus, A' 12, 88
Fast Forward 197, 207−18, 239
'Fast Forward' 211, 212−13
father *see* Porter, William
fear 22, 85, 149
First World War 33, 233, 244; *see
    also* war
'Five Generations' 11, 99, 253
'Flock and the Star, The' 209−10
'For John Clare from London' 69
'Forefathers' View of Failure' 12, 13,
    67, 78, 85, 173
'Fossil Gathering' 124, 125
Freud, Sigmund 76, 143, 151, 206,
    231, 235
    Civilization and its Discontents
    96−7, 225, 247−8
'Frogs at Lago di Bolsena' 169
'Future, The' 191
'Futurity' 20

gardens 8, 37, 108, 146−7, 158, 171,
    204
Garden of Eden 8, 146−7, 164, 189,
    241
'Ghosts' 5, 6, 10, 67, 182
'Giant Refreshed, A' 38, 39
'Going to Parties' 54
'Good Ghost, Gaunt Ghost' 188
'Good Vibes' 22
Gray, Robert 37, 183, 199
'Great Cow Journeys On, The' 131
Green, Dorothy 21, 26, 66, 194, 202
Grigson, Geoffrey 119
Group Anthology, A 61
Group, the xi, 18, 52−9, 62, 64, 87,
    97, 112, 198
'Grub Street *versus* Academe'
    119−20
'Guide to the Gods, A' 211
guilt xiv, 95, 149, 152, 160, 170

Hamilton, Ian 69, 99, 152, 170, 180
'Hand in Hand' 257
'Happening at Sordid Creek' 95−6
Hardy, Thomas 5, 103, 105, 128, 157,
    160, 161, 164, 200
Haydn, Joseph 96, 97, 105, 192
'He Would, Wouldn't He' 240−1
'Headland Near Adelaide, A' 248

Heaney, Seamus xiii, 130, 142, 183, 198
heaven xiv, 209, 216, 256; *see also* paradise
hell xiv, 1, 9, 15, 20, 37, 38, 196, 209, 216, 221, 242, 256
Henry, (Shirley) Jannice *see* Porter, Jannice
Hill, Geoffrey 183, 198, 201
'Hint from Ariosto, A' 128
'Historians Call Up Pain, The' 219
historicism, 92–8 *passim*, 103, 112, 133, 140, 166, 179, 216, 219, 220, 235, 252
'History of Music' *see* 'Three Poems for Music'
Hobsbaum, Philip 18, 52–3, 55, 61
Hoff, Ursula xvii, 134–5, 228
'Homage to Gaetano Donizetti' 38, 95
home xii, xiv, 29, 51, 77, 156, 157, 175–6, 215, 227
'Homily in the English Reading Rooms' 96, 97
Hope, A. D. xiii, 38, 76, 86, 198, 255
'Australia' 85
'How Important is Sex?' 183
'How to Get a Girl Friend' 95
Hughes, Ted xiii, 53, 57, 84, 103, 125, 141, 151, 183, 193, 198, 201, 218, 226
humour xiv, 19, 20, 45, 77, 121, 126, 144, 158, 162, 183, 187, 201, 209, 221, 231, 247–8, 255
Hungary invasion 55

Ibsen, Henrik 5, 6, 192, 214
iconography 1, 4, 76, 90, 132, 165, 167, 235
'Imperfection of the World, The' 192–3
'"In the New World Happiness is Allowed"' 97, 175–6
'Ingrate's England, An' 241–2
innocence 155, 156, 232
'Inspector Christopher Smart Calls' 49
irony xiv, 7, 20, 21, 31, 63, 76, 77, 82, 85, 114, 125, 126, 135, 209, 231, 239, 241, 247–8, 251
'Isle of Ink, The' 122
Israel 125–7

Italy xvi, 113, 118, 149, 153, 162, 167–9, 215, 227

'James Joyce Sings "Il mio tesoro"' 127
James, Clive xi, 19, 48, 68, 119, 122, 174–5, 207, 210, 219–20, 241
Jewish Quarterly 89
'John Marston Advises Anger' 60, 63, 69, 70–2, 84, 155
Jolley, Elizabeth 39
Jonah 61, 118, 132–4, 135
Jones, Evan 181, 196, 201, 214, 217
Joseph, Rosemary 53
journalism, as practised by Porter 97, 118–19, 120, 137, 139, 253–6
journeys 92, 100, 171, 192, 242, 250
'Jumping to Conclusions' 217

'Killing Ground, The' 187–8
'King of Limerick's Army, The' 128
'King of the Cats is Dead, The' 186

Lady and the Unicorn, The 118, 132, 133, 134–6
'Lament for a Proprietor' 73
'Landscape with Orpheus' 4, 42, 193, 196, 208, 229
Lane Cove River 4, 5
language 134, 177
plain and baroque 22, 224
Porter's fascination with 25, 26, 38, 178–9, 180, 218, 227, 243–4
Larkin, Philip xiii, 52, 54, 84, 183, 198
Last of England, The xi, 62, 109–16, 124, 202, 228
'Last of England, The' 111, 130
'Last of the Dinosaurs, The' 99–100, 101, 112
Lawrence, D. H. 56, 85, 139, 141
Lawson, Sylvia 66–7, 102, 145
Leavis, F. R. 53, 54, 56, 119, 163
'Legs on Wheels' 206–7
Lehmann, Geoffrey 146, 170, 205
Lehmann, Sally *see* McInerney, Sally
Listener 68, 89
'Little Buddha' 257
'Little Harmonic Labyrinth' 217
Living in a Calm Country 118, 135, 139–48

'Living in a Calm Country' 140
London 29, 44−5, 60
   'exile' (1951−74) 25
   'moral map' 51−2, 65, 136
   poems 69, 72, 75, 77, 81, 88, 94,
      96, 112, 240, 241
London Magazine 89
Losing Chance, The 48−9, 216
loss, sense of 1, 5, 7, 72, 95, 131, 149,
   154, 160, 161, 164, 167, 180
love 25, 82, 100, 108, 135, 161, 162,
   164, 170, 177, 183, 184, 185, 188,
   215, 257
Lowell, Robert 11−12, 152, 193, 201
Lucie-Smith, Edward 18, 53, 56, 61,
   87
Lumsdaine, David 48, 50, 132
'Lying Art, The' xiv, 177, 179
lyricism 57, 94, 106, 107, 109, 121,
   135, 136, 183, 231, 249, 255, 256

MacBeth, George 18, 53, 54, 95, 98
McInerney, Sally 146, 149, 169, 170,
   171, 184, 203, 204
'Made in Heaven' 69, 73, 82, 88
Main, Marion *see* Porter, Marion
Main, Mark 12, 33
Malouf, David 2, 29−31, 32, 168,
   181, 215, 255
'Marianne North's Submission'
   215−16
marriage *see* Porter, Jannice
Mars 197, 204, 213, 228, 233−6, 242
Martial 83, 111, 121, 122
Marx, Karl 96, 125, 247
Marxism 46, 206
Meanjin 36
Melbourne 35, 36, 38, 86
   University of 197
'Melbourne General Cemetery, The'
   227−8
memory 10, 42, 136−7, 142, 156,
   159, 160, 252−3
Mendelssohn, Felix 129
'Metamorphoses' 68
metamorphosis 74, 126−7, 132, 133,
   158, 229, 234
'Metamorphosis' 60, 69, 73−4, 78,
   147
metropolitanism xv, 5, 18, 29, 35, 56,
   69, 119, 144−6, 195, 230, 250; *see*

   also provincialism
Micklem, Neil 155
Milton, John 105, 188, 189, 243
'Missionary Position, The' 208−9
Modernism xv, 76, 132, 181, 200, 239
'Moral Tale has a Moral End, A' 242
Moraes, Dom 61, 86
'Mort aux Chats' 68, 128, 186−7
mother *see* Porter, Marion
Movement, the xi, 54, 103, 112, 119
Mozart, W. A. 38, 56, 69, 77, 96, 105
'Mr Roberts' 16−17, 18−20, 23, 234
Murray, Les A. xiii, 13, 143−6, 175,
   181, 183, 198, 218, 231−2
   critic of Porter 57, 90, 143−6, 199
   criticized by Porter 139, 143−6,
      230, 242, 251, 254−5
   nationalist poet xvi, 167, 205
   'Towards the Imminent Days'
      170−1, 177
music 4, 29, 35, 48, 50, 76, 96, 97,
   104−7 *passim*, 113, 132, 140, 184,
   196, 238, 256, 258
   and religion 77, 104, 142, 160−1
   see also names of composers;
   opera
'My Late T'ang Phase' 101−2, 113
'My Old Cat Dances' 185, 187
'Myopia' 193
mythology
   classical xiv, 5, 20, 134, 149, 188,
      196, 226, 230, 235−6
   of Australia 67, 92
   personal 1, 10, 83, 137

Narcissus xvi, 49, 187, 197, 228−32,
   235, 257
nationalism 137, 145, 185, 242
nature, response to 97, 137, 167, 169,
   172, 185, 249
'Need for Foreplay, The' 183
Neo-Romanticism 45, 54, 103, 133,
   193
Neville, Jill 43−6, 50, 51, 68, 70, 72,
   150, 151, 153, 193
   Fall-Girl 43−7, 52, 117
Neville, Richard 68
'New Mandeville, The' 239
New Poems 1962 61
New Poems 1971−1972 118
New Poetry 110

New Poetry, The xi, 61, 84
New Review, The 54, 119
New Statesman 68, 89, 91, 96, 111
'Next to Nothing' 246
'Nine Points of Law' 95
Nolan, Sidney 39, 90, 250
'Non Piangere, Liù' 164, 165—6, 188
nostalgia 39, 40, 60, 128, 136, 141, 149
'Nothing to Declare' 51

Observer 68, 118, 181, 197
'Occam's Razor' 192
'Ode to Afternoon' 141, 143, 255
Olson, Charles 110, 112, 138
'On First Looking into Chapman's Hesiod' 127, 143, 144—5, 242, 255
'On the Train between Wellington and Shrewsbury' 22, 124
'On This Day I Complete My Fortieth Year' 113, 114—16, 117
Once Bitten, Twice Bitten xi, 4, 39, 61, 62—86, 102, 197, 225
'Open-Air Theatre, Regent's Park' 257
opera 56, 122, 162, 165, 168, 195, 238
'Orchard in E-Flat, The' 258—9
'Orchid in the Rock, The' 172—3
Osborne, Charles 35, 125
outcast, outsider xiii, 15, 21, 102, 121, 122; *see also* exile
Owen, Margaret 53, 58—9, 64, 81

pain xii, 1, 7, 118, 155, 162, 167, 177, 180, 214
'Painter's Banquet, The' 229, 231
painting 76, 105, 132, 143, 168, 196, 209, 239
'Pantoum of the Opera' 221
'Paradis Artificiel' 209, 228, 229
paradise xiv, 1, 126, 168, 171, 196, 228, 246, 256
'Paradise Park' 219, 223, 233
'Party Line' 73
patriotism 34, 94, 145, 242—4; *see also* nationalism
Penguin Modern Poets 2 xi, xiii, 61, 67, 86
Penton, Brian 30
Peter Porter Reads from His Own Work 118

'Phar Lap in the Melbourne Museum' xiii, 67, 84—5, 86
Phillips, A. A. 26, 39
Phillips, Diana, *see* Watson-Taylor, Diana
'Philosopher of Captions, A' 177, 179—80
place, sense of 122, 128, 160, 166, 167, 189, 248, 250
Plath, Sylvia 151
'Poem to End Poems, The' 247
Poems Ancient and Modern 61, 86, 94—7, 121
poetic forms 20, 63, 93, 112, 160, 165, 166, 188, 214, 221—2, 240
'Poetry' 104
politics 34, 54—5, 233, 241, 242—3
'Pontormo's Sister' 229—30
Pope, Alexander 73, 83, 103, 105, 120, 138, 200, 239
Porter Folio, A 61, 86, 94, 97—108, 121, 217, 221
Porter Selected, A 24, 84, 98, 197, 228
'Porter Song Book, The' 94, 104, 106—7, 258
'Porter's Retreat', 251—2
Porter, Jannice 149—55, 156—69 *passim*
  death xii, 118, 139, 213, 214, 229
  inspiration for poetry 188—91, 215, 257
  marriage 24, 61, 65, 70, 114
Porter, Katherine and Jane 14, 24, 61, 94, 150, 152, 153, 170, 204
Porter, Marion (Marion Main), mother 2, 5—6, 7, 11, 12, 33, 95, 169, 191, 192—3, 204
  death 1, 9, 29, 32, 72, 117, 135—6, 149, 153, 204
Porter, Robert, grandfather 11, 32
Porter, William, father 1, 8—11, 29—31, 32, 33, 41, 88, 94, 95, 105, 107, 108, 153, 191, 203—4
Possible Worlds 50, 197, 237—59
postmodernism 71, 200, 217, 221, 222, 223
poststructuralism 217, 222, 223
Preaching to the Converted 118, 121—31, 135

'Preaching to the Converted' 124
'Prince of Anachronisms, The' 83,
    234
protests 60, 204, 212
provincialism 5, 18, 67, 131, 195; *see
    also* metropolitanism
psychoanalysis 151—2, 205—6, 225,
    231, 248; *see also* Freud, Sigmund

'Reading *MND* in Form 4B' 23—4, 67
'Real People' 111
'Recipe, The' 98, 113
reconciliation 148, 149, 170, 184,
    190, 204
'Redback' 220—1
Redgrove, Peter 53, 56—7, 58, 87,
    102, 149, 226
'Reflections on my own Name' 69
religion xv, 14, 34, 104, 142, 164,
    208, 209—11, 218
'Requiem for Mrs Hammelswang' 38
responsibility 6, 21, 193, 194
'Return of Inspector Christopher
    Smart, The' 49
'Returning' 191—2
Richardson, Henry Handel 12, 26, 39,
    43
'River Run' xvi
Roberts, H. E. headmaster, 16
Roloff Beny in Italy 118
Rolph, John, *see* Scorpion Press
roman fleuve xiv, 208
'Roman Incident' 149, 169, 171—2,
    204, 229
romanticism 5, 69, 162, 169, 193
Roper, Susanna 150, 151, 154, 205
Rosenthal, T. G. 132—3

'Sadness of the Creatures, The'
    113—14, 117, 141, 186, 201
'St Cecilia's Day Epigrams' 104
'St Cecilia's Day, 1710' 105, 107
'Sanitized Sonnets' 103, 110, 113
'Santa Cecilia in Trastevere' 215
'Santa Lucia' 28
satire 61, 63, 67, 68, 70, 82—3, 108,
    113, 145, 239, 245
Scarlatti, Domenico 127, 187
school, *see* education
Schubert, Franz 4, 96, 97
Scorpion Press xi, 61, 62—3

'Scream and Variations' 164—5
'Seahorses' 100, 125
'Seaside Resort' xvi, 124, 128—31,
    216
Second World War 55, 79; *see also*
    war
'Secret House, The' 28
self, image of xiv, xvii, 19, 22, 36, 64,
    78, 83, 92, 95, 98, 106, 150, 201,
    229, 231, 233, 244
self-pity 130, 219
sex 16, 36, 46, 122—3, 183—5, 188,
    232, 235
'Sex and the Over Forties' 123—4
Shakespeare, William 23—4, 103,
    104, 105, 120, 156, 188, 189, 200
Shaw, G. B. xv, 46, 104
Shoalhaven River xvi, 136, 229, 231
'Shopping Scenes' 86
'Short Story' 111, 112
'Sick Man's Jewel' 51
Siege of Munster, The 118
'Sins of the Fathers, The' 95
Skinner, Mollie 139
Smart, Christopher xiv, 57, 187
'Smell on the Landing, The' 69
social poet xii, 59, 61, 63, 67, 68, 71,
    109, 202, 239, 245
socialism 35, 46, 88, 91, 175
'Some Questions for Thomas à
    Kempis' 210
'Somme and Flanders' 33, 67, 233
'Sonata Form: The Australian
    Magpie' 183—5, 191
'South of the Duodenum' 82
Spender, Stephen xiii, 198, 212
'Spiderwise' 13, 137, 219—20
Stein, Gertrude 24, 138, 173, 174
Stephensen, P. R. ('Inky') 26, 30
Stevens, Wallace xiii, 78, 112, 120,
    138, 179, 238
'Sticking to the Text' 222, 247
'Storm, The' 139—40, 143
'Story of my Conversion, The' 96,
    142—3
'Story of U, The' 4, 225
Stow, Randolph xii, 37, 147
'Stratagems of the Spirit' xiv, 238—9,
    256
Stravinsky, Igor 56, 127
'Stroking the Chin' 112